Nanoscale Semiconductor Devices

This book covers 2D materials, transistors, and low-power transistor technologies. It also discusses a wide range of semiconductor devices, including transistors, LEDs, energy harvesting devices, and spintronics devices.

- Covers nanoscale semiconductor materials, low-power transistors, and applications in integrated circuits.
- Offers comprehensive coverage of design, simulation, and experimentation for a range of field-effect transistors.
- Highlights the importance of using piezoelectric nanogenerators for energy applications.
- Showcases the developments in nanoscale device simulation and opportunities in electronic fabrication.
- Explains the latest nanoscale sensors, micro- to nanoscale energy-harvesting devices, and nano energy storage devices.

The book is primarily written for senior undergraduates, graduate students, and academic researchers in electrical engineering, electronics, and communications engineering.

Electronic Materials, Devices, Systems and Manufacturing
Series Editors:
Jean-Pierre Leburton
University of Illinois, Urbana-Champaign, Illinois, USA
Yongbing Xu
The University of York, UK

Nanoscale Semiconductor Devices: Technology and Applications
Edited by D. Nirmal

For more information about this series, please visit:
https://www.routledge.com/Electronic-Materials-and-Devices-Series/book-series/ELEMATDEVSER

Nanoscale Semiconductor Devices

Technology and Applications

Edited by

D. Nirmal

CRC Press
Taylor & Francis Group
Boca Raton London New York

CRC Press is an imprint of the
Taylor & Francis Group, an **informa** business

First edition published 2027
by CRC Press
2385 NW Executive Center Drive, Suite 320, Boca Raton FL 33431

and by CRC Press
4 Park Square, Milton Park, Abingdon, Oxon, OX14 4RN

CRC Press is an imprint of Taylor & Francis Group, LLC

First edition published by CRC Press 2027

ISBN: 978-1-041-06371-1 (hbk)
ISBN: 978-1-041-06372-8 (pbk)
ISBN: 978-1-003-63512-3 (ebk)

DOI: 10.1201/9781003635123

Typeset in Sabon
by SPi Technologies India Pvt Ltd (Straive)

Contents

Editor biography

D. Nirmal currently serves as Dean of Academic Affairs at Karunya Institute of Technology and Sciences, India. Dr. Nirmal has played a key role in driving academic reforms, promoting outcome-based competency-focused curriculum frameworks, and facilitating national accreditation initiatives. His research focuses on Modeling and TCAD simulation of power semiconductor devices, GaN HEMT devices, design, fabrication, and reliability analysis and circuit applications. Dr. Nirmal has successfully secured ₹21.1 million in research funding from DRDO, ISRO, the Ministry of Electronics and Information Technology, and AICTE, along with product development grants from the IIT Tirupati Navavishkar I-Hub Foundation (IITTNiF) and Analog Devices India Pvt. Ltd. (ADI). He is the founding chair of the IEEE Electron Devices Society (EDS) Coimbatore Chapter and currently serves as the IEEE EDS Region 10 SRC Chair. Additionally, he is an active volunteer on several IEEE committees and a technical committee member for EDS Compound Semiconductor Devices and Circuits. His outstanding contributions have earned him numerous accolades, including the IEI Young Engineer Award, the IETE Smt. Manorama Rathore Memorial Award (2022), and the Young Scientist Award (2019) from the Academy of Sciences. A prolific researcher, he has authored over 180 peer-reviewed publications and holds three patents. He is a Fellow of IETE and a Senior Member of IEEE. Dr. Nirmal is also a sought-after speaker, delivering keynote addresses and lectures at national and international conferences.

Contributors list

J. Ajayan
Department of Electronics and Communication Engineering
SR University
Warangal, Telangana, India

Sunny Anand
HCL Technologies

N. Aruna Kumari
Wipro Limited
Bangalore, Karnataka, India

Navjeet Bagga
School of Electrical and Computer Sciences
IIT Bhubaneswar
Kansapada, Odisha, India

Suresh Balanethiram
J.C. Bose Dept. of Electronics and Communication Engineering
National Institute of Technology Puducherry
Karaikal, Puducherry, India

Vidhya Bhojan
Division of Physical Sciences
Karunya Institute of Technology and Sciences
Coimbatore, Tamil Nadu, India

Madhuparna Chakraborty
School of Electrical Engineering and Computer Science
Ohio University
Athens, Ohio, USA

J. Charles Pravin
Department of ECE
Kalasalingam Academy of Research and Education
Virudhunagar District, Tamil Nadu, India

Siva Chidambaram
Department of Physics and Nanotechnology
SRM Institute of Science and Technology
Chennai, Tamil Nadu, India

Komatreddy Damodar Reddy
Department of Physics, Stanley College of Engineering and Technology for Women, Abids
Hyderabad-500001, India

Sourav Das
Solid State Electronics and VLSI Lab
Department of ECE
NIT Jamshedpur
Jamshedpur, Jharkhand, India

G. Dhivyasri
Department of Computer Science and Engineering
Sai Vidya Institute of Technology
Bangalore, Karnataka, India

Flavia P.N. Inbanathan
Electrical and Electronics Engineering Technology, CATS
Kent State University at Trumbull
Warren, Ohio, USA

Satyabrata Jit
Department of Electronics Engineering
IIT (BHU)
Varanasi, Uttar Pradesh, India

Naveen Kumar
University of Glasgow
Glasgow, Scotland

Pradeep Kumar
Amity University
Noida, Uttar Pradesh, India

Sandeep Kumar
IIT Bhubaneswar
Kansapada, Odisha, India

Bipul Kumar Mahato
NSSE, NIE
NTU
Singapore

Velusamy Maheskumar
School of Chemistry
Southwest Jiaotong University
Chengdu, Sichuan, China

R. Manaswini
Department of Electronics and Communication Engineering
Presidency University
Bangalore, Karnataka, India

M. Manikandan
Department of Electronics and Communication Engineering
Presidency University
Bangalore, Karnataka, India

B. Mounika
Department of ECE
SR University
Warangal, Telangana, India

Chhandak Mukherjee
IMS Laboratory
University of Bordeaux
Bordeaux, France

Raju Nandhakumar
Division of Physical Sciences
Karunya Institute of Technology and Sciences
Coimbatore, Tamil Nadu, India

K.J. Niveditha
Division of Physical Sciences
Karunya Institute of Technology and Sciences
Coimbatore, Tamil Nadu, India

Eswaran Parthasarathy
Department of Electronics and Communication Engineering
SRM Institute of Science and Technology
Chennai, Tamil Nadu, India

Naini Pavan Kumar
Department of Sciences and Humanities
Matrusri Engineering College
Hyderabad, Telangana, India

K.P. Pradhan
Department of Electronics and Communication
IIITDM Kancheepuram
Chennai, Tamil Nadu, India

R. Priyadharshni
J.C. Bose Department of Electronics and Communication Engineering
National Institute of Technology Puducherry
Karaikal, Puducherry, India

Shiromani Balmukund Rahi
University School of Information and Communication Technology
Gautam Buddha University
Greater Noida, Uttar Pradesh, India

E. Ramola
Department of ECE
Kalasalingam Academy of Research and Education
Virudhunagar, Tamil Nadu, India

Rajdeep Singh Rawat
NSSE, NIE
NTU
Singapore

M. Saravanan
Department of Electronics and Communication Engineering
Sri Eshwar College of Engineering
Coimbatore, Tamil Nadu, India

M. Sathvika
Department of Electronics and Communication
IIITDM Kancheepuram
Chennai, Tamil Nadu, India

Anany Saxena
Amity University
Noida, Uttar Pradesh, India

Kunal Singh
Solid State Electronics and VLSI Lab
Department of ECE
NIT Jamshedpur
Jamshedpur, Jharkhand, India

Prabhat Singh
IIIT Senapati
Imphal, Manipur, India

S. Sreejith
New Horizon College of Engineering
Bangalore, Karnataka, India

S. Srimaan
S.A. Engineering College
Chennai, Tamil Nadu, India

Rajesh Swaminathan
Division of Physical Sciences
Karunya Institute of Technology and Sciences
Coimbatore, Tamil Nadu, India

Samuel Tensingh
Western Sydney University
Sydney, New South Wales, Australia

Sresta Valasa
Department of Electronics and Communication Engineering
Manipal Institute of Technology
Bangalore, Karnataka, India

K. Vijaya Kumar
J.C. Bose Dept. of Electronics and Communication Engineering
National Institute of Technology Puducherry
Karaikal, Puducherry, India

Weiqiang Yuan
School of Electrical Engineering and Computer Science
Ohio University
Athens, Ohio, USA

Preface

The rapid evolution of semiconductor technology continues to drive innovation across modern electronics, enabling breakthroughs in computation, communication, renewable energy, and nanotechnology. As conventional CMOS scaling approaches its physical limits, emerging device concepts and advanced materials have become essential for achieving low-power, high-speed, and energy-efficient performance. This book presents a comprehensive overview of recent advancements in semiconductor and nanoscale device technologies, covering topics such as emerging transistors for digital and analog integrated circuits, tunnel field-effect transistors as sustainable alternatives, and two-dimensional field-effect transistors with exceptional electrical properties. It also explores the design and optimization of thin-film solar cells, nano energy harvesting devices, and piezoelectric nanogenerators for self-powered systems. Further discussions on storage devices, memory technologies, and recent developments in light-emitting diodes illustrate the convergence of materials science and device engineering. The chapters on lithography, nanoscale device modeling, and simulation highlight the fabrication and analytical tools critical for realizing next-generation electronics, while the study of molecular electronics showcases the potential of atomic-scale systems. Collectively, these topics reflect the interdisciplinary advances shaping the future of sustainable, high-performance, and intelligent semiconductor technologies.

Chapter 1

Introduction of emerging semiconductor devices for low-power and high-speed digital/analog integrated circuit applications

J. Ajayan and N. Aruna Kumari

1.1 INTRODUCTION

From the 22 nm technology node to the 5 nm technology node, fin field effect transistors (FinFETs) ruled the semiconductor industry; nonetheless, the architecture approached its end of life. The constraints of fin scaling (spacing and fin height) impose increasing external resistance and capacitance, and the dissipated power is more than that of similar nanosheet field effect transistors (NSFETs), which indicates the end of the era of FinFETs. A gate-all-around (GAA) architecture is necessary to completely enclose the transistor channel with a gate in order to overcome the FinFET constraints. A breakthrough transistor architecture to carry on complementary metal oxide semiconductor (CMOS) technology scaling beyond the 5 nm technology node is the NSFET. The nanosheet's wider channel offers superior channel carrier mobility than nanowire field effect transistors (FETs), which suffer from self-heating and mobility deterioration brought on by interface roughness. The greatest advantage of the NS design is that it produces high-quality lateral interfaces through epitaxial fabrication, minimizing the channel's mobility reduction resulting from interface roughness. To provide an adequate driving current, GAA transistors must also be stacked vertically to form parallel channels. Samsung has accepted the GAA-NSFET design for mass manufacturing of the chips, considering its ability to provide a high drive current and low device characteristic variability (3 nm Process Technology) [1–4].

1.2 PERFORMANCE OVERVIEW OF NSFETs

Transistor size reduction is a crucial factor for the semiconductor industry as it boosts integrated circuit speed. The development of a smaller transistor that operates at a high frequency and consume less power is the main goal of design engineers in the chip industry. Large volume and fast static random access memory (SRAM) occupy a significant amount of the die space in many computing chips and aids in boosting system performance. Thus,

DOI: 10.1201/9781003635123-1

one of the most significant challenges to continue the scaling of CMOS technology in past decades is the development of more dense SRAM with smaller bitcell size and stronger driving ability within. For design flexibility with different nanosheet channel widths, the performance of SRAM may be enhanced by the use of vertically stacked GAA-NSFETs. Silicon GAA-NSFETs that are vertically stacked are becoming more and more popular. This is because, in comparison to FinFETs, NSFETs provide better electrostatic channel control within the same footprint, making them perfect for scaling down to sub-3 nm technology nodes. For the sub-3-nm technology node and beyond, stacked GAA-NSFETs are generally considered to be the most promising FinFET replacement to continue the CMOS scaling. This sophisticated design greatly enhances current drivability, short-channel effect immunity, and gate electrostatic control. However, as device dimensions continue to shrink, more parasitic components are brought in by the vertical stacking design and more complex 3-D geometrics, potentially declining the performance of both dynamic and static devices. The CMOS performance enhancement of ultrascaled NSFETs is anticipated to be significantly influenced by the reduction of parasitics [5–16]. There are several techniques available for mitigating subchannel leakage, such as partial/complete bottom dielectric isolation (BDI) and punch-through stoppers (PTSs). In 2024, Chunlei Wu et al. [7] introduced, for the first time, a reverse-biased p–i–n subtunnel FET (TFET) channel to efficiently reduce parasitic subchannel leakage in a unique GAA-NSFET with bottom p–i–n isolation.

NSFETs mainly have two drawbacks: 1) poor heat dissipation due to the thinner channel, and 2) high punch-through current in the PTS. In 2024, Sanguk Lee et al. [9] suggested that NSFETs with trench gates in the PTS area and shortened NS channels can effectively address these issues. Shobhit Srivastava et al. [14] demonstrated that a source/drain extension region can be an effective technique for the mitigation of self-heating effects in NSFETs. However, this extension of source and drain regions leads to the deterioration of drive current and transconductance due to increased parasitics. Reducing the 2-D cross-sectional geometry of NSFETs while simultaneously increasing the channel length is necessary for maintaining good electrostatics. When channel length shrinks, drift-diffusive transport gives way to ballistic transport. At the same time, substantial quantum confinement effects occur by an ultrascaled channel cross-section [17–28]. 2-D materials with atomically thin structures and surfaces devoid of dangling bonds, including WSe_2, MoS_2, and InSe, have emerged as the most promising materials to further advance the technological node [21, 29–48]. The architectures of emerging transistors (NSFET, CombFET, and TreeFET) are depicted in Figure 1.1. The CombFET and TreeFETs have inter-bridge (IB) channels compared with the NSFET. The CombFET and TreeFET offer superior on current (I_{ON}) and ON/OFF current ratio (I_{ON}/I_{OFF}) (Figure 1.2). This is due to the presence of the IB channel. However, the CombFET and TreeFET also exhibit higher I_{OFF} and drain induced barrier lowering (DIBL) due to the increased parasitic arises from IB channel. Therefore, NSFET is the best performer at lower technology nodes. At lower technology nodes, NSFET

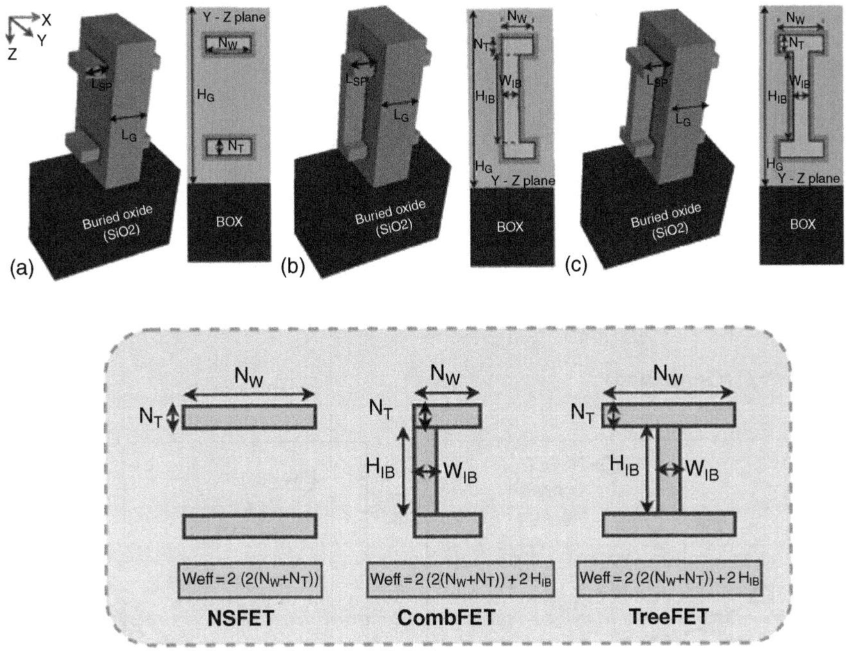

Figure 1.1 Architectures of (a) NSFET, (b) CombFET, and (c) TreeFET.

(Aruna Kumari et al. [38].)

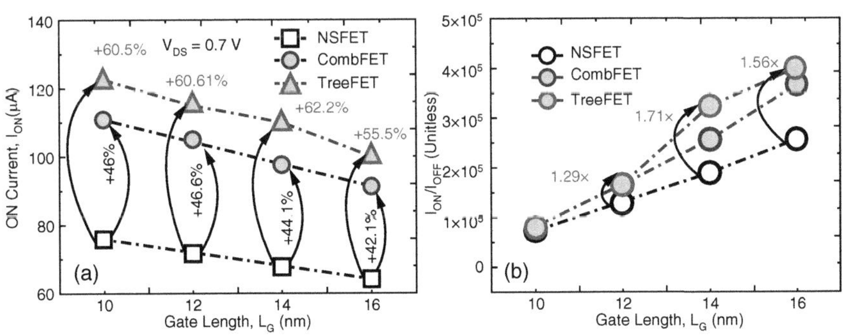

Figure 1.2 (a–d) L_G (gate length) scaling effects on NSFET, CombFET, and TreeFET.

(Aruna Kumari et al. [38].)

(Continued)

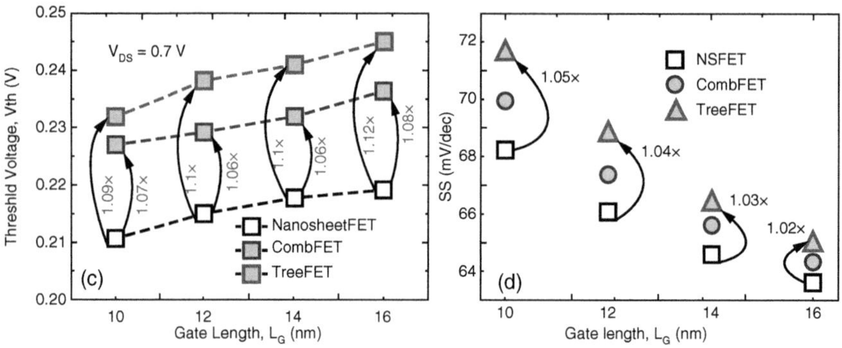

Figure 1.2 (Continued)

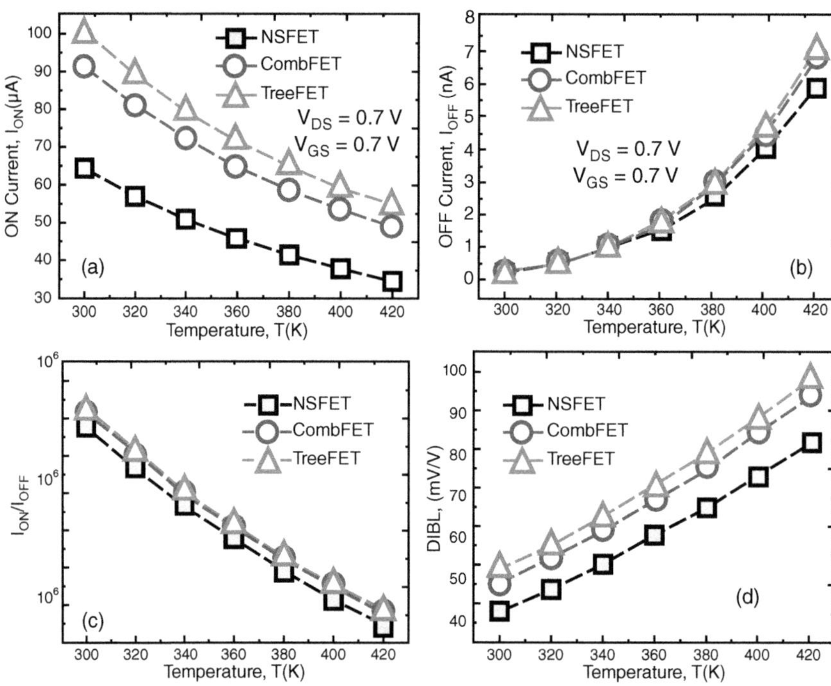

Figure 1.3 (a–d) Thermal variation effects in NSFET, CombFET, and TreeFET. **(Aruna Kumari et al. [38].)**

offers better SS (subthreshold swing) and lower V_{th} (Threshold voltage) compared to its counterparts.

The I_{ON} and I_{ON}/I_{OFF} declines I_{OFF} and DIBL increases with an elevation in temperature due to different phenomena like Lattice and phonon scatterings (Figure 1.3). The TreeFET exhibits larger I_{OFF} and DIBL compared with the combFET and NSFET at higher temperatures [38]. The spacer material also

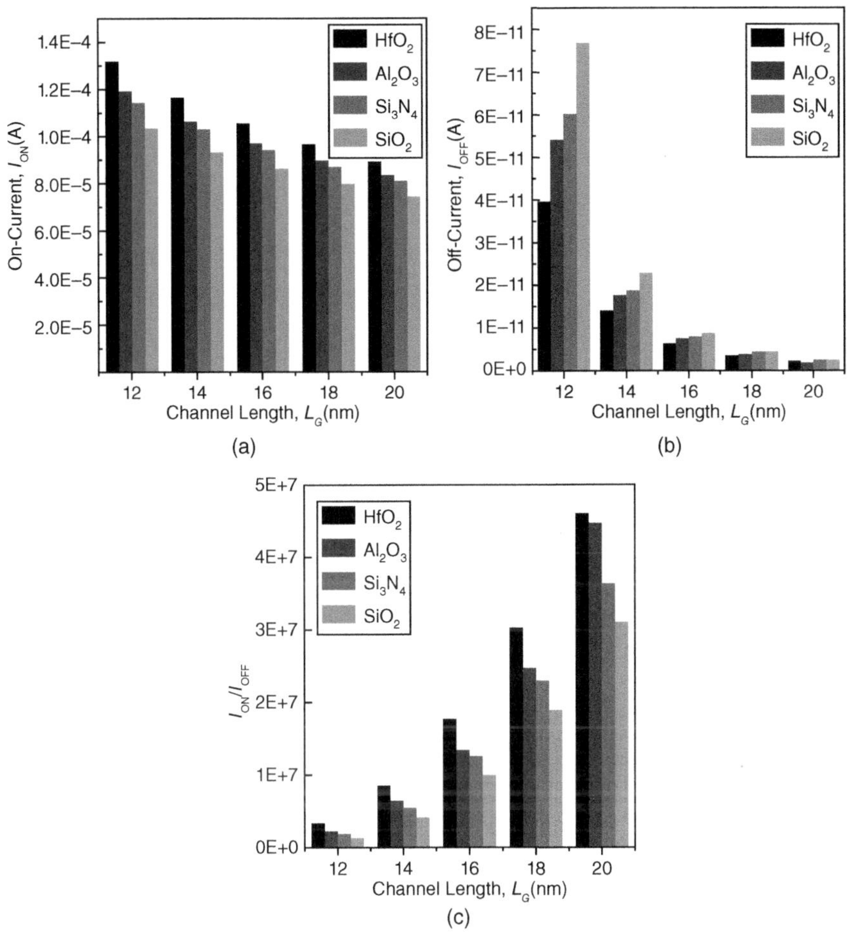

Figure 1.4 Impact of spacer materials and L_G scaling on (a) I_{ON}, (b) $I_{OFF,}$ and (c) I_{ON}/I_{OFF} of TreeFET.

(Reprinted from Parvathi and Prithvi [40] with permission from Elsevier.)

significantly affects the performance of these emerging transistors (Figures 1.4 and 1.5). The TreeFET with high-K spacer material offers superior transconductance, I_{ON}, $I_{OFF,}$ and I_{ON}/I_{OFF} characteristics. However, SS and DIBL degrade with the usage of high-K spacers. Finally, the radio frequency (RF) performance of TreeFET degrades with the usage of high-K spacers due to the increase of output conductance and gate capacitance.

In 2020, Yunho Choi et al. [46] demonstrated the effect of parasitic channel height (H_{ch}) on the performance of GAA-NSFETs (Figures 1.6–1.8). A large H_{ch} significantly enhances I_{ON}/I_{OFF} and reduces SS and DIBL of GAA-NSFETs. On the other hand, GAA-NSFETs with smaller H_{ch} suffer from

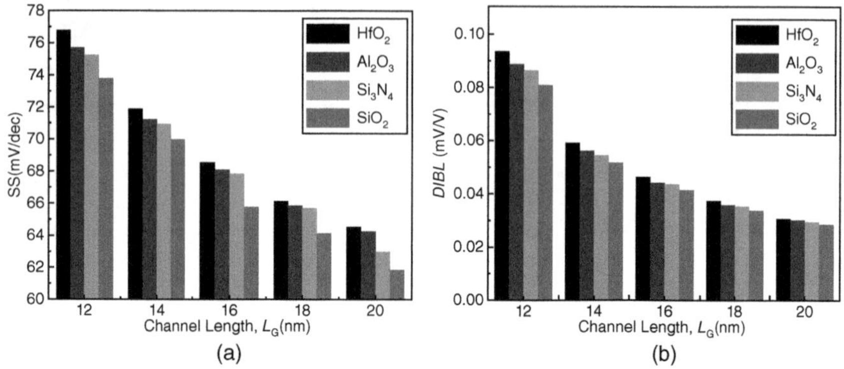

Figure 1.5 Impact of spacer materials and L_G scaling on (a) SS and (b) DIBL of TreeFET.

(Reprinted from Parvathi and Prithvi [40] with permission from Elsevier.)

severe short-channel effects (SCEs) and I_{ON}/I_{OFF} degradation. However, extension of H_{ch} leads to an increase in gate capacitance and RC delay in GAA-NSFETs.

The channel length (L_G), thickness of nanosheet (N_T), and width of nanosheet (N_W) significantly affect the DC and RF traits of NSFETs. Both I_{ON} and I_{OFF} of NSFETs increase with an increase in N_W (Figure 1.9). The rise in I_{OFF} due to the increase of N_W is because of the poor gate control over the

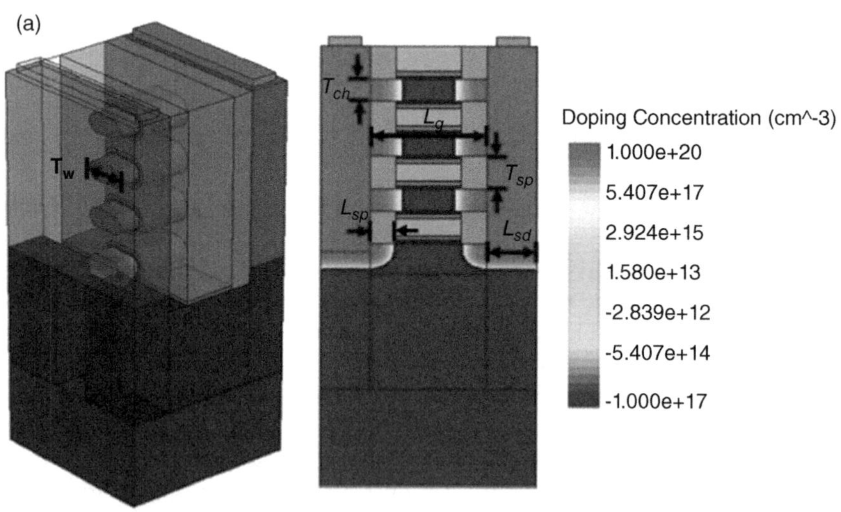

Figure 1.6 (a and b) Cross-sectional view of vertically stacked GAA-NSFET.

(Reprinted from Choi et al. [46] with permission from Elsevier.)

(*Continued*)

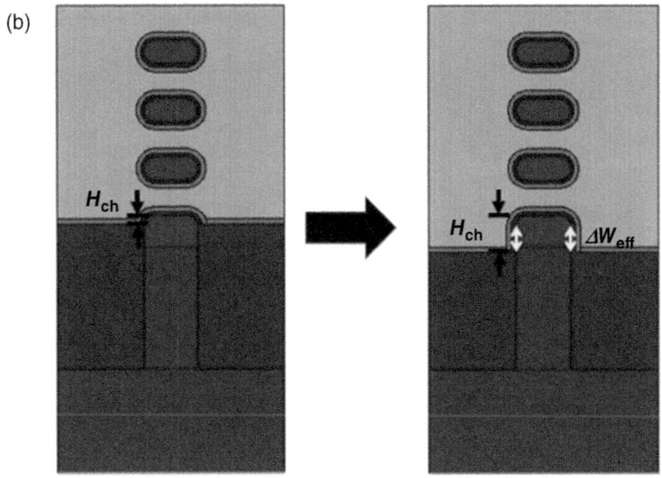

Figure 1.6 (Continued)

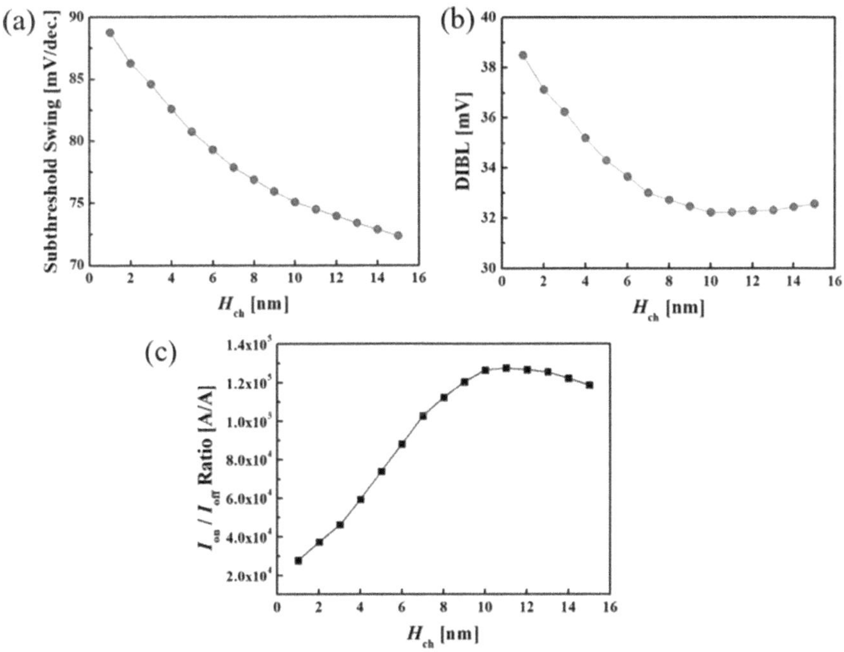

Figure 1.7 Effect of H_{ch} on (a) SS, (b) DIBL, and (c) I_{ON}/I_{OFF} of vertically stacked GAA-NSFET.

(Reprinted from Choi et al. [46] with permission from Elsevier.)

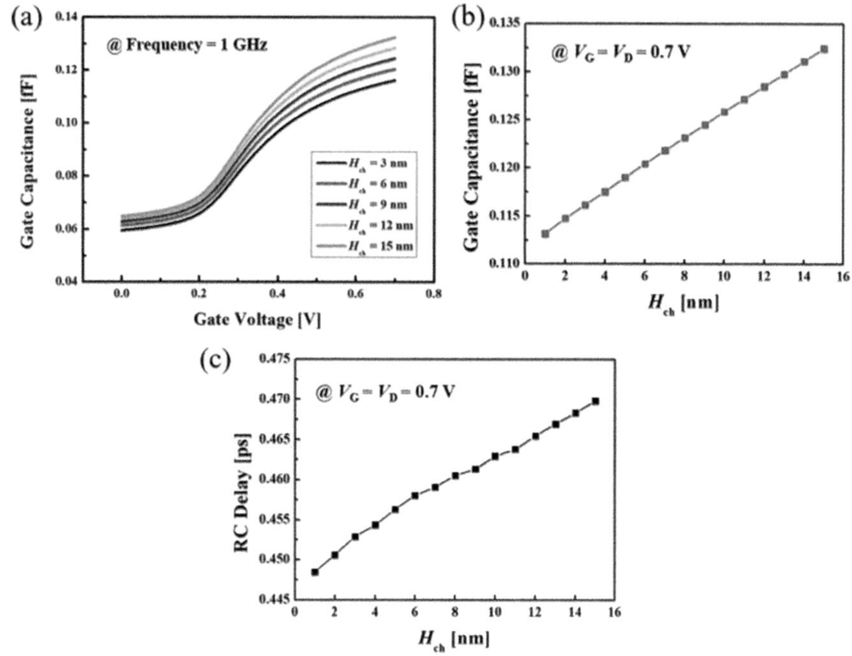

Figure 1.8 Effect of H_{ch} on (a and b) gate capacitance and (c) RC delay of vertically stacked GAA-NSFET.

(Reprinted from Choi et al. [46] with permission from Elsevier.)

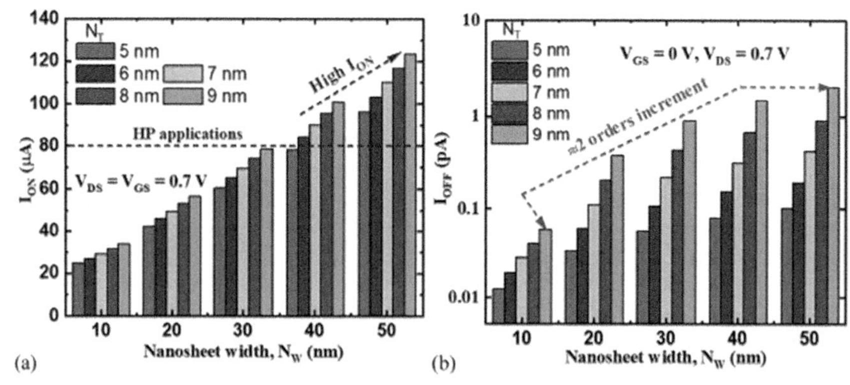

Figure 1.9 Impact of N_W and N_T on (a) I_{ON} and (b) I_{OFF} of NSFETs for different N_T.

(Aruna Kumari et al. [49].)

nanosheet channel. The I_{ON} and I_{OFF} of NSFETs also significantly increase with an increase in N_T, and the degradation in I_{OFF} at higher N_T is due to the poor gate control over the nanosheet channel, as in the case of N_W. Lower N_W and N_T are essential for achieving a good I_{ON}/I_{OFF} ratio due to the better gate control over nanosheet channels (Figure 1.10). The threshold voltage (V_{th}) of GAA-NSFET is inversely proportional to the N_W and N_T. That is, when N_W and N_T goes up, the V_{th} declines, which boosts the drive current of NSFETs. The GAA-NSFETs with vertically stacked nanosheet channels offer an I_{ON}/I_{OFF} of over 10^9. GAA-NSFETs with large N_W and N_T values lead to the degradation of short-channel traits in terms of SS and DIBL. That is, GAA-NSFETs with reduced N_W and N_T are required for attaining superior short-channel traits (Figure 1.11).

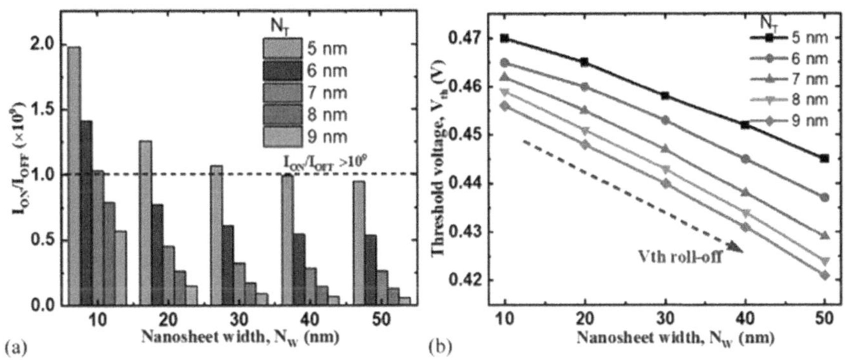

Figure 1.10 Impact of N_W and N_T on (a) I_{ON}/I_{OFF} and (b) V_{th} of NSFETs for different N_T.

(Aruna Kumari et al. [49].)

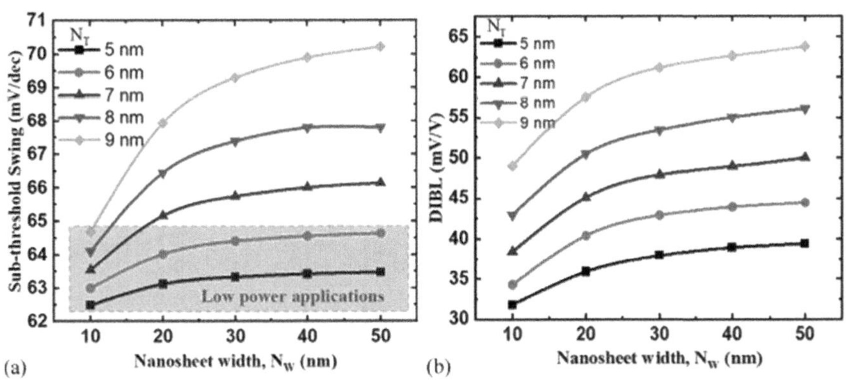

Figure 1.11 Impact of N_W and N_T on (a) SS and (b) DIBL of NSFETs for different N_T.

(Aruna Kumari et al. [49].)

The g_m and g_{ds} of GAA-NSFETs goes up with increment in N_W and N_T (Figures 1.12(a) and 1.13(a)). However, the transconductance generation factor (TGF) and maximum gain of GAA-NSFETs declines with elevation in N_W and N_T (Figures 1.12(b) and 1.13(b)). The RF traits of GAA-NSFETs can be measured in terms of cut-off frequency (f_T). The gate capacitance and C_{gd} of GAA-NSFETs also goes up with increment in N_W and N_T (Figure 1.14). The f_T of GAA-NSFETs declines with increase in N_W and N_T (Figure 1.15), due to the enhancement of g_m. The intrinsic time delay (τ) declines with increase in N_W and N_T (Figure 1.15).

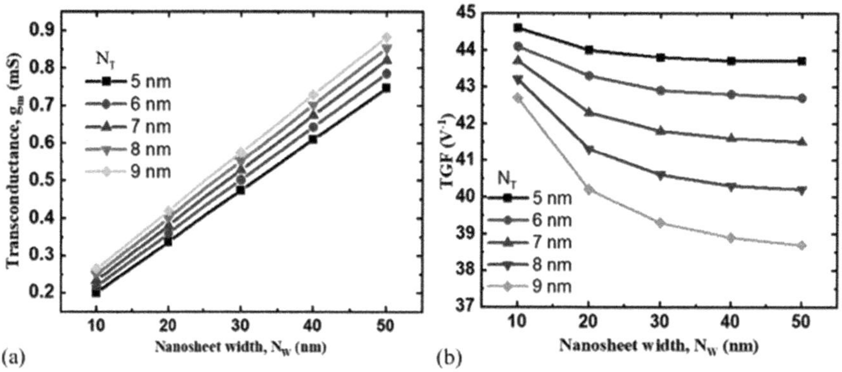

Figure 1.12 Impact of N_W and N_T on (a) transconductance (g_m) and (b) TGF of NSFETs for different N_T.

(Aruna Kumari et al. [49].)

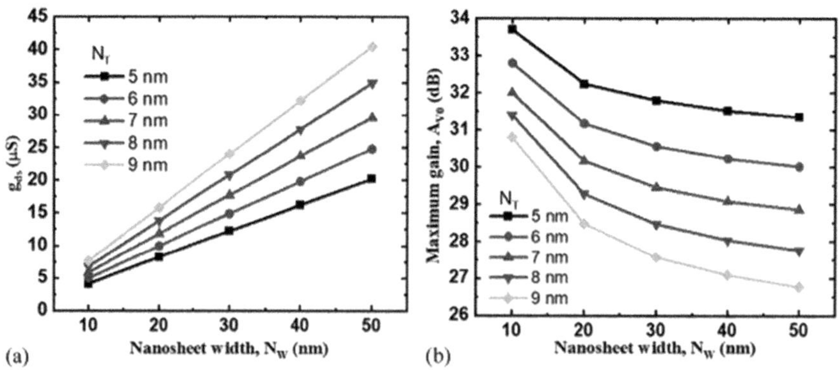

Figure 1.13 Impact of N_W and N_T on (a) g_{ds} and (b) gain of NSFETs for different N_T.

(Aruna Kumari et al. [49].)

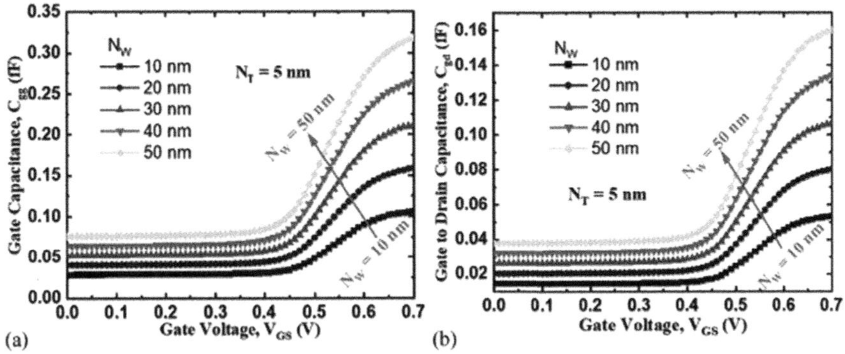

Figure 1.14 Impact of N_W and N_T on (a) C_{gg} and (b) C_{gd} of NSFETs for different N_T.

(Aruna Kumari et al. [49].)

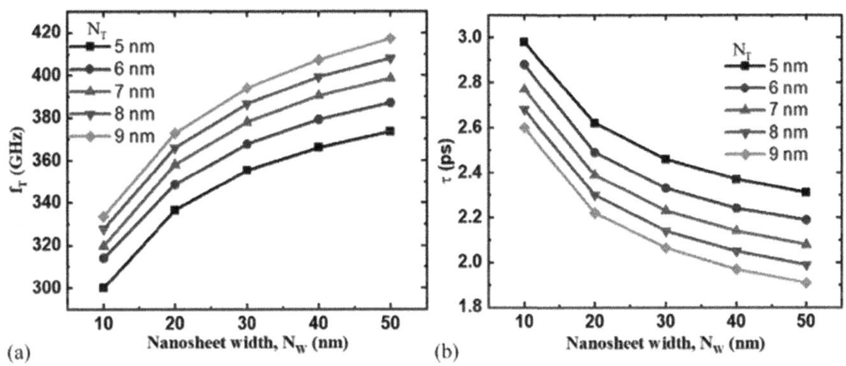

Figure 1.15 Impact of N_W and N_T on (a) f_T and (b) intrinsic time delay (τ) of NSFETs for different N_T.

(Aruna Kumari et al. [49].)

In 2023, Sresta Valasa et al. [50] developed a dual metal (DM) gate junctionless (JL) GAA-NSFET (Figure 1.16), which performs better than single-metal-gate-based GAA-NSFETs. The DM-JL-GAA-NSFETs exhibit higher electric field in the channel than single-metal-gate-based GAA-NSFETs, which in turn boost the electron velocity of carriers in the nanosheet channel. This also causes the enhancement in drive current. When the work function difference (Δw) between the two metal gates (M1 and M2) increases, the g_m declines and g_{ds} increases. The C_{gg} and C_{gd} of DM-JL-GAA-NSFET goes up with rise in Δw. Therefore, f_T, fmax, and gain of DM-JL-GAA-NSFETs declines with an increase in Δw [50]. The interface trap concentration (N_{it}) also affects the DC and RF traits of GAA-NSFETs [51–65]. I_{ON} and I_{ON}/I_{OFF} declines with rise in N_{it} of GAA-NSFETs. The rise in N_{it} also degrades the I_{OFF} and SS of GAA-NSFETs.

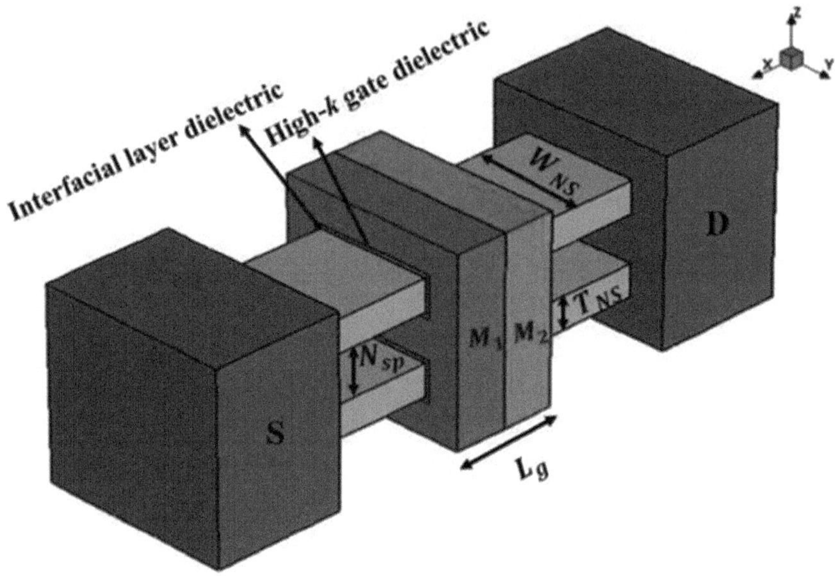

Figure 1.16 Dual metal gate junctionless GAA-NSFET.

(Reprinted from Valasa et al. [50] with permission from Elsevier.)

1.3 FORKSHEET FETs

Forksheet FETs have the ability to scale beyond the 3-nm technological node. An innovative Forksheet design can significantly reduce logic cell height at sub-3-nm nodes by providing a dielectric wall between the pMOS and nMOS regions. The device incorporates vertically stacked lateral pFET and nFET nanosheets on opposing sides of a dielectric wall, controlled by a forked gate structure. NSFET and Forksheet FET (Figure 1.17)

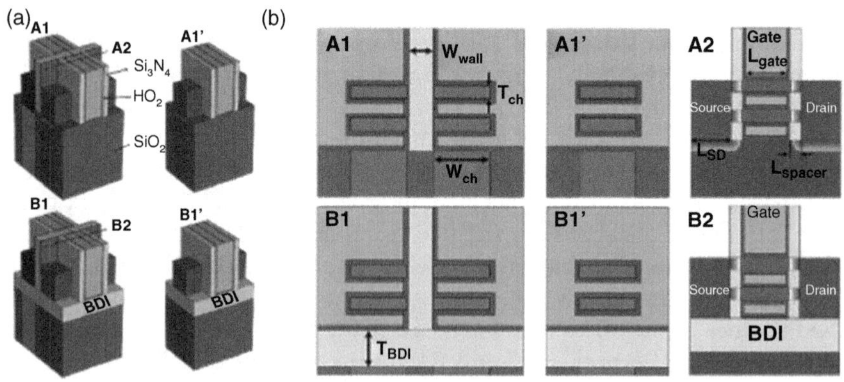

Figure 1.17 (a) Forksheet FET and (b) NSFET cross-sections.

(Choi and Jeon [66].)

vary structurally in that the former has the gate completely covering the nanosheet channel on both sides, while the latter does not have the gate enclosing the channel in the direction of the dielectric wall. The Forksheet FET has comparatively lower gate controllability as a result. This declined gate controllability makes it difficult to regulate the generated electron–hole pair (EHP), which causes the Forksheet FET's drain current to be comparatively greater (Figure 1.18) [66–75]. The performance of Forksheet FETs at lower technology nodes can be further improved by adopting JL and negative-capacitance designs.

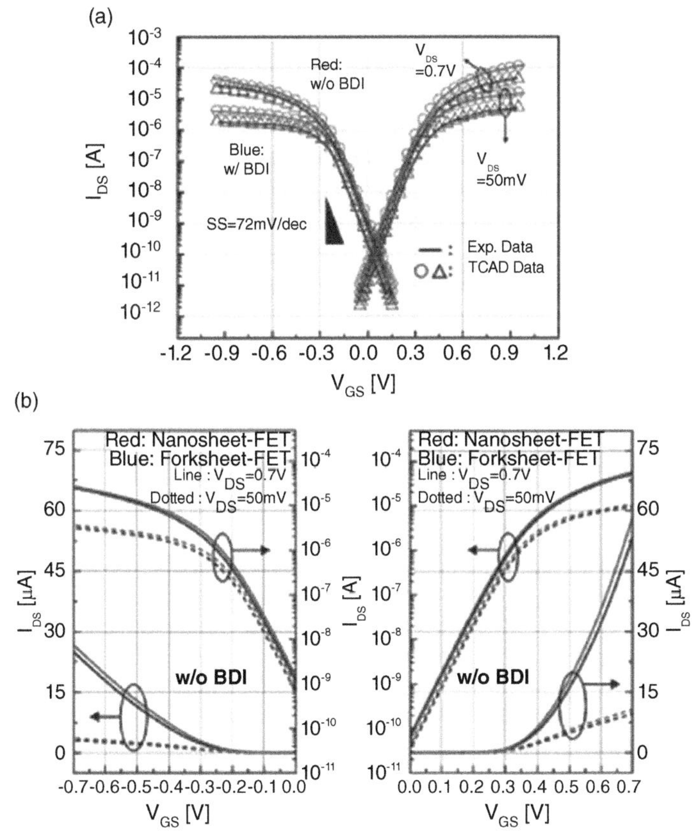

Figure 1.18 (a–c) Performance comparison of Forksheet FET and NSFET. **(Choi and Jeon [66].)**

(Continued)

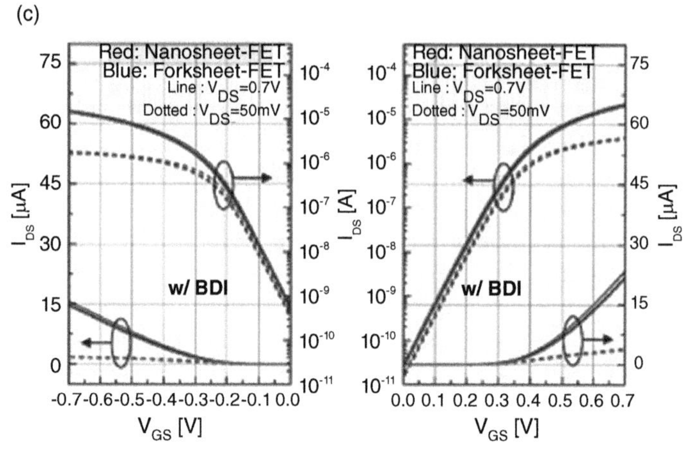

Figure 1.18 (Continued)

1.4 SUMMARY

The Forksheet FET has recently emerged as an attractive choice for next-generation applications due to its seamless integration of pFET and nFET with a minimum n–p gap. However, the presence of interface traps is one of the major issues that degrade the reliability of Forksheet FETs. In the coming years, JL and negative-capacitance designs can be adopted in these emerging transistors to further continue the scaling beyond the 2-nm node. This book has covered the electrical traits of emerging transistors such as NSFETs, CombFETs, TreeFETs, and Forksheet FETs.

REFERENCES

1. M. G. K. Alabdullah et al., "Scaling Challenges of Nanosheet Field-Effect Transistors Into Sub-2 nm Nodes," in *IEEE Journal of the Electron Devices Society*, vol. 12, pp. 479–485, 2024.
2. R. Kaur and N. R. Mohapatra, "Exploring Sheet Thickness Scaling and Substrate Orientation for Maximizing Nanosheet pFET Performance," in *IEEE Transactions on Electron Devices*, vol. 71, no. 9, pp. 5754–5760, 2024.
3. Y. S. Song, H. Kim and J. Hyun Kim, "Improvement of Thermal Characteristics and On-Current in Vertically Stacked Nanosheet FET by Parasitic Channel Height Engineering," in *IEEE Access*, vol. 12, pp. 105878–105886, 2024.
4. V. B. Sreenivasulu et al., "Analysis of Novel Core-Shell Junctionless Nanosheet FET for CMOS Logic Applications," in *IEEE Access*, vol. 12, pp. 144479–144488, 2024.
5. S. Zhang et al., "An NEGF-Based Comprehensive Analysis of Parasitic Effects in Stacked Nanosheet FETs at 1.5-nm Technology Node," in *IEEE Transactions on Electron Devices*, vol. 71, no. 10, pp. 5815–5823, 2024.

6. M. Bang et al., "Mitigation of Single Event Upset Effects in Nanosheet FET 6T SRAM Cell," in *IEEE Access*, vol. 12, pp. 130347–130355, 2024.
7. C. Wu et al., "Bottom p-i-n Isolation Enabling GAA Nanosheet Transistor for Low-Power Applications," in *IEEE Transactions on Electron Devices*, vol. 71, no. 11, pp. 6493–6498, 2024.
8. A. Alam et al., "Lattice Matched Germanium-on-AlAs Nanosheet-FETs for More-Moore CMOS," in *IEEE Transactions on Electron Devices*, vol. 71, no. 12, pp. 7874–7881, 2024.
9. S. Lee, J. Jeong and R.-H. Baek, "Channel Trimming Process to Improve Electro-Thermal Characteristics for Sub-3-nm Node Si Nanosheet FETs," in *IEEE Transactions on Electron Devices*, vol. 71, no. 11, pp. 7184–7191, 2024.
10. S. Sharma, S. Sahay and R. Dey, "Source-Drain Series Resistance Model for N-Stack Nanosheet FETs Using Transmission Line Matrix Method," in *IEEE Transactions on Electron Devices*, vol. 71, no. 11, pp. 6520–6526, 2024.
11. Shubham and R. K. Pandey, "Role of Oxygen Vacancy in the Performance Variability and Lattice Temperature of the Stacked Nanosheet FET," in *IEEE Access*, vol. 12, pp. 162565–162580, 2024.
12. S. Lee et al., "Multi-Task Learning for Real-Time BSIM-CMG Parameter Extraction of NSFETs With Multiple Structural Variations," in *IEEE Access*, vol. 12, pp. 184619–184628, 2024.
13. J. Jeong, S. Lee and R.-H. Baek, "Optimization of Asymmetries in Source/Drain Configurations and Tapered Channels for Vertical-Transport Silicon Nanosheet FETs," in *IEEE Transactions on Electron Devices*, vol. 72, no. 1, pp. 75–82, 2025.
14. S. Srivastava et al., "Influences of Source/Drain Extension Region on Thermal Behavior of Stacked Nanosheet FET," in *IEEE Transactions on Electron Devices*, vol. 71, no. 3, pp. 2171–2176, 2024.
15. W.-H. Hsieh et al., "Interfacial-Layer-Free Ge0.95Si0.05 Nanosheet FeFETs," in *IEEE Transactions on Electron Devices*, vol. 71, no. 3, pp. 1758–1763, 2024.
16. X. Zhang et al., "Hybrid Integration of Gate-All-Around Stacked Si Nanosheet FET and Si/SiGe Super-Lattice FinFET to Optimize 6T-SRAM for N3 Node and Beyond," in *IEEE Transactions on Electron Devices*, vol. 71, no. 3, pp. 1776–1783, 2024.
17. Y. Zhao et al., "A Comprehensive Comparison of Different Wafer/Channel Orientations for Ultrascaled Nanosheet FETs," in *IEEE Transactions on Electron Devices*, vol. 71, no. 3, pp. 1784–1791, 2024.
18. A. K. Panigrahy et al., "Analysis of GAA Junction Less NS FET Towards Analog and RF Applications at 30 nm Regime," in *IEEE Open Journal of Nanotechnology*, vol. 5, pp. 1–8, 2024.
19. D. Wang et al., "Design Technology Co-Optimization for Gate-All-Around Nanosheet Transistors Considering Source/Drain Confinement and Post-Gate Single Diffusion Break," in *IEEE Transactions on Electron Devices*, vol. 71, no. 7, pp. 4021–4027, 2024.
20. Y. Lyu, W. Chen and L. Cai, "Electrothermal Small-Signal Model of Nanosheet FETs With Zero-Temperature-Coefficient Based Parameters Extraction Method," in *IEEE Transactions on Electron Devices*, vol. 71, no. 7, pp. 4153–4159, 2024.

21. A. K. Panigrahy et al., "Spacer Dielectric Analysis of Multi-Channel Nanosheet FET for Nanoscale Applications," in *IEEE Access*, vol. 12, pp. 73160–73168, 2024.
22. M. Balasubbareddy and K. Sivasankaran, "Impact of Device-to-Device Thermal Interference Due to Self-Heating on the Performance of Stacked Nanosheet FETs," in *IEEE Access*, vol. 12, pp. 26401–26409, 2024.
23. D. Yakimets et al., "Inflection Points in GAA NS-FET to C-FET Scaling Considering Impact of DTCO Boosters," in *IEEE Transactions on Electron Devices*, vol. 71, no. 4, pp. 2309–2314, 2024.
24. W. Kwon, C. Yoo and J. Jeon, "Electrothermal Modeling of Multi-Nanosheet FETs With Various Layouts," in *IEEE Transactions on Electron Devices*, vol. 71, no. 4, pp. 2592–2597, 2024.
25. J. Hattori, K. Fukuda, T. Ikegami and Y. Hayashi, "Effects of Electron Quantum Confinement on Velocity Overshoot in Si Nanosheet Gate-All-Around FETs," in *IEEE Transactions on Electron Devices*, vol. 71, no. 6, pp. 3498–3503, 2024.
26. U. Gowthami et al., "Performance Improvement of Spacer-Engineered N-Type Tree Shaped NSFET Toward Advanced Technology Nodes," in *IEEE Access*, vol. 12, pp. 59716–59725, 2024.
27. A. Rawat and B. Rawat, "Multichannel Two-Dimensional MoS_2 Nanosheet MOSFET for Future Technology Node," in *IEEE Transactions on Electron Devices*, vol. 71, no. 6, pp. 3945–3951, 2024.
28. A. Singh, O. Maheshwari and N. R. Mohapatra, "Parasitic Capacitance in Nanosheet FETs: Extraction of Different Components and Their Analytical Modeling," in *IEEE Transactions on Electron Devices*, vol. 71, no. 5, pp. 2894–2900, 2024.
29. J. You, H. Kim and D. Kwon, "Analysis of Nanosheet Field-Effect Transistor With Local Bottom Isolation," in *IEEE Transactions on Electron Devices*, vol. 71, no. 5, pp. 2844–2848, 2024.
30. W.-H. Hsieh et al., "Monolithic 3-D Self-Aligned Heterogeneous Nanosheet Channel Complementary FETs With Matched VT by Band Alignments of Individual Channels," in *IEEE Transactions on Electron Devices*, vol. 71, no. 5, pp. 3383–3389, 2024.
31. K. Hikake et al., "A Nanosheet Oxide Semiconductor FET Using ALD InGaOx Channel for 3-D Integrated Devices," in *IEEE Transactions on Electron Devices*, vol. 71, no. 4, pp. 2373–2379, 2024.
32. R. Kaur and N. R. Mohapatra, "Comprehensive Analysis of Sheet Thickness Scaling on the Performance of Nanosheet nFETs," in *IEEE Transactions on Electron Devices*, vol. 71, no. 5, pp. 2856–2862, 2024.
33. S. R. Kola and Y. Li, "Electrical Characteristic and Power Fluctuations of GAA Si NS CFETs by Simultaneously Considering Six Process Variation Factors," in *IEEE Open Journal of Nanotechnology*, vol. 4, pp. 229–238, 2023.
34. Y.-S. Yang, Y. Li and S. R. R. Kola, "A Physical-Based Artificial Neural Networks Compact Modeling Framework for Emerging FETs," in *IEEE Transactions on Electron Devices*, vol. 71, no. 1, pp. 223–230, 2024.
35. S. Sharma, S. Sahay and R. Dey, "Parasitic Capacitance Model for Stacked Gate-All-Around Nanosheet FETs," in *IEEE Transactions on Electron Devices*, vol. 71, no. 1, pp. 37–45, 2024.

36. W.-L. Sung, Y. Li and M.-H. Chuang, "Nanosized-Metal-Grain-Pattern-Dependent Threshold-Voltage Models for the Vertically Stacked Multichannel Gate-All-Around Si Nanosheet MOSFETs and Their Applications in Circuit Simulation," in *IEEE Transactions on Electron Devices*, vol. 71, no. 1, pp. 350–358, 2024.
37. S. K. Kim et al., "Heterogeneous 3-D Sequential CFETs With Ge (110) Nanosheet p-FETs on Si (100) Bulk n-FETs," in *IEEE Transactions on Electron Devices*, vol. 71, no. 1, pp. 393–399, 2024.
38. N. A. Kumari et al., "Performance Comparison of Nanosheet FET, CombFET, and TreeFET: Device and Circuit Perspective," in *IEEE Access*, vol. 12, pp. 9563–9571, 2024.
39. M. M. Jatin and M. Shrivastava, "On the Multifinger Turn-On Instability in Drain Extended Vertically Stacked Nanosheet FETs Under ESD Stress Conditions," in *IEEE Transactions on Electron Devices*, vol. 71, no. 2, pp. 983–990, 2024.
40. D. Parvathi and P. Prithvi, "Innovative Spacer Material Integration in Tree-FETs for Enhanced Performance Across Variable Channel Lengths," in *Micro and Nanostructures*," vol. 195, p. 207974, 2024.
41. Y.-W. Zhang, K.-H. Gao and Z.-Q. Li, "Oxygen Sensitivity of Ga_2O_3 Nanosheet Field-Effect Transistor," *Applied Surface Science*, vol. 688, p. 162391, 2025.
42. W. Chen et al., "Statistical Simulation of Self-Heating Induced Variability and Reliability with Application to Nanosheet-FETs based SRAM," in *Microelectronics Reliability*, vol. 98, pp. 63–68, 2019.
43. A. Veloso et al., "Ryckaert, Nanowire & Nanosheet FETs for Ultra-scaled, High-density Logic and Memory Applications," in *Solid-State Electronics*, vol. 168, p. 107736, 2020.
44. G. Darbandy et al., "Performance Analysis of Parallel Array of Nanowires and a Nanosheet in SG, DG and GAA FETs," in *Solid-State Electronics*, vol. 162, p. 107641, 2019.
45. T.-K. Chiang, Nanosheet FET: A New Subthreshold Current Model Caused by Interface-Trapped-Charge and Its Application for Evaluation of Subthreshold Logic Gate," in *Microelectronics Journal*, vol. 104, p. 104893, 2020.
46. Y. Choi et al., "Simulation of the Effect of Parasitic Channel Height on Characteristics of Stacked Gate-all-around Nanosheet FET," in *Solid-State Electronics*, vol. 164, p. 107686, 2020.
47. V. C. P. Silva et al., Evaluation of n-Type Gate-all-around Vertically-stacked Nanosheet FETs from 473 K Down to 173 K for Analog Applications," in *Solid-State Electronics*, vol. 208, p. 108729, 2023.
48. S. Srivastava et al., "Understanding the Impact of Extension Region on Stacked Nanosheet FET: Analog Design Perspective," in *Solid-State Electronics*, vol. 208, p. 108758, 2023.
49. N. Aruna Kumari and P. Prithvi, "Device and Circuit-level Performance Comparison of GAA Nanosheet FET with Varied Geometrical Parameters," in *Microelectronics Journal*, vol. 125, p. 105432, 2022.
50. S. Valasa, S. Tayal and L. R. Thoutam, "Performance Analysis of Metal Gate Engineered Junctionless Nanosheet FET with a ft/fmax of 224/342 GHz for Beyond 5g (b5g) Applications," in *Micro and Nanostructures*, vol. 179, p. 207582, 2023.

51. S. Rathore, R. K. Jaisawal, P. N. Kondekar and N. Bagga, "Trap and Self-heating Effect Based Reliability Analysis to Reveal Early Aging Effect in Nanosheet FET," in *Solid-State Electronics*, vol. 200, p. 108546, 2023.
52. V. Bheemudu, D. Vaithiyanathan and B. Kaur, "Design Insights into a Junctionless Nanosheet FET (JL-NSFET) for Switching and Analog/RF Applications: Device to Circuit Level Assessment," in *Microelectronics Journal*, vol. 149, p. 106220, 2024.
53. S. Valasa et al., "A Critical Review on Performance, Reliability, and Fabrication Challenges in Nanosheet FET for Future Analog/Digital IC Applications," in *Micro and Nanostructures*, vol. 170, p. 207374, 2022.
54. F. Zhao et al., "Hybrid Integrated Si Nanosheet GAA-FET and Stacked SiGe/Si FinFET Using Selective Channel Release Strategy," in *Microelectronic Engineering*, vol. 275, p. 111993, 2023.
55. S. Valasa, V. R. Kotha and N. Vadthiya, "Interface Traps in the Sub-3 nm Technology Node: A Comprehensive Analysis and Benchmarking of Negative Capacitance FinFET and Nanosheet FETs - A Reliability Perspective from Device to Circuit Level," in *Microelectronics Reliability*, vol. 160, p. 115479, 2024.
56. S. Rathore, R. K. Jaisawal, P. N. Kondekar and N. Bagga, "Investigation of Analog/RF and Linearity Performance with Self-heating Effect in Nanosheet FET," in *Microelectronics Journal*, vol. 139, p. 105904, 2023.
57. J. G. Fernandez et al., "An Accurate Machine Learning Model to Study the Impact of Realistic Metal Grain Granularity on Nanosheet FETs," in *Solid-State Electronics*, vol. 207, p. 108710, 2023.
58. M. Balasubbareddy, K. Sivasankaran, A. E. Atamuratov and M. M. Khalilloev, "Optimization of Vertically Stacked Nanosheet FET Immune to Self-heating," in *Micro and Nanostructures*, vol. 182, p. 207633, 2023.
59. J. Park et al., "In Man Kang, Statistical Analysis of Vertically Stacked Nanosheet Complementary FET Based on Polycrystalline Silicon with Multiple Grain Boundaries," in *Results in Physics*, vol. 63, p. 107873, 2024.
60. M. Balasubbareddy and K. Sivasankaran, "Optimization of Junctionless Stacked Nanosheet FET – RF Stability Perspective," in *Microelectronics Journal*, vol. 145, p. 106123, 2024.
61. R. Kaur and N. R. Mohapatra, "Exploring the Impact of Sheet Thickness Scaling on Nanosheet FET Gate Electrostatics Using k.p Based Simulations," in *Microelectronics Journal*, vol. 156, p. 106480, 2025.
62. M. P. Gogoi, R. Saha, and S. Baishya, "Investigation on Electrical Parameters in Nanowire FET and Nanosheet FET Including Trap Charges and Its Circuit Applications," in *Micro and Nanostructures*, vol. 198, p. 208068, 2025.
63. I. Myeong, Q. Nguyen-Gia and I. Song, "Self-Heating-Based Channel Thermal Noise of Advanced Sub-5-nm-Node Nanosheet FET," in *Results in Physics*, vol. 67, p. 108059, 2024.
64. S. M. Majd et al., "Design of a Novel Aptamer/Molecularly Imprinted Polymer Hybrid Modified Ag–Au@Insulin Nanoclusters/Au-gate-Based MoS2 Nanosheet Field-Effect Transistor for Attomolar Detection of BRCA1 Gene," in *Talanta*, vol. 257, p. 124394, 2023.
65. J. Ajayan et al., "Nanosheet Field Effect Transistors – A Next Generation Device to Keep Moore's Law Alive: An Intensive Study," in *Microelectronics Journal*, vol. 114, p. 105141, 2021.

66. G. Choi and J. Jeon, "Radiation Effects on Multi-channel Forksheet-FET and Nanosheet-FET Considering the Bottom Dielectric Isolation Scheme," in *Nuclear Engineering and Technology*, vol. 56, no. 11, pp. 4679–4687, 2024.
67. G. Choi, S. Kim and J. Jeon, "Investigation of Radiation Effects on Multichannel Nanosheet-FETs, Forksheet-FETs, and Complementary-FETs," in *IEEE Transactions on Electron Devices*, vol. 71, no. 12, pp. 7263–7269, 2024.
68. Y. Arafat Pritom, H. Biswas and M. Hossain, "Analysis of Negative Capacitance Effect in Sub-3-nm Forksheet FETs," in *IEEE Transactions on Electron Devices*, vol. 72, no. 1, pp. 83–89, 2025.
69. G. Ryali et al., "Interface Trap Characterization in Junctionless Forksheet FET at Sub-3 nm Technology Node: A Reliability Assessment on Digital, Analog/RF, and Circuit Applications," in *IEEE Transactions on Device and Materials Reliability*, doi:10.1109/TDMR.2024.3516084
70. D. Gurre et al., "Spacer Design Strategies at sub-5 nm technology node for Junctionless Forksheet FET: Bridging Device Optimization and Circuit Efficacy – A Dielectric Perspective," in *IEEE Transactions on Dielectrics and Electrical Insulation*, doi:10.1109/TDEI.2024.3497877
71. J. Lee, J.-S. Yoon, J. Lim and R.-H. Baek, "Electrical Coupling Effect of Forksheet FET for Power, Performance, and Area Analysis," in *IEEE Transactions on Electron Devices*, vol. 69, no. 12, pp. 7096–7101, 2022.
72. J. Lim et al., "Investigation of Electrothermal Characteristics in Silicon Forksheet FETs for Sub-3-nm Node," in *IEEE Transactions on Electron Devices*, vol. 70, no. 12, pp. 6132–6137, 2023.
73. M. Vandemaele et al., "Trapping of Hot Carriers in the Forksheet FET Wall: A TCAD Study," in *IEEE Electron Device Letters*, vol. 44, no. 2, pp. 197–200, 2023.
74. S. Sharma, S. Sahay and R. Dey, "Parasitic Gate Capacitance Model for N-Stack Forksheet FETs," in *IEEE Transactions on Electron Devices*, vol. 71, no. 3, pp. 1728–1736, 2024.
75. D. Li et al., "Controllable Carrier Transfer Modulation of Ambipolar van der Waals Semiconductors Toward Forksheet FETs," in *Applied Physics Letters*, vol. 124, no. 21, p. 213501, 2024. doi:10.1063/5.0197266

Chapter 2

Tunnel FETs

A sustainable energy-efficient innovation for modern applications

Sourav Das, Satyabrata Jit, and Kunal Singh

2.1 INTRODUCTION

In the last decade of the century, MOSFETs played a crucial role in the progress of most integrated circuits by enabling their size reduction to the nanometer scale. MOSFETs (Metal-Oxide-Semiconductor Field-Effect Transistor) are versatile devices that, depending on their configuration and design, can be used in both low-power and high-power applications. The current in a conventional MOSFET is primarily driven by the thermionic emission of carriers across a potential barrier. Therefore, one of the critical challenges that conventional MOSFETs face at advanced technological nodes is keeping power consumption within an acceptable range. T. Baba developed the tunnel transistor in 1992 as one of the most promising alternatives to traditional MOSFETs, based on several performance metrics, as shown below:

- Sub 60 mV/decade Subthreshold Swing (SS).
- Low short channel effects.
- Reduced leakage currents.
- The fabrication method is similar to that of MOSFETs.
- Increased I_{OFF}/I_{ON} current ratio [1].

Considering all the aforementioned factors, a prospective tunnel field-effect transistors (TFETs) substitute for the MOSFET, which could be used for high-speed, ultra-low power, and energy-efficient integrated circuit applications. The potential of TFETs in analog and mixed-signal applications, where minimizing leakage current and achieving high signal fidelity are essential, is also being investigated. They are also promising candidates for biosensing applications due to their sensitivity to low voltages, enabling accurate and energy-efficient detection of biological molecules. The goal of ongoing research is to maximize TFET performance, explore novel materials, and develop fabrication methods to fully harness their potential in next-generation electronics, contributing to the creation of superior and environmentally friendly electronic products. Since tunneling is directly controlled by the gate voltage, the V-TFET has a steeper subthreshold slope than the

 DOI: 10.1201/9781003635123-2

lateral TFET, among other benefits. Operating at low power and with a low OFF-current is made possible by the steep subthreshold slope [2, 3].

2.2 FUNDAMENTAL PRINCIPLES OF TFET OPERATION

We proceed on discussing the design and operation of TFETs in this chapter. We cover the history and development of a few TFET structures and identify their electrical properties. We provide an estimate of the tunneling current and the electrical and physical characteristics of TFETs. This will make it easier for readers to adjust their device configuration to optimize the electrical characteristics of TFETs. We examine a few TFET applications and proposed vertical-based TFET device structures, considering their benefits and drawbacks in analog applications.

2.2.1 TFET device framework

The basic operation of n-type and p-type TFETs is shown in Figure 2.1. An n-type TFET has a p+ doped source and an n+ doped drain, while a p-type

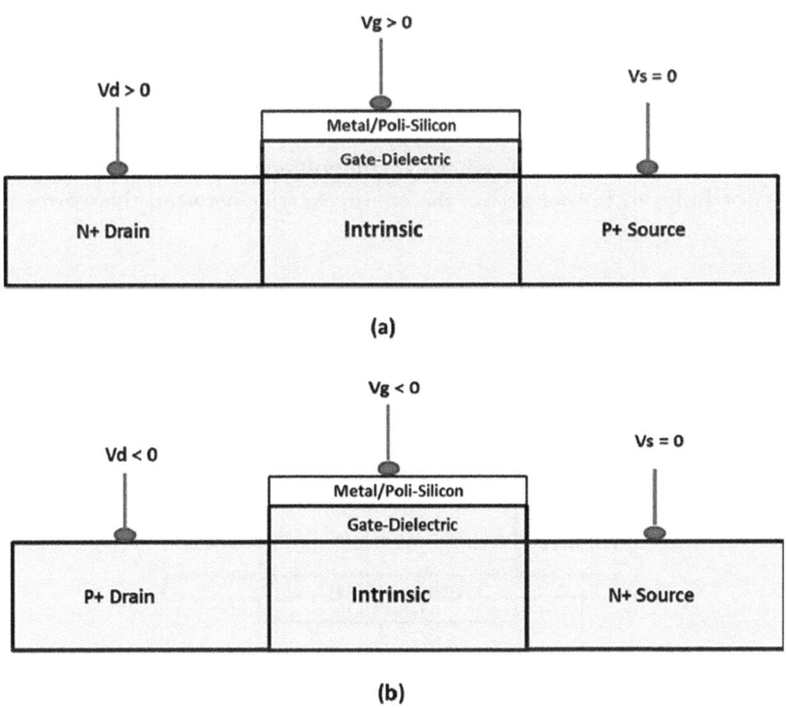

Figure 2.1 A simple configuration of a TFET: (a) n-type TFET. (b) p-type TFET. **(Saurabh and Kumar [4].)**

TFET has an n+ doped source and a p+ doped drain. The channel is formed by an intrinsic or lightly doped p-type or n-type semiconductor. Similar to a conventional MOSFET, a dielectric layer separates the channel from the gate electrode. In an n-type TFET, the source is grounded, while a positive voltage is applied to the drain and gate electrodes.

The source remains at ground and a negative voltage is applied to the gate and drain contacts in a p-type. Electrons are the major carriers in the channel of an n-type TFET, whereas holes in the channel are the dominant carriers of a p-type TFET. In an n-type TFET, electrons enter the channel through the source and flow through the drain. A p-type TFET's holes flow into the channel through the source and out of the channel through the drain respectively. Through a process known as band-to-band tunneling (BTBT), the dominant carriers enter the channel region. A very small ON-current is the most notable of the many design shortcomings in the TFET designs.

Hence, researchers have designed and developed TFETs in several structural configurations to overcome the limitations of the basic TFET design. One such advanced TFET structure is the DG-TFET, as shown in Figure 2.2, which presents a cross-sectional view of an n-type DG-TFET [4]. Since DG-TFETs can operate at lower voltages while maintaining excellent switching efficiency, they are highly beneficial for low-power and ultra-low-power applications. It should be noted that many researchers have used different variants of DG-TFET in their theoretical and experimental work. [5–8].

2.2.2 Operation

TFETs operate using BTBT, where a gate voltage is applied to induce electrons or holes to tunnel across the energy barrier between the source and

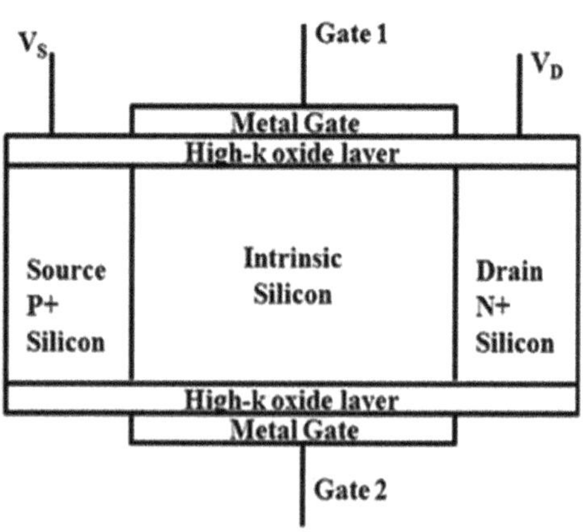

Figure 2.2 2-D sectional view of a DG-TFET.

(Saurabh and Kumar [4].)

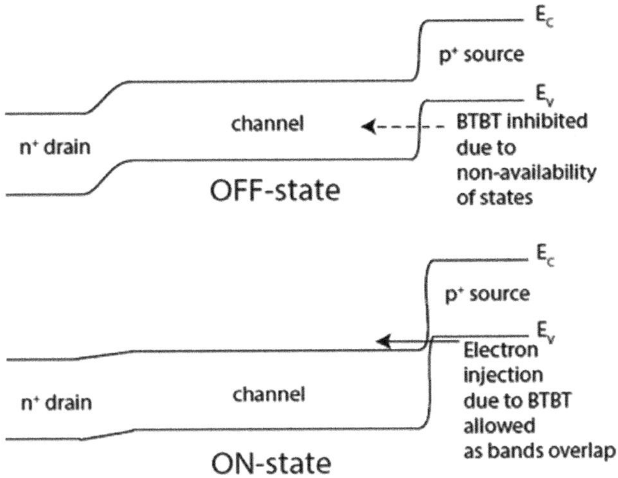

Figure 2.3 (a) OFF-state (b) ON-state of n-type TFET.

(Saurabh and Kumar [4].)

the channel. The ON and OFF band diagrams of an n-type TFET are shown in Figure 2.3. When the gate voltage is zero, the TFET is considered to be in the OFF state. Here, the conduction band of the channel is positioned above the valence band of the source. A higher gate voltage increases the carrier density and shifts the conduction band in the channel downward.

When a relatively high voltage is applied, band bending occurs at the gate, aligning the valence band of the source with the conduction band of the channel, as shown in Figure 2.3. The positive drain bias triggers electrons to tunnel into the channel and reach the drain terminal. This is the basic fundamental operation of an n-type TFET.

2.2.3 Transfer characteristics of TFET

Three regions can be used to typically classify the TFET transfer characteristics shown in Figure 2.4.

- **OFF-state** ($0 < V_{GS} < V_{OFF}$): In this region, BTBT decreases, and the TFET's current becomes extremely low. The drain current begins to flow at the gate voltage, also known as V_{OFF}.
- **Subthreshold area** ($V_{OFF} < V_{GS} < V_T$): The drain current increases significantly in this area. Here, V_T is the threshold voltage of the DG-TFET.
- **Super-threshold region** ($V_{GS} > V_T$): Here, as the gate voltage decreases, the drain current increases correspondingly.

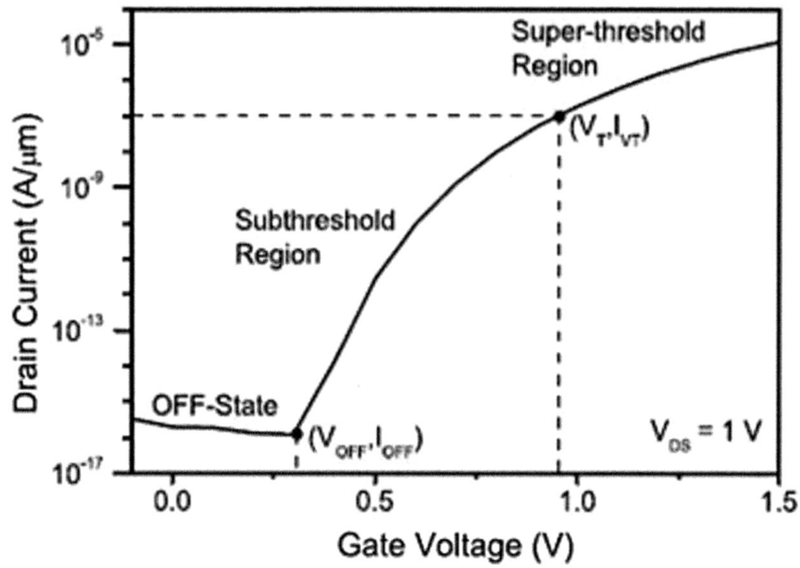

Figure 2.4 Simulated transfer characteristics of the TFET.

(Saurabh and Kumar [4].)

2.2.4 Point versus line tunneling

In a TFET, tunneling can be categorized into two types: point tunneling and line tunneling. These depend on how the applied gate potential shapes the electric field in the tunneling direction. Figure 2.5 illustrates the typical flow of point and line tunneling. The concept of point tunneling arises when the channel is fully covered by the gate. The concept of "point tunneling" refers to a curved tunneling pathway that forms at the interface between the gate metal and the gate dielectric above the tunnel junction [9]. As illustrated in Figure 2.5(b), line tunneling may result from substantial overlap of the gate over the source region [9].

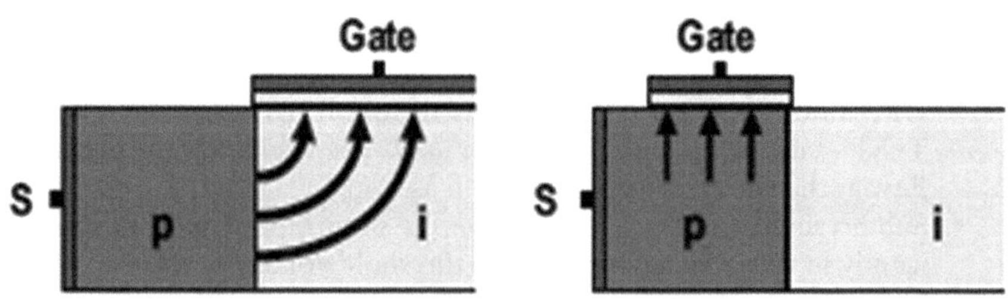

Figure 2.5 (a) Point and (b) line tunneling TFET arrangement.

(Verreck et al. [9].)

In point tunneling, the tunnel channels are ideally perpendicular to the gate dielectric when no parasitic paths are present [10]. To improve the current effectiveness of electronic devices, researchers are constantly concentrating on TFET technology. To increase the I_{ON}/I_{OFF} ratio, several device topologies have been investigated, including vertical, source-pocket-designed, and L-shaped/U-shaped designs [11]. Additionally, once the device turns on, the tunneling area and electric field at the top of the tunneling junction are increased by the lateral gate–pocket overlap [9].

2.3 DIFFERENT TYPES OF TFETs

TFETs come in various forms, each with unique characteristics that make them advantageous depending on their intended use, designs, materials, and manufacturability. To enhance performance (low power consumption and steep SS) and address various challenges, several TFET architectures have been developed. This chapter discusses the benefits and drawbacks of various types of TFETs with different configurations:

1. Homojunction TFETs [8, 12]
   - **Description**: The source and drain of these TFETs are formed from the same semiconductor material.
   - **Materials used**: Silicon and germanium.
   - **Features**: Since only a single material is used, the fabrication process is simpler.
   - **Challenges**: Owing to the intrinsic properties of the material, performance gains are limited.
2. Heterojunction TFETs [13–15]
   - **Description**: These TFETs form a heterojunction by using different semiconductor materials at the source and drain locations.
   - **Materials used**: Silicon–Germanium (SiGe) and III–V compounds (such as InGaAs).
   - **Advantages**: Increased ON-state current due to improved tunneling efficiency and better band alignment.
   - **Features**: More complex fabrication and potential issues with material compatibility.
3. III–V TFETs [13, 16]
   - **Description**: These TFETs utilize III–V compound semiconductors, known for their favorable bandgap characteristics and high electron mobility.
   - **Materials used**: InGaAs, InAs, and GaSb.
   - **Advantages**: Improved performance due to increased ON-state current and enhanced tunneling efficiency.
   - **Features**: Higher production costs and challenges in integration with existing silicon technology.

4. Vertical TFETs [14, 17]
   - **Description**: These TFETs feature a vertical structure, with the source and drain stacked on top of each other.
   - **Advantages**: Enhanced electrostatic control and potentially higher driving current.
   - **Features**: A more complex manufacturing process and challenges in scaling down the size.
5. Graphene nanoribbon TFETs (GNR-TFETs) [18, 19]
   - **Description**: These TFETs utilize thin graphene strips as the channel material.
   - **Advantages**: Exceptional carrier mobility and enhanced electrostatic modulation.
   - **Challenges**: Bandgap engineering in graphene is challenging, and fabrication techniques are still evolving.

2.4 APPLICATIONS OF TFETs

According to several theoretical research studies, TFET-based circuits may perform better at low supply voltages than conventional CMOS (Complementary Metal-Oxide-Semiconductor) circuits. A number of applications, such as memory, sensors, analog circuits, conventional logic gates, and more, are being investigated for the use of TFETs. These results have prompted researchers to investigate circuit designs that utilize the special properties of TFETs to minimize power and energy consumption.

2.4.1 Electrical characteristics

A number of electrical characteristics give TFETs an advantage over conventional MOSFETs. The following features are crucial for their use in circuit design.

- **Unidirectional conduction**: Compared with a conventional MOSFET, a TFET allows current to flow in only one direction due to its asymmetric p–i–n configuration. For a given gate voltage, a significant current flow in an n-type TFET only when the V_{DS} is positive or the p–i–n structure is reverse-biased [20–23]. The unidirectional properties of the TFET can be leveraged to enhance the efficiency of RF (Radio Frequency) rectifier circuits by blocking reverse leakage paths.
- **Subthreshold swing**: A crucial factor for evaluating the performance of transistors, particularly TFETs, is SS. It evaluates the transistor's ability to transition between the OFF and ON states, particularly in the subthreshold region, where the gate voltage and current change very quickly. Compared to a conventional MOSFET, TFETs can exhibit SS of less than 60 mV/decade at room temperature. Voltage scaling is possible without affecting TFET operation due to their minimal SS.

- **Gate-to-drain Miller capacitance:** At all bias configurations, the gate-to-drain capacitance (C_{GD}) in a TFET exceeds the total gate capacitance (C_{GG}). The channel-to-drain resistance in an OFF-state TFET is substantially lower than the channel-to-source resistance, and C_{GD} is significantly greater than C_{GS}. When the TFET is in the ON-state, a nominal potential drop exists between the channel and the drain, leading to a high C_{GD} [24]. This differs from a standard MOSFET, where the gate-to-drain capacitance (C_{GD}) contributes almost equally in the linear region, while the gate-to-source capacitance (C_{GS}) dominates in the saturation region [24]. In circuits, a TFET's large C_{GD} creates strong coupling between its input pin (the gate) and output pin (the drain). This coupling, influenced by the Miller effect, significantly impacts the circuit's behavior.
- **Identical properties of N-type TFETs and P-type TFETs:** The integration of TFETs in CMOS circuit design requires that the characteristics of an N-TFET and a P-TFET be identical. By properly adjusting the width (W) to length (L) ratios of the transistors, the drive characteristics of NMOS and PMOS transistors can be optimized in a conventional MOSFET. However, there are also differences between an N-TFET and a P-TFET in terms of electrical characteristics, such as SS, and mismatch in driving strengths [25].

2.4.2 Memories and digital circuits

The key characteristics of matrices adapted to evaluate standard digital circuits include delay properties, specifically noise adaptability, circuit area, rise/fall times, power dissipation, and propagation delay [26]. These characteristics are analyzed in TFET-based circuits and compared with conventional CMOS circuits to determine the potential advantages of TFETs over MOSFETs. Since inverters are the most fundamental type of digital circuit, researchers are placing greater emphasis on the properties of inverters implemented with TFETs. TFETs are being deal with for possible applications in digital circuits beyond inverters [27, 28]. For SRAM (Static Random-Access Memory) development, TFETs have been extensively studied due to their exceptional SS and extremely low leakage current [29, 30].

2.4.3 Analog circuits

Considering their special qualities, especially their capacity to operate efficiently at extremely low voltages and achieve low power consumption, TFETs are attracting interest in analog circuit design. A significant number of research studies have focused on TFETs for digital applications and as an alternative to traditional MOSFETs in logic circuits. However, researchers have recently explored the advantages of using TFETs in various analog circuit applications [31–35]. In addition, these studies have shown that TFETs are well-suited for low-power analog circuits due to their low SS and excellent output characteristics. The implementation of Si TFETs has recently been compared with a CMOS design for the same circuit [34]. Amplifier circuits using conventional MOSFETs can be similar to those incorporating TFETs. By using TFETs to reduce the positive drain bias, a Schmitt-trigger inverter can be implemented [36].

2.4.4 Future-oriented TFETs circuit observation

Most studies have shown that TFETs provide significant advantages over conventional MOSFETs in terms of power conservation at low to medium frequencies. When TFETs operate below the quantum capacitance limit, their switching rates decrease. The switching rates of TFETs decrease when they operate below the quantum capacitance limit [37]. Additionally, TFETs show promise for applications in SRAMs and ultra-low-power analog circuits. However, several studies have shown that TFETs can achieve higher energy efficiency than CMOS only at low supply voltages.

2.4.5 Over view of literature

Shilpi Gupta [38] used the TCAD (Technology Computer-Aided Design) simulation tool to study a vertical TFET device with an n + SiGe-doped layer, employing a gate-stacking method. It should be noted that VTFETs are denser, easier to fabricate, and less prone to trap-related issues than conventional transistors [39–41]. A study by Das et al. examined low-frequency flicker noise on dual dielectric pockets with three distinct gate-source underlap lengths and thicknesses. The proposed device exhibits a high current ratio and a noticeably steep slope [42]. For high-frequency applications, Seema [43] recommended utilizing a hetero-dielectric buried oxide (HDB) and a double-gate vertical TFET linearity analysis. Tunneling width is decreased, and drain current is increased when low bandgap material is used. However, Gonzalez [44] indicates that staggered heterojunction transistors (TFETs) based on III–V materials exhibit a trade-off between performance and power consumption, making them suitable for low-voltage operation.

2.5 SCHEMATIC OF DOUBLE-GATE HETEROSTRUCTURE V-TFET

2.5.1 Sample of device framework

The two-dimensional cross-sectional design of the DG-Hetero-Pocket-V-TFET is shown in Figure 2.6.

Auxiliary gates have a higher work function, which helps regulate the OFF-current (I_{OFF}), while tunnel gates have a lower work function, which

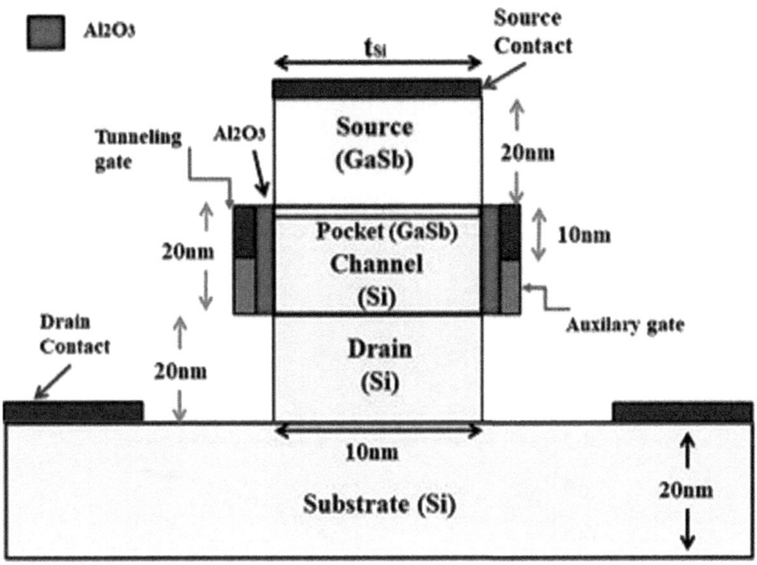

Figure 2.6 Schematic cross-sectional view of the dual-metal double-gate Hetero Pocket V-TFET.

controls the ON-current (I_{ON}). To enable energy band shifting, a BTBT nonlocal model has been utilized in TCAD simulation. The different device parameters are shown in Table 2.1.

This simulation incorporates several additional models, including Shockley-Read-Hall generation-recombination, Fermi-Dirac, field-dependent mobility, Auger recombination, and band gap-narrowing. This plots the drain current against the gate voltage, which varies from 0 to 1 V. High-k (~ 9.3) Al_2O_3 has been utilized in the gate oxide to enhance the drain current and fabrication reliability. After simulating a Hetero-Pocket-V-TFET with the 2-D ATLAS device simulator (version 5.26.1.R), we generated a sample output file.

Table 2.1 Device constraints for the DG-Hetero-Pocket-V-TFET

Parameters	*Values (Hetero-Pocket-V-TFET)*
Al_2O_3 thickness (t_{Al2O3})	2 nm
Pocket length (L_p)	3 nm
Channel doping (n-type)	1×10^{15} cm^{-3}
Drain doping (n-type)	5×10^{19} cm^{-3}
Pocket doping (n-type)	6×10^{18} cm^{-3}
Source doping (p-type)	1×10^{20} cm^{-3}
Gate work function Φ_{Tunnl}/Φ_{Aux}	4.1/4.7

2.5.2 Output file

The drain current properties of the proposed pocket hetero-V-TFET are shown in Figure 2.7(a). Drain voltage (V_{DS}) remains constant at 0.9 V, while V_{GS} fluctuates between 0 V and 1 V in the drain current vs. gate voltage characteristics. Depending on the tool used, the tunneling model may need to be recalibrated, and other tool-specific parameters may require adjustment. The resulting parameters of the proposed Hetero-Pocket-V-TFET are shown in Table 2.2.

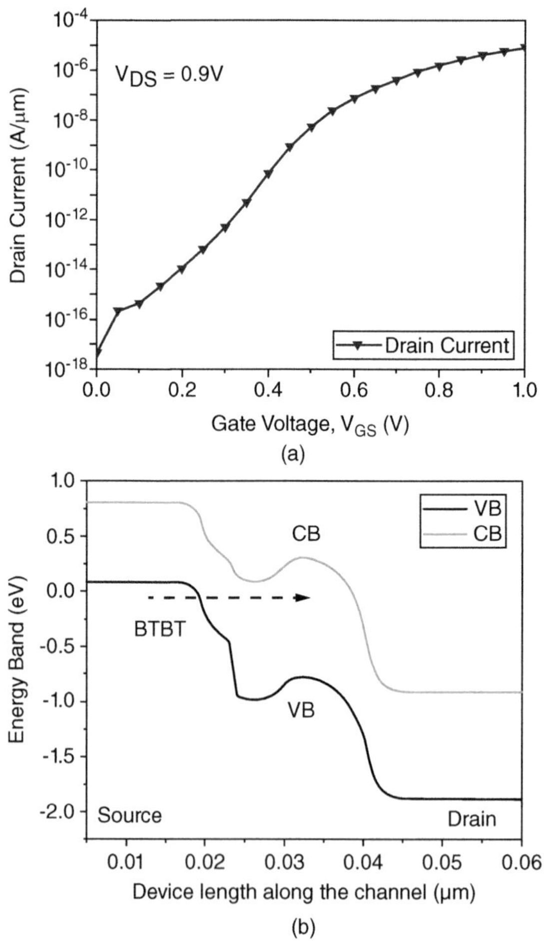

Figure 2.7 (a) I_D–V_{GS} characteristics. (b) Energy band diagram of DG-Hetero-V-TFET.

Table 2.2 Different parameters of Hetero-Pocket-V-TFET

Parameters	*Values (Hetero-Pocket-V-TFET)*
Subthreshold swing (SS)	29.75 mV/decade
Threshold voltage (V_{th})	0.3768 V
OFF-state current (I_{OFF})	4.56×10^{-18} A/μm
ON-state current (I_{ON})	8.38×10^{-6} A/μm
Ratio (I_{ON}/I_{OFF})	1.84×10^{12}

2.6 CHALLENGES IN FABRICATING TFETS FOR LOW-POWER APPLICATIONS AND CURRENT STATUS OF TFET TECHNOLOGY NODES

TFETs have demanding design requirements and unique operating principles, which make them difficult to fabricate for low-power applications. Here is an overview of the primary fabrication concerns:

- **Steep doping gradients at source–channel junction**
  - **The significant role**: BTBT, which is essential to TFET operation, is highly sensitive to the electric field at the junction.
  - **Issue**: It is challenging to achieve abrupt and high-gradient doping profiles using conventional ion implantation and thermal annealing (e.g., a significantly doped source with a lightly doped or intrinsic channel).
  - **Impact**: Gradual connections increase SS and power consumption while decreasing tunneling efficiency.
- **Material integration (e.g., III–V semiconductors)**
  - **The significant role**: Materials having low effective masses and very small band gaps, such as InAs and GaSb, are required for high tunneling currents.
  - **Issue**: These include problems such as lattice mismatch, thermal expansion mismatch, interdiffusion, and flaws that are not inherent to silicon platforms.
  - **Impact**: Lowers performance while increasing fabrication complexity and expense.
- **Precise gate alignment**
  - **The significant role**: Gate placement has an impact on TFET performance, particularly in vertical or multi-gate architectures.
  - **Issue**: Variability, higher leakage, and inadequate electrostatic control are all triggered by misalignment.
  - **Impact**: Reduces the energy efficiency and scalability of the device.
- **Scalability and variability**
  - **The significant role**: At sub-10 nm nodes, TFETs are competing with FinFETs and GAAFETs [45].

- **Issue:** TFET performance is more negatively impacted by process variability (such as gate length, doping, and material quality).
- **Impact:** Uniform low-power behavior in large-scale integration is challenging to attain.

- **Complex architectures** (e.g., **vertical** and **triple-gate**) [46]
  - **The significant role:** Control and tunneling direction are enhanced by advanced geometries.
  - **Issue:** It is difficult to fabricate 3D objects at the nanoscale with low defect density and great precision.
  - **Impact:** Lowers dependability and raises costs in commercial manufacturing.

2.6.1 Current status of TFET technology nodes

Technology node	*Current state of TFET research*	*Notes*
14–10 nm [47]	Demonstrated in labs	For proof-of-concept research, simple TFET models have been developed at these nodes.
7–5 nm [48]	Active research	Experiments and device simulations continue to be in progress. The focus is on vertical TFETs and gate-all-around (GAA).
<5 nm [49]	Exploratory research	The possibility of TFETs to function effectively at these extreme scales is being evaluated, although commercial integration has not yet been achieved.
Beyond CMOS/ IRDS 2031+ [50]	Road-mapped candidate	In long-term roadmaps such as IRDS, TFETs are taken into consideration for wearable electronics, IoT, and ultra-low-power logic.

2.7 CONCLUSION

This chapter has examined the characteristics of modern TFETs and their cost advantages over MOSFETs. Exploration continues to focus on the economic advantages of TFETs over traditional MOSFETs for analog applications, as well as their use in various types of analog circuits. We will discuss the significance of TFET's benefits and drawbacks in analog applications, along with examples of their implementation in recently proposed analog circuits. As we have seen, three main strategies for improving I_{ON} in TFETs are enhancing the gate structure, modifying the tunneling junction, and incorporating new material types. Technological advancements have made TFETs the most effective alternative for low-power industrial applications.

REFERENCES

1. Aswathy M, Biju N M and Komaragiri R. 2013. Comparison of a 30nm tunnel field effect transistor and CMOS inverter characteristics. *Proceedings – 2013 Third International Conference on Advances in Computing and Communications*, 149–52.
2. Han Z F, Ru G P and Ruan G. 2011. A simulation study of vertical tunnel field effect transistors. *Proceedings – 2011 9th IEEE International Conference on ASIC*, 665–8.
3. Das S, Kumar B, Bundela P and Singh K. 2024. Performance assessment of Si based dual metal double gate vertical TFET biosensor. *Micro and Nanostructures* **191**, 207864.
4. Saurabh S and Kumar M J. 2016. Fundamentals of tunnel field-effect transistors. In S Saurabh, M J Kumar (eds.), *Fundamentals of Tunnel Field-Effect Transistors*, CRC Press, 1–291.
5. Toh E H, Wang G H, Samudra G and Yeo Y C. 2007. Device physics and design of double-gate tunneling field-effect transistor by silicon film thickness optimization. *Appl. Phys. Lett.* **90**, 263507.
6. Toh E H, Wang G H, Chan L, Samudra G and Yeo Y C. 2007. Device physics and guiding principles for the design of double-gate tunneling field effect transistor with silicon-germanium source heterojunction. *Appl. Phys. Lett.* **91**, 243505.
7. Verhulst A S, Vandenberghe W G, Maex K and Groeseneken G. 2007. Tunnel field-effect transistor without gate-drain overlap. *Appl. Phys. Lett.* **91**, 053102.
8. Alian A, Franco J, Vandooren A, Mols Y, Verhulst A, El Kazzi S, Rooyackers R, Verreck D, Smets Q, Mocuta A, Collaert N, Lin D and Thean A. 2015. Record performance InGaAs homo-junction TFET with superior SS reliability over MOSFET. *2015 IEEE International Electron Devices Meeting (IEDM)*, 31.7.1–31.7.4.
9. Verreck D, Verhulst A S and Groeseneken G. 2016. The tunnel field-effect transistor. In J Webster (ed.), *Wiley Encyclopedia of Electrical and Electronics Engineering*, Wiley, 1–24.
10. Li W, Liu H, Wang S, Chen S and Yang Z. 2017. Design of high performance Si/SiGe heterojunction tunneling FETs with a T-shaped gate. *Nanoscale Res. Lett.* **12**, 1–8.
11. Tripathy M R, Singh A K, Samad A, Chander S, Baral K, Singh P K and Jit S. 2020. Device and circuit-level assessment of GaSb/Si heterojunction vertical tunnel-FET for low-power applications. *IEEE Trans. Electron Devices* **67**, 1285–92.
12. Vandooren A, Leonelli D, Rooyackers R, Hikavyy A, Devriendt K, Demand M, Loo R, Groeseneken G and Huyghebaert C. 2013. Analysis of trap-assisted tunneling in vertical Si homo-junction and SiGe hetero-junction Tunnel-FETs. *Solid-State Electron.* **83**, 50–5.

13. Convertino C, Zota C B, Schmid H, Ionescu A M and Moselund K E. 2018. III–V heterostructure tunnel field-effect transistor. *J. Phys. Condens. Matter* **30**, 264005.
14. Vanlalawpuia K and Bhowmick B. 2019. Investigation of a Ge-source vertical TFET with delta-doped layer. *IEEE Trans. Electron Devices* **66**, 4439–45.
15. Knoch J and Appenzeller J. 2010. Modeling of high-performance p-type IIIV heterojunction tunnel FETs. *IEEE Electron Device Lett.* **31**, 305–7.
16. Moselund K E, Cutaia D, Schmid H, Riel H, Sant S and Schenk A. 2016. Complementary III–V heterostructure tunnel FETs. *European Solid-State Device Research Conference 2016*, 403–7.
17. Chen F, Ilatikhameneh H, Tan Y, Klimeck G and Rahman R. 2018. Switching mechanism and the scalability of vertical-TFETs. *IEEE Trans. Electron Devices* **65**, 3065–8.
18. Zhang Q, Fang T, Xing H, Seabaugh A and Jena D. 2008. Graphene nanoribbon tunnel transistors. *IEEE Electron Device Lett.* **29**, 1344–6.
19. Aghanejad Ahmadchally A and Gholipour M. 2021. Investigation of 6-armchair graphene nanoribbon tunnel FETs. *J. Comput. Electron.* **20**, 1114–24.
20. Yang X and Mohanram K. 2011. Robust 6T Si tunneling transistor SRAM design. *Proceedings – 2011 Design, Automation & Test in Europe*, 740–5.
21. Solomon P M, Lauer I, Majumdar A, Teherani J T, Luisier M, Cai J and Koester S J. 2011. Effect of uniaxial strain on the drain current of a heterojunction tunneling field-effect transistor. *IEEE Electron Device Lett.* **32**, 464–6.
22. Mukundrajan R, Cotter M, Saripalli V, Irwin M J, Datta S and Narayanan V. 2012. Ultra low power circuit design using tunnel FETs. *Proceedings – 2012 IEEE Computer Society Annual Symposium on VLSI*, 153–8.
23. Lee Y, Kim D, Cai J, Lauer I, Chang L, Koester S J, Blaauw D and Sylvester D. 2013. Low-power circuit analysis and design based on heterojunction tunneling transistors (HETTs). *IEEE Trans. Very Large Scale Integr. Syst.* **21**, 1632–43.
24. Mookerjea S, Krishnan R, Datta S and Narayanan V. 2009. On enhanced miller capacitance effect □ in interband tunnel transistors. *IEEE Electron Device Lett.* **30**, 1102–4.
25. Wu C, Huang R, Huang Q, Wang J and Wang Y. 2016. Design guideline for complementary heterostructure tunnel FETs with steep slope and improved output behavior. *IEEE Electron Device Lett.* **37**, 20–3.
26. Wang S, Pan A, Chui C O and Gupta P. 2016. PROCEED: A pareto optimization-based circuit-level evaluator for emerging devices. *IEEE Trans. Very Large Scale Integr. Syst.* **24**, 192–205.
27. Zhao Q T, Richter S, Schulte-Braucks C, Knoll L, Blaeser S, Luong G V, Trellenkamp S, Schafer A, Tiedemann A, Hartmann J M, Bourdelle K and Mantl S. 2015. Strained Si and SiGe nanowire tunnel FETs for logic and analog applications. *IEEE J. Electron Devices Soc.* **3**, 103–14.
28. Morris D H, Avci U E, Rios R and Young I A. 2014. Design of low voltage tunneling-FET logic circuits considering asymmetric conduction characteristics. *IEEE J. Emerg. Sel. Top. Circuits Syst.* **4**, 380–8.
29. Singh J, Ramakrishnan K, Mookerjea S, Datta S, Vijaykrishnan N and Pradhan D. 2010. A novel Si-tunnel FET based SRAM design for ultra low-power 0.3V V DD applications. *Proceedings – 2010 15th Asia and South Pacific Design Automation Conference (ASP-DAC)*, 181–6.

30. Saripalli V, Mohata D K, Mookerjea S, Datta S and Narayanan V. 2010. Low power loadless 4T SRAM cell based on degenerately doped source (DDS) In0.53Ga0.47As tunnel FETs. *68th Device Research Conference*, 101–2.
31. Fulde M, Heigl A, Wachutka G, Knoblinger G and Schmitt-Landsiedel D. 2009. Complementary multi-gate tunnelling FETs: Fabrication, optimisation and application aspects. *Int. J. Nanotechnol.* **6**, 628–39.
32. Guo P F, Yang Y, Samudra G, Heng C H and Yeo Y C. 2010. Temperature independent current biasing employing TFET. *Electron. Lett.* **46**, 786–7.
33. Senale-Rodriguez B, Yeqing Lu, Fay P, Jena D, Seabaugh A, Xing H G, Barboni L and Silveira F. 2012. Perspectives of TFETs for low power analog ICs. *2012 IEEE Subthreshold Microelectronics Conference (SubVT)*.
34. Mallik A and Chattopadhyay A. 2012. Tunnel field-effect transistors for analog/mixed-signal system-on-chip applications. *IEEE Trans. Electron Devices* **59**, 888–94.
35. Trivedi A R, Carlo S and Mukhopadhyay S. 2013. Exploring tunnel-FET for ultra low power analog applications: A case study on operational transconductance amplifier. *Proceedings – 2013 50th ACM/EDAC/IEEE Design Automation Conference*.
36. Sedighi B, Hu X S, Liu H, Nahas J J and Niemier M. 2015. Analog circuit design using tunnel-FETs. *IEEE Trans. Circuits Syst. I Regul. Pap.* **62**, 39–48.
37. Koswatta S O, Lundstrom M S and Nikonov D E. 2009. Performance comparison between p-i-n tunneling transistors and conventional MOSFETs. *IEEE Trans. Electron Devices* **56**, 456–65.
38. Gupta S, Wairya S and Singh S. 2022. Design and analysis of triple metal vertical TFET gate stacked with N-type SiGe delta-doped layer. *Silicon* **14**, 4217–25.
39. Chander S, Baishya S, Sinha S K, Kumar S, Singh P K, Baral K, Tripathy M R, Singh A K and Jit S. 2019. Two-dimensional analytical modeling for electrical characteristics of Ge/Si SOI-tunnel FinFETs. *Superlattices Microstruct.* **131**, 30–9.
40. Neves F S, Agopian P G D, Martino J A, Cretu B, Rooyackers R, Vandooren A, Simoen E, Thean A V Y and Claeys C. 2016. Low-frequency noise analysis and modeling in vertical tunnel FETs with Ge source. *IEEE Trans. Electron Devices* **63**, 1658–65.
41. Singh S and Raj B. 2020. Two-dimensional analytical modeling of the surface potential and drain current of a double-gate vertical t-shaped tunnel field-effect transistor. *J. Comput. Electron.* **19**, 1154–63.
42. Das D and Chakraborty U. 2021. A study on dual dielectric pocket heterojunction SOI tunnel FET performance and flicker noise analysis in presence of interface Traps. *Silicon* **13**, 787–98.
43. Seema N and Chauhan S S. 2021. Linearity performance analysis of double gate (DG) VTFET using HDB for RF applications. *Silicon* **13**, 1121–5.
44. Gonzalez T. 2015. Introduction to special issue on noise modeling. *J. Comput. Electron.* **14**, 1–3.
45. Sharma A, Goud A A and Roy K. 2015. Sub-10 nm FinFETs and tunnel-FETs: From devices to systems. *Proceedings – 2015 Design, Automation & Test in Europe Conference & Exhibition*, 1443–8.
46. Singh K, Mishra R A and Priyadarshani K N. 2025. Design and analysis of trigate Ge pocket vertical tunnel FET for enhanced drive current. *Mater. Sci. Eng. B* **315**, 118096.

47. Kim M S, Cane-Wissing W, Li X, Sampson J, Datta S, Gupta S K and Narayanan V. 2016. Comparative area and parasitics analysis in FinFET and heterojunction vertical TFET standard cells. *ACM J. Emerg. Technol. Comput. Syst.* **12**, 1–23.
48. Park J. 2016. Comparison study of the future logic device candidates for under 7nm era. *Proceeding EDISON Challenge.* 295–8.
49. Radamson H H, Miao Y, Zhou Z, Wu Z, Kong Z, Gao J, Yang H, Ren Y, Zhang Y, Shi J, Xiang J, Cui H, Lu B, Li J, Liu J, Lin H, Xu H, Li M, Cao J, He C, Duan X, Zhao X, Su J, Du Y, Yu J, Wu Y, Jiang M, Liang D, Li B, Dong Y and Wang G. 2024. CMOS Scaling for the 5 nm node and beyond: Device, process and technology. *Nanomaterials* **14**, 837.
50. Zhang Z, Zhang N and Zhang Z. 2025. High-performance carbon nanotube electronic devices: Progress and challenges. *Micromachines* **16**, 554.

Chapter 3

Nanoscale 2D field-effect transistors

Suresh Balanethiram, Chhandak Mukherjee, R. Priyadharshni, and K. Vijaya Kumar

3.1 FROM BULK TO NANOSCALE TRANSISTORS

3.1.1 Introduction

The field-effect transistor (FET) has long served as the backbone of semiconductor electronics, enabling remarkable progress in circuit density and performance through continual scaling, as envisioned by Moore's Law [1]. Over the past five decades, traditional bulk materials such as silicon and III–V semiconductors have been scaled down to nanometer dimensions, driving the success of modern microprocessors [2]. However, as devices approach the sub-10 nm technology node, numerous physical limitations emerge, which include short-channel effects, increased leakage currents, and thermal management issues posing fundamental obstacles to further miniaturization [3]. The origin of these effects lies in poor gate control and increased scattering from surface roughness and interface states, especially as body thickness drops below 5 nm. Even advanced architectures like FinFETs and Semiconductor on Insulator (SOI) devices, although effective up to a certain extent, suffer from mobility degradation due to thickness fluctuations and surface defects. These challenges underscore the urgent need for alternative materials and device architectures that retain performance and electrostatic integrity at atomic dimensions. Two-dimensional (2D) semiconductors offer notable advantages, such as atomically flat, defect-free surfaces and a reduced rate of mobility degradation at scaled thicknesses, positioning them as strong candidates for future technology nodes [4].

3.1.2 Need for new channel materials: beyond silicon

With continued transistor scaling, traditional silicon-based MOSFETs are increasingly challenged by limitations in mobility, electrostatic integrity, and quantum tunneling. As gate lengths approach sub-5 nm, short-channel effects become severe, and the source-drain tunneling current rises to unacceptable levels, compromising switching performance [3]. Efforts to incorporate high-mobility materials such as germanium and III–V semiconductors

DOI: 10.1201/9781003635123-3

have shown promise, yet challenges like lattice mismatch and incompatibility with Complementary Metal Oxide Semiconductor (CMOS) fabrication persist [1]. Moreover, scaling bulk materials into ultra-thin-body (UTB) form introduces surface roughness and dangling bonds that scatter carriers and degrade performance [2]. The physical thickness constraints of silicon impose a hard limit on further scaling, even with sophisticated structures like FinFETs or gate-all-around transistors [5]. 2D semiconductors present an elegant solution: their atomically thin nature eliminates dangling bonds and allows excellent electrostatic control, enabling lower gate lengths without significant mobility loss [4]. Thus, the transition toward alternative materials is not only desirable but essential for sustaining Moore's Law in the post-silicon era.

3.1.3 Overview of 2D materials and carbon nanotubes for FETs

The emergence of graphene in 2004 marked the beginning of an intense global exploration of 2D materials for electronic applications [3]. Although initially hailed as a revolutionary material due to its exceptional carrier mobility, graphene's intrinsic lack of a bandgap limits its use in conventional logic FETs, where high on/off ratios are essential [1]. This realization prompted a shift toward alternative 2D semiconductors, such as transition metal dichalcogenides (TMDs), phosphorene, silicene, and h-BN, many of which possess bandgaps in the 1–2 eV range and exhibit superior gate control at the atomic scale [2, 5]. These materials feature dangling-bond-free surfaces, improving interface quality and enabling precise electrostatic modulation [4]. Meanwhile, carbon nanotubes (CNTs), with their one-dimensional (1D) structure and high mobility, also offer compelling characteristics for FET applications that include near-ballistic transport and immunity to short-channel effects [6]. Together, 2D materials and CNTs represent two complementary classes of emerging channels that hold promise for sub-10 nm transistor technology.

3.1.4 Scope and structure of the chapter

This chapter provides a comprehensive review of nanoscale FETs built from 2D materials and CNTs, highlighting their potential as successors to conventional silicon-based devices. Beginning with an exploration of the limitations of bulk and UTB silicon transistors, we examine the motivation behind the search for new materials and introduce the family of 2D semiconductors and CNTs. The subsequent sections detail the physics of device operation, focusing on electrostatics, scaling, and contact engineering. We then evaluate various modeling strategies, including analytical, drift-diffusion, and quantum transport frameworks. This is followed by a benchmarking of key performance metrics and a discussion of their relevance in practical applications such as flexible electronics, neuromorphic computing, and biosensing. This chapter concludes by addressing current challenges in fabrication, integration, and reliability. Also, offers recommendations for advancing these technologies toward industrial adoption.

3.2 OVERVIEW OF KEY 2D MATERIALS AND CNTs FOR FETs

3.2.1 Transition metal dichalcogenides (TMDs): MoS_2, WS_2

TMDs such as MoS_2 and WS_2 have emerged as leading contenders in nanoscale transistor research owing to their unique structural and electronic characteristics. Monolayer MoS_2 exhibits a direct bandgap of approximately 1.8 eV, in contrast to its bulk form, which has an indirect bandgap, making it ideal for low-power switching and optoelectronic applications [7]. The atomically thin body ensures excellent gate electrostatics, enabling superior subthreshold swings (SSs) approaching the Boltzmann limit [7, 8]. Furthermore, the absence of dangling bonds on TMD surfaces results in pristine dielectric interfaces, which are critical for device scaling [9]. Experimental demonstrations have shown successful MoS_2 transistor operation at gate lengths as small as 15 nm, with promising suppression of short-channel effects (SCEs), as shown in Figure 3.1 [9]. The figure compares the transfer characteristics $\left(I_{DS} - V_{Gate}\right)$ of monolayer MoS_2, $MoTe_2$, WSe_2, and WTe_2 transistors, highlighting that WTe_2 delivers the highest ON-current and steepest subthreshold slope, followed by WSe_2 and $MoTe_2$, with MoS_2 exhibiting the lowest current drive.

The structural phase plays a significant role in determining the electrical behavior of TMDs. 2H–MoS_2 is semiconducting, whereas its 1T counterpart exhibits metallic properties. Additionally, photoluminescence and nonlinear optical responses are layer-number dependent, opening new avenues in valleytronics and optoelectronics. WS_2, sharing similar traits with MoS_2, offers comparable carrier mobility but superior photoluminescence yield. When benchmarked for analog and RF applications, MoS_2-FETs demonstrate intrinsic cutoff frequencies up to 42 GHz, whereas WSe_2-FETs show even greater potential due to higher drive currents and intrinsic gain. A quantitative comparison of electronic parameters such as bandgap, effective carrier masses, and dielectric constants for various TMDs is summarized in Table 3.1. Despite these advantages, challenges like contact resistance, Schottky barriers (SB), and variability in layer-dependent behavior affect performance consistency [8]. Still, with advancements in synthesis and interface engineering, monolayer TMDs are poised to play a critical role in ultra-scaled digital and RF transistor technologies [10–12].

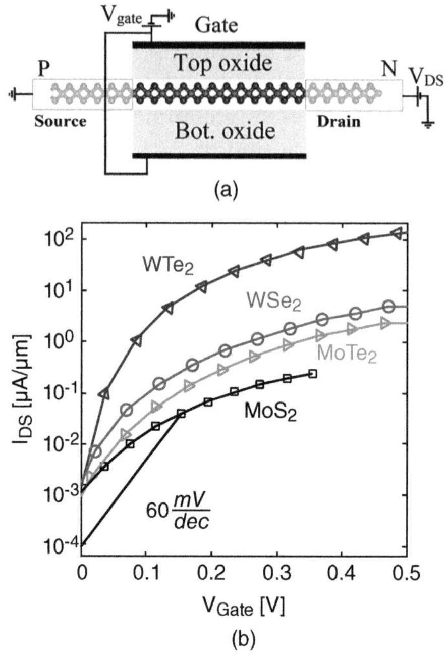

Figure 3.1 (a) Schematic illustration of a monolayer transition metal dichalcogenide (TMD) field-effect transistor (FET) structure with top and bottom oxides. (b) Transfer characteristics $(I_{DS} - V_{Gate})$ of different monolayer TMD FETs, showing that WTe_2 achieves the highest ON-current and the steepest subthreshold slope compared to WSe_2, $MoTe_2$, and MoS_2.

(Adapted with permission from [9].)

Table 3.1 Electronic properties of various TMD materials

Parameters	MoS_2	WSe_2	$MoTe_2$	WTe_2
E_g (eV)	1.68	1.56	1.085	0.75
$m_e^*[m_0]$	0.52	0.36	0.57	0.37
$m_h^*[m_0]$	0.64	0.5	0.75	0.3
ε_r^{in}	4.2	4.5	8	5.7
ε_r^{out}	2.8	2.9	4.4	3.3

3.2.2 Graphene: high-mobility but zero-bandgap challenge

Graphene, a monolayer of carbon atoms arranged in a honeycomb lattice, offers extraordinary charge carrier properties, including mobility values as high as 200,000 $cm^2/V{\cdot}s$ and saturation velocities up to 4×10^7 cm/s.

These attributes make it highly appealing for high-speed transistors and RF circuits. However, graphene's inherent limitation lies in its zero bandgap, which restricts the I_{ON} / I_{OFF} current ratio to values less than 10 even under optimized conditions, making it unsuitable for digital logic applications that require distinct switching behavior [13]. To address this, several bandgap engineering strategies have been explored. Among them, bilayer graphene (BLG) has shown promise due to its ability to open a bandgap of up to ~250 meV under an applied perpendicular electric field [13]. This tunable bandgap enables modest switching characteristics while retaining some of graphene's mobility advantages. Nevertheless, even with BLG, the induced gap is insufficient to fully suppress leakage currents in the OFF-state, limiting its application in ultra-low-power CMOS logic. An alternative and more scalable solution is the use of graphene nanoribbons (GNRs), where quantum confinement and edge effects can induce significant bandgaps depending on ribbon width and edge orientation, as illustrated in Ref. [14]. Despite these digital limitations, graphene-based FETs, especially those leveraging bilayer structures, have demonstrated cutoff frequencies exceeding 100 GHz, positioning them as strong candidates for analog and RF applications [15]. In such domains, where strict I_{ON} / I_{OFF} ratios are not mandatory, graphene's near-ballistic transport and high-speed response can be effectively utilized.

3.2.3 Carbon nanotubes (CNTs): one-dimensional electrostatics

CNTs have emerged as strong contenders for post-silicon electronics, driven by their exceptional electrical and mechanical properties. Notably, CNTs exhibit extremely high carrier mobility due to weak electron–phonon interactions, along with robust mechanical and thermal stability. These properties, combined with their inherent 1D geometry, allow for enhanced electrostatic control over the channel, crucial for FET scaling. In carbon nanotube field-effect transistors (CNT-FETs), the electrostatics differ significantly from conventional silicon-based devices. Their 1D density of states (DOS) leads to sharper band edges and affects capacitance profiles, making quantum mechanical modeling essential for accurate predictions. These effects not only impact the gate control but also determine threshold behavior and switching efficiency. Studies have shown that densely packed CNT arrays on planar insulators suffer reduced capacitance due to electrostatic screening between adjacent nanotubes [16]. Another major advantage of CNTs is their intrinsic nature. CNT channels typically remain undoped, and charge carriers are injected directly from metal contacts, thereby preserving the defect-free sp^2 carbon lattice. This doping-free operation enables near-ballistic transport and has been key to achieving high-performance symmetric CMOS devices [17]. Due to their cylindrical geometry and strong lateral confinement, CNTs behave as quasi-1D nanowires, where the gate control becomes more efficient compared to planar structures [18]. The structural

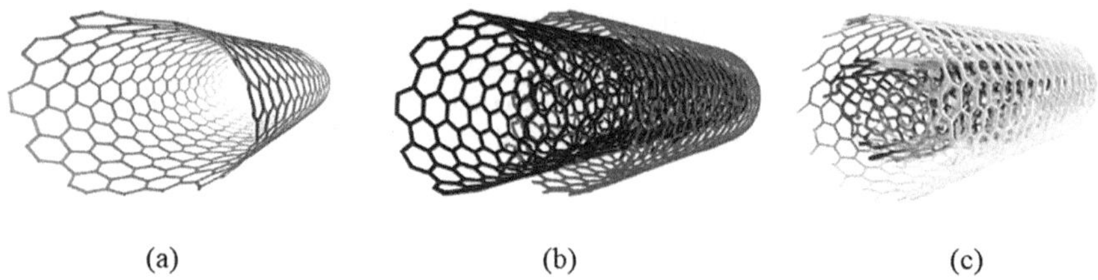

Figure 3.2 Structural description of (a) single-walled carbon nanotube (SWCNT), (b) double-walled carbon nanotube (DWCNT), and (c) multi-walled carbon nanotube (MWCNT).

variation between single-walled, double-walled, and multi-walled CNTs is illustrated in Figure 3.2 and significantly affects their electronic properties and suitability for different applications. Recent studies employing self-consistent Poisson-Schrödinger solvers further highlight the necessity of accounting for quantum confinement effects to accurately model the carrier distribution and electrostatics in such scaled devices [18, 19]. These electrostatic advantages are clearly exemplified in the early CNT-FET demonstration in Ref. [20]. In conclusion, the unique electrostatics of CNTs driven by their one-dimensionality and ballistic transport characteristics form the foundation for their integration in next-generation FET architectures. Their superior control over SCEs, reduced power consumption, and potential for CMOS compatibility make CNTs a promising platform for future nanoelectronic devices.

3.3 DEVICE DESIGN AND ELECTROSTATIC CONSIDERATIONS

3.3.1 Short-channel effects in nanoscale FETs

As transistor dimensions scale below 10 nm, SCEs like drain-induced barrier lowering (DIBL), threshold voltage roll-off, and leakage become increasingly severe in conventional FETs [3]. These effects limit gate control and require new materials and structures for continued scaling. 2D materials such as MoS_2 offer improved immunity to SCEs due to their atomic thickness and excellent electrostatic integrity [15]. Simulations comparing Si, InGaAs, and MoS_2 show that 2D FETs can retain better control at reduced dimensions [21]. Device architectures like double-gate and gate-all-around further enhance this control, enabling operation at dimensions even below 5 nm [22]. Experimental studies have demonstrated MoS_2 transistors with SSs near the theoretical 60 mV/dec limit, confirming their robustness under extreme scaling. Classical models become inadequate at these dimensions due to quantum confinement effects, necessitating quantum transport simulations to capture carrier behavior accurately [23]. Overall, 2D FETs show clear advantages in suppressing SCEs, making them strong candidates for future ultra-scaled technologies.

3.3.2 Gate control and electrostatics in 2D material-based FETs

Effective gate control is critical for ensuring strong electrostatic modulation in nanoscale transistors. In conventional FETs, scaling often weakens this control, whereas 2D materials offer a key advantage due to their atomically thin channels. This geometry allows the gate field to penetrate more effectively, resulting in improved SS and reduced leakage. Studies on MoS_2 and WS_2 FETs reveal enhanced electrostatic coupling in top-gate and gate-all-around configurations, even at short gate lengths. The lack of bulk body and minimal dielectric interface traps contributes to superior gate-channel coupling [24, 25]. Electrostatic integrity also improves with high-κ dielectrics and dual-gate designs, enabling steep-slope behavior and better threshold control [26]. Simulation results confirm that 2D-FETs can achieve ideal subthreshold slopes and strong gate modulation over a wide bias range. These features make them highly scalable and energy-efficient for future low-power logic nodes [26, 27].

3.3.3 CNT-FET electrostatics and channel modulation

CNT-FETs offer exceptional electrostatic control due to their quasi-1D structure, which confines carriers along the tube axis and minimizes SCEs [28]. The cylindrical geometry ensures excellent gate-to-channel coupling, especially in gate-all-around architectures [29]. CNTs exhibit high carrier mobility and ballistic transport over short lengths, leading to fast switching and minimal power dissipation [30]. The absence of surface roughness and reduced scattering enhances their electrostatics compared to silicon and 2D planar devices. Furthermore, channel modulation is strongly influenced by the *chirality* and *diameter* of the CNT, which determine its bandgap and threshold voltage [31]. Advanced simulations and experimental studies show that optimizing gate alignment, dielectric interfaces, and doping profiles can further improve electrostatic integrity and ON/OFF ratios in CNT-FETs [32]. These advantages make CNT-FETs highly attractive for high-speed, low-power applications.

3.3.4 Contact engineering and interface control

At nanoscale dimensions, contact resistance becomes a dominant performance limiter in both 2D and CNT-FETs. In 2D transistors, the absence of dangling bonds reduces interface states; however, forming low-resistance ohmic contacts remains challenging due to Fermi-level pinning and SB [33,

34]. Techniques such as phase engineering, doping near contacts, and metal work function tuning are employed to reduce barrier heights and enhance current injection [35]. Contact geometry also plays a critical role—edge contacts in graphene and other 2D materials often outperform top contacts in terms of carrier injection efficiency [36]. For CNT-FETs, contact alignment and the quality of the tube-metal interface largely dictate device resistance. High-performance devices often use metals such as palladium (Pd) and scandium (Sc), which align well with CNT energy levels and reduce injection barriers [37]. Controlled interface engineering, combined with clean fabrication techniques, is essential for achieving consistent and scalable performance across both 2D and CNT-based technologies.

3.4 MODELING AND SIMULATION OF 2D AND CNT-FETs

3.4.1 Analytical and compact models for 2D FETs

With the emergence of 2D materials such as graphene and BLG, conventional transistor modeling approaches have undergone significant adaptations to reflect their distinctive electrostatic and transport properties. Unlike bulk silicon, 2D semiconductors exhibit quantum confinement, weak screening, and symmetry-driven band structures that demand physics-based formulations. The operating principle of a graphene field-effect transistor (GFET) is illustrated schematically in Figure 3.3. As such, ambipolar conduction is observed when the gate bias is swept from a negative to a positive value, showcasing a pure hole conduction (Figure 3.3(a)) branch for a negative gate bias and a pure electron conduction (Figure 3.3(b)) branch for a positive bias. Around the Dirac point, the channel showcases a transition between p- and n-type conduction (Figures 3.3(c) and (d)), highlighting a minimum net carrier density at the charge neutrality point. The cross-sectional view of a BLG FET (Figure 3.4(a)) highlights the dual-gate configuration that enables electrostatic bandgap tuning. Analytical models for 2D FETs are typically derived from drift-diffusion or ballistic transport theory and incorporate material-dependent parameters like the effective mass, quantum capacitance, and DOS [39, 40]. These models are instrumental in understanding how gate bias, channel doping, and energy band engineering affect the carrier transport across the channel. Electrostatic modeling through equivalent circuits (as shown in Figure 3.4(b)) helps bridge quantum models with classical charge control. In contrast, compact models are designed to be computationally efficient while retaining essential physics for circuit simulation. They are often implemented in Verilog-A for integration into SPICE-level design environments. Notably, Ref. [39] developed a large-signal compact model for bilayer GFETs that includes quantum capacitance, access resistance modeling based on contact barrier heights, and flat-band voltage shifts

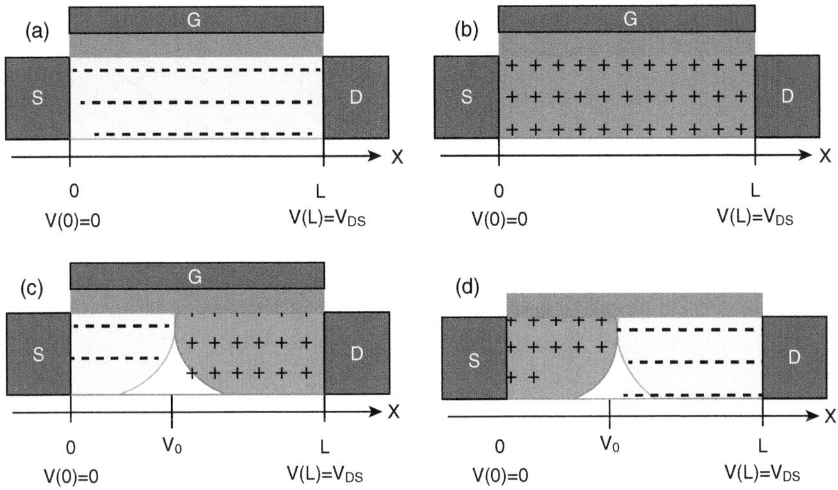

Figure 3.3 Schematic illustration of the operating principle of a graphene field-effect transistor (GFET). (a) Pure hole conduction under negative gate bias, (b) pure electron conduction under positive gate bias, (c) transition from hole to electron conduction near the charge neutrality (Dirac) point, and (d) complementary transition with mixed carrier transport. Ambipolar behavior is observed as the gate bias is swept from negative to positive values, highlighting the minimum carrier density at the Dirac point.

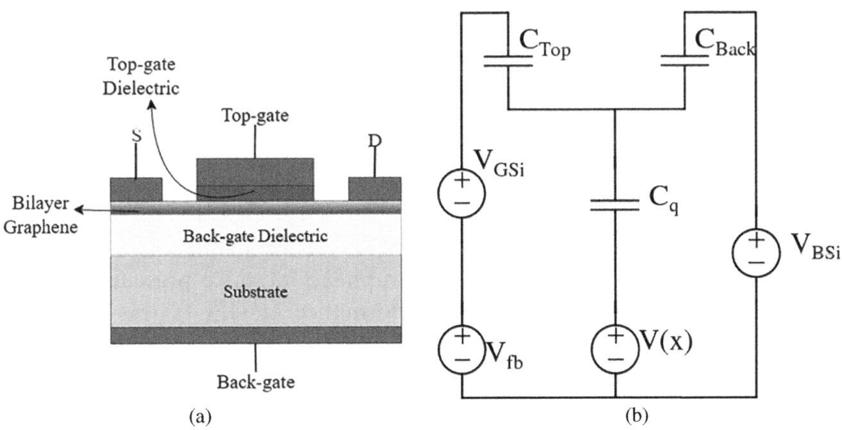

Figure 3.4 (a) Device structure, (b) capacitive model, (c) transfer characteristics, and (d) output characteristics of the bilayer GFET.

(Post-processed from Aguirre-Morales et al. [38].)

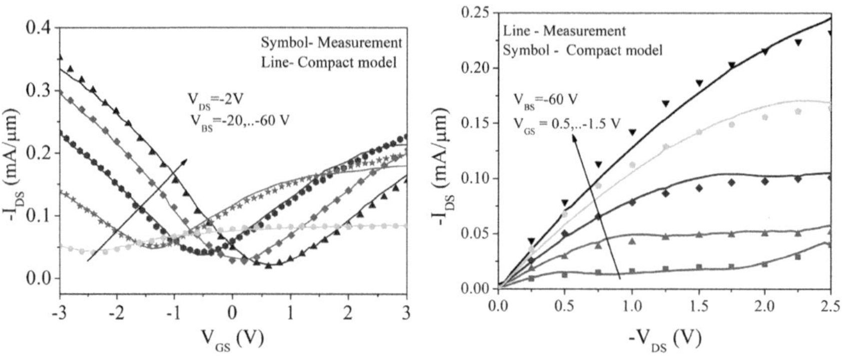

(c) (d)

Figure 3.4 (Continued)

as a function of back-gate bias. These models have been validated against experimental transfer and output characteristics (Figures 3.4(c) and (d)), demonstrating strong agreement across different gate and drain biasing conditions. Overall, both analytical and compact modeling approaches have proven essential in enabling predictive design and performance evaluation of 2D FETs, particularly for high-frequency and low-power applications.

3.4.2 Modeling of contact resistance and threshold voltage shifts

The accurate modeling of contact resistance and threshold voltage shifts is essential for predicting the performance of GFETs, particularly in nanoscale and high-frequency applications. Due to the unique electronic properties of monolayer and BLG, such as linear dispersion relations and tunable bandgaps, the electrical behavior of GFETs deviates significantly from that of traditional semiconductor devices.

3.4.2.1 *Contact resistance modeling*

Contact resistance (R_C) at the metal–graphene interface presents a major bottleneck in achieving high device performance. In GFETs, this resistance arises from the formation of a potential barrier at the metal contact, which depends on local electrostatics and the type of carrier involved. Unlike silicon-based devices, GFETs exhibit ambipolar transport, necessitating the modeling of contact resistance as a function of both electron and hole injection barriers. In compact models proposed for both monolayer and bilayer GFETs [39, 41], the source and drain access resistances $(R_S \text{ and } R_D)$ are treated as bias-dependent parameters. These resistances are formulated using an exponential relationship with the effective barrier height, which itself varies linearly with the applied back-gate voltage:

$$R_{S/D}\left(V_{BG}\right) = R_0 + r \cdot \exp\left(\frac{q\Phi_B\left(V_{BG}\right)}{kT}\right) \tag{3.1}$$

This expression captures the modulation of the Schottky-like barrier at the metal–graphene interface. The barrier height, Φ_B, is influenced by the doping level and the electrostatic environment, particularly in dual-gated structures. The use of this formulation enables the compact model to accurately replicate experimental $I-V$ characteristics across a wide bias range. The effectiveness of this modeling approach is demonstrated in Figures 3.4(c) and (d), where the simulated and measured output characteristics closely align, validating the role of contact resistance in shaping the current response at different gate biases.

3.4.2.2 Threshold voltage modeling

In GFETs, defining and modeling the threshold voltage $\left(V_{TH}\right)$ is challenging due to the absence of a well-defined bandgap in monolayer graphene and its intrinsic ambipolar conduction mechanism. Nevertheless, a pseudo-threshold voltage is often defined as the gate voltage corresponding to the minimum drain current, also known as the Dirac point. In bilayer graphene FETs (BGFETs), an electric field applied perpendicular to the graphene layers can open a bandgap, enabling more pronounced switching behavior and allowing for a more practical definition of threshold voltage. In these devices, the threshold voltage is influenced by the flat-band voltage $\left(V_{FB}\right)$, which is itself modulated by the back-gate voltage $\left(V_{BG}\right)$ and interface properties. The compact model captures this dependence using an empirical exponential relationship:

$$V_{FB}\left(V_{BG}\right) = V_1 + V_2 \cdot \exp\left(\frac{V_{BG}}{V_3}\right) \tag{3.2}$$

where $V_1, V_2,$and V_3 are fitting parameters. This expression accounts for the exponential screening and electrostatic coupling effects that arise in bilayer graphene structures, especially under varying back-gate bias. As a result, the model can accurately reproduce the threshold voltage shifts observed in experimental data. The impact of V_{BG} on the threshold behavior is illustrated in Figure 3.4, which shows the shift in the transfer characteristics as the back-gate voltage is varied.

3.4.3 Transport models: drift-diffusion and ballistic approaches

Accurate transport modeling is essential for describing the electrical characteristics of GFETs, especially under varying bias and geometry conditions. Two main regimes, drift-diffusion and ballistic transport, are typically used in analytical and compact models of monolayer and bilayer GFETs. The drift-diffusion approach is widely applied in compact models due to its compatibility with standard EDA tools and its ability to represent ambipolar behavior. It assumes that carriers undergo frequent scattering, and current is computed using the local field and carrier concentration. Ref. [39, 41] also implemented this approach in their Verilog-A models for both monolayer and bilayer GFETs. The total drain current is often separated into electron and hole components to account for graphene's ambipolarity:

$$I_{\mathrm{DS}} = I_n + I_p = \frac{\mu_n Q_n}{1 + \mu_n Q_n / v_{\mathrm{sat},n}} + \frac{\mu_p Q_p}{1 + \mu_p Q_p / v_{\mathrm{sat},p}} \tag{3.3}$$

where, Q_n and Q_p are the carrier densities, μ are mobilities, and v_{sat} are saturation velocities. The charge and capacitance expressions are also derived based on the density of states of graphene. The current saturation behavior in the output characteristics of bilayer GFETs (Figure 3.4(d)) reflects the transition toward quasi-ballistic regimes, especially at high V_{DS}.

To bridge these regimes, modern compact models often use velocity saturation terms and semiempirical fitting to smoothly transition between drift-limited and ballistic-like behavior. The use of bias-dependent mobility and contact resistance expressions, as implemented in Ref. [40, 41], enables the models to capture real GFET behavior across a wide range of biasing conditions.

3.4.3.1 Circuit application example based on large-signal GFET models

The intrinsic GFET large-signal compact model in Ref. [38] includes a current source for the drain current described in eq. (3.3) along with the access resistances at the source and the drain sides as well as two current branches that account for the gate and the drain charges, that is, the intrinsic device capacitances (Figure 3.5(a)). In addition, external current sources are added to the intrinsic large-signal circuit to account for the parasitic elements of the test structures and the bias-tee (Figure 3.5(b)), enabling accurate capture of the experimental conditions.

To extract the values of these external parasitic elements, which were considered in the AC simulations at high frequencies, electromagnetic (EM) simulations of the measurement environment were carried out. In particular, specific test structures (OPEN-PAD, MUTE, and SHORT), which were

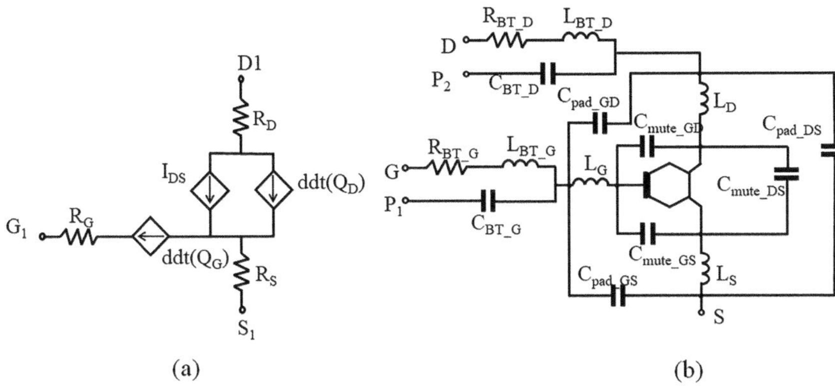

Figure 3.5 (a) Intrinsic large-signal equivalent circuit. (b) Large-signal equivalent circuit showing external elements.

(Aguirre-Morales et al. [38].)

available from the process, were reproduced in the EM simulation environment, leveraging the Graphic Data System (GDS) layout of the fabricated devices. Once the values of these parasitic elements are determined from EM simulations, they are used together with the intrinsic device parameter set for the validation of the complete GFET compact model against both DC and AC experimental data.

3.4.3.2 GFET balun circuit

Following the transistor-level validation of the EM-SPICE co-simulation approach, the developed compact model was leveraged to assess the performance of GFET-based circuits, specifically a balun circuit architecture presented in Ref. [38] (Figure 3.6(a)). The operating principle of the circuit is to convert a single-ended signal into a differential signal using two GFETs. In this example, two GFET active loads are used for the two load resistances R_1 and R_2 While the gates and the sources of the GFETs are connected together, the differential output signals $\left(V_{DS1}\text{ and } V_{DS2}\right)$ are retrieved at the drain terminals referenced to the common source. The supply voltage V_{DD} is set to 2.9 V, while V_{SS} is kept at 5 V. A sinusoidal input signal of 0.2 V amplitude and a frequency of 1 GHz was then applied to the common gate terminal, and the two differential outputs were recorded. The simulated *S*-parameters are shown in Figure 3.6(b), which exhibited a 180°-phase shift between Ports 2 and 3 (Figure 3.6(c)). Figure 3.6(d) shows the two output signals, showcasing a nearly perfect amplitude balance for the two output signals. This study underlines the benefits of an accurate large-signal SPICE-EM co-simulation framework for GFET-based technologies, allowing optimal RF circuit design exploration.

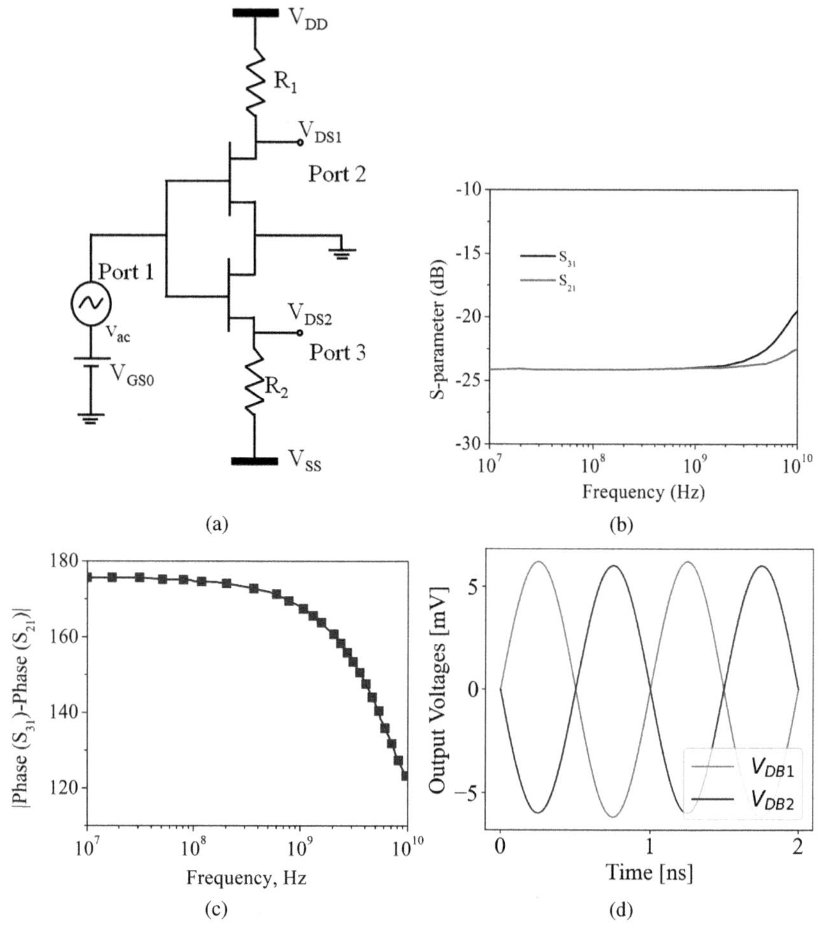

Figure 3.6 (a) Schematic circuit diagram and (b) S-parameter and (c) phase shift between Ports 2 and 3. (d) Balun's output voltages V_{DS1} and V_{DS2} for an input voltage of 0.2 V at f = 1GHz.

(Post-processed from Aguirre-Morales et al. [38])

A conventional CNT-FET operates on the principle of ballistic transport, with heavily doped source/drain regions providing barrier height modulation through gate control [42]. While other architectures, such as SB CNT-FETs, are among the possible choices that exhibit ambipolar conduction, the conventional type has the advantages of unipolar transport, reduced OFF-state leakage, and higher ON-current due to the absence of SB. It also provides better scalability, thereby making them more interesting from a design standpoint. The ballistic transport and electron confinement lead to a constant current through the channel, which is calculated at the top of the energy barrier, while the helicity and the radius of the nanotube warrant discretization of the energy sub-band structure [43]. Among the CNT-FET compact models presented in Ref. [44], the DG-CNT-FET model was used to explore the performance of a dynamically reconfigurable eight-function logic gate, reporting on its power consumption and transient parameters based on the first layout and simulations.

3.4.3.3 Dynamically reconfigurable 8-function logic gate

CNT-FETs have been shown to support CMOS-compatible logic circuits [45], especially in multi-value logic [46] and single-transistor logic [47], leveraging specific properties of the CNT-FETs. In Ref. [44], the authors demonstrated a novel dynamically reconfigurable logic design utilizing the ambipolarity of DG-CNT-FETs.

The dynamically reconfigurable eight-function logic cell proposed in Ref. [44] (Figure 3.7) comprises 7 CNT-FETs arranged for specific parts of the cell, such as the logic function or the inverter/follower stages. The overall assembly can be configured to operate as eight basic binary functions depending on the applied bias polarity. Table. 3.2 shows the back-gate bias combinations to control the polarities of the CNT-FET, denoted as TC1,TC2,and TC3 (Figure 3.7) in order to realize the said eight basic logic functions. In total, the cell contains seven inputs and one output: two

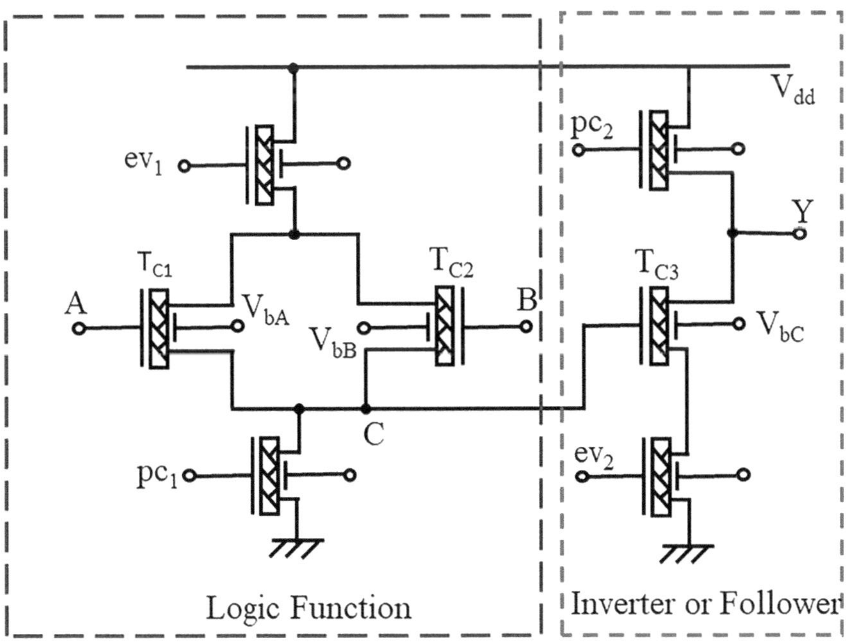

Figure 3.7 Dynamically reconfigurable 8-function logic gate.

(O'Connor et al. [44].)

Table 3.2 Power consumption for different logic functions at 20 GHz and 250 MHz

				Power (nW)	*Power (nW)*
V_{bA}	V_{bB}	V_{bC}	Y	*@20 GHz*	*@250 MHz*
+V	+V	+V	$\overline{A+B}$	895.7	32.2
+V	+V	−V	$A+B$	848.6	31.29
−V	−V	+V	$A{\cdot}B$	725.7	25.89
−V	−V	−V	$\overline{A\cdot B}$	686.0	24.98
+V	−V	+V	$\bar{A}B$	874.6	32.26
+V	−V	−V	$A+\bar{B}$	834.8	31.33
−V	+V	+V	$A\bar{B}$	828.3	31.93
−V	+V	−V	$B+\bar{A}$	788.1	31.10

Boolean inputs, *A* and *B*, as the data operands; three control inputs, V_{bA}, V_{bB}, and V_{bC}, which can be used to switch the polarity from *n*- to *p*-type and vice versa; a set of clocking signals consisting of two pre-charge inputs, pc_1 and pc_2, and two evaluation inputs, ev_1 and ev_2; and finally, the circuit output *Y*. The table also includes the power consumption for each logic configuration evaluated using the compact model of the CNT-FETs. In perspective, these design efforts show particular promise for CMOS add-on technologies, as classical CMOS-only systems would require circuitry that are manifold complex, thus inducing significant power penalties to achieve equivalent circuit functionality.

3.5 RELIABILITY ASPECTS

RF figures of merit of modern transistors are key metrics that impact the performance of analog circuits such as mixers, oscillators, or low-noise amplifiers. These figures of merit, such as device transconductance, output conductance, and maximum oscillation frequencies, are significantly influenced by reliability issues such as noise or long-term wear-off (linked with technology lifetime), which are further accentuated by device miniaturization. This section addresses some of these critical reliability concerns in 2D material-based FET technologies.

3.5.1 Low-frequency noise

With technology downscaling, low-frequency flicker noise has become a dominant reliability concern in electronic circuits owing to the deterioration of the signal-to-noise ratio and unwanted phase noise in circuit applications [48]. It is therefore crucial to investigate the low-frequency noise behavior of modern electronic devices in order to assess their process quality and for defect spectroscopy. Like all electronic devices, 2D material-based FETs are also susceptible to the ubiquitous flicker noise that can impact the performance of circuits based on these technologies. Naturally, noise spectroscopy is a popular technique that has been widely used as a primary reliability indicator for various graphene FET technologies, such as exfoliated graphene on Si/SiO_2 substrate [49], exfoliated BLG FETs [50], GNR [51], and Chemical Vapour Deposition (CVD)-grown graphene FETs [52], as well as graphene layers obtained from thermal decomposition of SiC substrates [53]. While most studies have reported a $1/f$ type flicker noise spectra in the 1 Hz to 10 kHz range, there are some reports of random generation–recombination noise (exhibiting Lorentzian humps in the noise spectra), indicating a nonuniform defect distribution in the oxide [50]. In Ref. [54], particular attention was given to isolating the noise contributions of the access resistances, which were shown to have a non-negligible impact on the overall noise. This study also compared noise response qualitatively across various graphene FET technologies fabricated using different methods, indicating the highest noise level in exfoliated graphene compared to SiC- or CVD-grown graphene devices. Multilayer graphene FETs were also identified as superior to monolayer devices in terms of noise performance. This study highlights, in particular, the importance of the fabrication method for 2D material-based technologies in order to improve device noise level, a precursor to the long-term reliability of the technology.

3.5.2 Long-term reliability

Long-term reliability of electronic devices is a prerequisite for designing high-performance circuit applications. While technology maturity plays a key role in variability and technological dispersion, improving device lifetime is an ever-evolving process that requires close interaction between designers and process engineers through continuous optimization cycles in the early stages of technology development. These technology improvement loops go hand in hand with reliability-aware design techniques that make use of architectural optimization without additional process improvement. While quite a few studies have reported reliability analysis of graphene FETs [55–58] through electrical bias stress measurements, Ref. [59] particularly sets the premise for reliability-aware design methodologies for 2D material FETs through the development of a compact model that can dynamically account for long-term degradation during circuit operation. The main idea is to study and understand the physical mechanisms of degradation under accelerated electrical bias stress in order to derive mathematical equations of long-term failure. The accelerated aging tests are practical from an experimental standpoint and can be used to derive the aging laws for lifetime assessment during the normal operation of the devices under test. The derived analytical equations are then implemented in the existing Verilog-A compact model for the GFETs. To study the effect of long-term stress during dynamic circuit operation, the implementation of the aging model should be compatible with existing SPICE software, ensuring time-invariance. Among typically reported degradation mechanisms, hot-carrier degradation (HCD) and bias-temperature instability (BTI) are commonly observed in modern FET technologies. While hot-carrier-induced damage is typically observed when both high gate and drain biases are applied, BTI can occur even under normal operating conditions when the gate bias is sufficiently large, even at very low drain biases. The latter is therefore the most recurrent degradation mechanism in modern transistors. BTI-induced degradation is typically observed through a shift of the device threshold voltage, along with mobility degradation in some cases, due to the formation or activation of defects within the Si/SiO_2 stack. Depending on whether the stressed devices exhibit recovery during a relaxation phase, one can determine the type of defects present within the device. For example, a non-recoverable degradation is typically due to the creation of new traps at the Si/SiO_2 interface due to the application of bias stress, whereas recoverable degradation is linked with charge trapping into preexisting defects that are formed during the fabrication process. In both cases, the created/activated defects lead to a change in the effective charge density, causing a shift of the threshold voltage. In some cases, a change in the channel resistance due to trap activity can lead to current/mobility degradation. It should be noted that charge trapping in preexisting defects alone has been reported to be insufficient to capture the full BTI kinetics, while models based on the reaction–diffusion (R-D) theory for interface trap generation have also been found to be inadequate, especially in describing recovery dynamics. In compact modeling practices, BTI is modeled through a combination of charge trapping into preexisting defects located in the gate insulator, which shifts the device characteristics according to trap polarity, and the creation of defects (interface states) [60], which leads to a change of slope of the current–voltage characteristics as well as mobility degradation. For example, a positive gate bias will lead to electron trapping in the dielectric, inducing a positive charge layer at the graphene/substrate interface and causing the threshold voltage shift in the direction of negative gate voltage. In the case of interface defect generation leading to current degradation, a positive gate stress bias creates negative interface defects that reduce the electric field in the graphene channel, resulting in current reduction along with an induced threshold voltage shift [61].

In Ref. [59], the physical mechanisms of the aging and stress-induced degradation under long-term bias stress are identified and implemented in Verilog-A for the GFET compact model developed in Ref. [13]. Experimental results of gate bias stress in different GFET technologies were reproduced using a stretched-exponential model for charge trapping during both stress and relaxation phases under dynamic stress conditions, as well as interface state creation as a function of the oxide field. In both cases, threshold voltage shift reaches a saturation after long stress periods, indicating that all broken bonds, which act as trap centers, are filled, leading to no further evolution of the device characteristics over time. A general observation is that at higher stress bias, the trapping time constants reduce, indicating a faster trap response due to increased emission of trap charges toward deep states in the bulk dielectric under a higher electric field [58]. The physics-based compact aging modeling approach presented in Ref. [59] offers the possibility for further exploration at the circuit level to study the impact of bias-induced degradation on circuit performances and lifetime. The methodology can be subsequently leveraged for reliability-aware design of graphene FET-based circuits.

3.6 EMERGING APPLICATIONS

3.6.1 Flexible and wearable electronics

2D materials have emerged as key enablers for flexible and wearable electronics due to their atomically thin form factor, mechanical resilience, and superior electrical properties [62, 63]. Recent studies highlight the integration of 2D semiconductors, such as MoS_2 and WS_2, into ultra-thin, flexible transistors that can maintain performance under strain and bending conditions. These transistors exhibit high ON/OFF ratios and stable carrier mobility under flexed states, making them ideal for applications such as skin-mounted sensors, health monitoring patches, and bendable displays [62]. Encapsulation techniques and substrate engineering have further improved thermal stability and strain tolerance, enabling reliable operation in dynamic environments [63]. The low-temperature fabrication of 2D FETs on flexible substrates also makes them compatible with roll-to-roll manufacturing processes. Overall, 2D materials present a scalable and reliable platform for next-generation wearable electronics, effectively bridging the gap between high performance and mechanical flexibility.

3.6.2 Neuromorphic computing with 2D FETs and CNT-FETs

Neuromorphic systems aim to replicate biological brain functions using hardware-efficient architectures. 2D materials and CNT-based transistors are promising candidates for this paradigm due to their low power consumption, excellent electrostatic control, and scalability [64, 65]. CNT-based ferroelectric transistors have demonstrated artificial synaptic behaviors with high tunability, making them suitable for edge neuromorphic platforms [64]. Similarly, 2D-TFETs exhibit ultra-low SSs, outperforming conventional Metal Oxide Semiconductor Field effect transistors (MOSFETs) in energy efficiency and enabling event-driven, spike-based computing [66]. Furthermore, recent reviews emphasize the potential of 2D materials in realizing multiple synaptic functions—such as potentiation, depression, and plasticity—within compact device footprints [65]. These features offer a promising path toward dense, brain-inspired networks with minimal thermal and energy overhead.

3.6.3 2D FETs and CNTs in biosensing applications

2D materials and CNTs have emerged as compelling platforms for biosensing applications due to their high surface-to-volume ratio, inherent biocompatibility, and exceptional electrical sensitivity [67, 68]. FETs based on TMDCs and graphene derivatives have demonstrated strong performance in detecting a wide range of biomarkers, including proteins, DNA, and small molecules. Notably, MoS_2-based biosensors have shown high selectivity for microRNA (miRNA) detection down to femtomolar concentrations, attributed to their tunable bandgap and ease of surface functionalization [24, 68]. CNT-based FET devices offer rapid response times and label-free detection capabilities, making them particularly effective for detecting cancer biomarkers and neurotransmitters [69]. These FET-based biosensors enable real-time, point-of-care diagnostics with compact form factors and low power consumption. Current research is directed toward improving stability, reproducibility, and integration with flexible substrates to advance wearable healthcare technologies.

3.6.4 Other potential areas: RF, energy-efficient logic

Beyond digital logic and sensing, 2D and CNT-based FETs are gaining traction in high-frequency and low-power domains. For RF applications, physics-based compact modeling of 2D FETs has shown promising large signal performance, achieving high f_T and f_{max} values due to superior electrostatics and reduced parasitics [38]. These devices exhibit tunable transconductance and linearity, making them suitable for flexible RF front-ends and reconfigurable circuits.

In energy-efficient logic, innovations such as quantum-engineered polarity reconfigurable transistors have been demonstrated using ultra-thin 2D materials, offering dual n-/p-mode behavior for area and power optimization [70]. Additionally, emerging 2D logic technologies emphasize steep subthreshold slopes and near-threshold operation to support ultra-low power computing in IoT and edge systems [71].

Collectively, these developments underline the versatility of 2D and CNT-FETs across diverse electronics sectors beyond conventional CMOS replacement.

3.7 CHALLENGES AND FUTURE DIRECTIONS

3.7.1 Material quality, contact engineering, and process variability

Despite the promise of 2D materials and CNTs, issues related to material quality and contact engineering remain primary bottlenecks to commercialization. Imperfections such as grain boundaries, vacancies, and surface contamination degrade device performance and lead to significant variability across fabricated samples [5, 72]. Achieving large-area uniformity in CVD-grown 2D materials remains a challenge, especially for monolayer TMDs, where small structural defects can drastically impact transport properties. Additionally, metal–semiconductor contacts suffer from SB and Fermi-level pinning, leading to high contact resistance and reduced ON-state performance [3]. While techniques such as phase engineering and edge contacts offer promise, they lack scalability and repeatability in manufacturing. CNT-FETs are similarly constrained by chirality variation and inconsistent alignment, which influence threshold voltage and drive current [73]. Moreover, even small perturbations in doping profiles or dielectric interfaces can cause major shifts in device behavior, raising concerns about process-induced variability [74]. Addressing these challenges calls for improved synthesis protocols, selective doping, and self-aligned contact schemes that preserve interface integrity while enabling reproducible electrical characteristics.

3.7.2 Modeling gaps and standardization of metrics

The field continues to lack unified modeling frameworks that can accurately predict the behavior of 2D and CNT-FETs across different scales and configurations. Traditional drift-diffusion approaches fall short in capturing quantum confinement, ballistic transport, and DOS peculiarities inherent to low-dimensional systems [75, 76]. Furthermore, compact models used in circuit design often ignore critical factors such as contact resistance modulation, interface traps, and gate-induced bandgap changes. While several efforts have been made to develop Verilog-A compatible models for BLG and monolayer MoS_2, their accuracy remains limited to narrow bias regimes [21]. Another major issue is the inconsistent benchmarking of device performance. Metrics like SS, mobility, and contact resistance are often reported under differing test conditions, substrate choices, or extraction techniques, hindering cross-comparison and progress assessment [5]. Recent literature calls for standardized testing protocols and open-access benchmarking datasets that reflect realistic operating environments. Moving forward, the integration of physics-informed machine learning tools with quantum transport models may offer a scalable solution to bridge this modeling gap and accelerate predictive device design.

3.7.3 Long-term reliability and environmental stability

For practical deployment, devices based on 2D materials and CNTs must demonstrate long-term operational stability under thermal, mechanical, and environmental stress. However, current studies indicate pronounced degradation under prolonged bias stress, ambient exposure, and high-humidity conditions [77, 78]. 2D semiconductors like MoS_2 and black phosphorus are particularly susceptible to surface oxidation and adsorption of atmospheric species, which deteriorate carrier mobility and threshold voltage stability [3]. CNTs, while chemically inert, also exhibit hysteresis in transfer curves due to moisture-induced charge trapping. The development of robust encapsulation techniques using Atomic Layer Deposition (ALD)-grown oxides or polymeric layers has partially mitigated these effects [5]. Still, there is a need for comprehensive reliability studies that simulate real-world operating conditions over extended periods. Investigations into BTI, time-dependent dielectric breakdown (TDDB), and electrostatic discharge (ESD) sensitivity remain scarce. Future research should also examine interfacial degradation mechanisms at atomic resolution to develop chemically passivated and environmentally resilient architectures.

3.7.4 Opportunities and recommendations for future work

Looking ahead, several opportunities exist to transition 2D and CNT-FETs from lab-scale prototypes to industry-ready technologies. First, hybrid integration with mature silicon CMOS platforms can facilitate near-term adoption, enabling heterogeneous systems with enhanced performance and functionality [15]. Second, the use of 2D materials in nontraditional computing paradigms, such as neuromorphic and quantum computing, can bypass digital logic constraints and exploit their intrinsic physical properties [72]. From a materials perspective, defect-tolerant growth techniques and in-situ doping strategies offer routes to minimize variability. On the modeling front, open-source, multi-scale simulation platforms that couple ab initio methods with compact models will accelerate design cycles. Finally, establishing standard roadmaps for performance metrics and fabrication protocols—as advocated by several recent reviews—can align global research efforts. Collaborations between academia, foundries, and EDA tool developers are essential to close the gap between exploratory research and large-scale adoption. Collectively, these initiatives will determine whether the extraordinary properties of 2D and CNT-FETs can translate into practical technological impact.

REFERENCES

1. Frank Schwierz, Jörg Pezoldt, and Ralf Granzner. "Two-dimensional materials and their prospects in transistor electronics." *Nanoscale* 7.18 (2015), pp. 8261–8283.
2. Manish Chhowalla, Debdeep Jena, and Hua Zhang. "Two-dimensional semiconductors for transistors." *Nature Reviews Materials* 1.11 (2016), pp. 1–15.
3. Yuan Liu et al. "Promises and prospects of two-dimensional transistors." *Nature* 591.7848 (2021), pp. 43–53.
4. Wei Cao et al. "2D semiconductor FETs—Projections and design for sub-10 nm VLSI." *IEEE Transactions on Electron Devices* 62.11 (2015), pp. 3459–3469.
5. Senfeng Zeng, Chunsen Liu, and Peng Zhou. "Transistor engineering based on 2D materials in the post-silicon era." *Nature Reviews Electrical Engineering* 1.5 (2024), pp. 335–348.
6. Cedric Huyghebaert et al. "2D materials: roadmap to CMOS integration." *2018 IEEE International Electron Devices Meeting (IEDM)*. IEEE. 2018, pp. 22–31.
7. Abu Sebastian et al. "Benchmarking monolayer MoS_2 and WS_2 field-effect transistors." *Nature Communications* 12.1 (2021), p. 693. DOI: https://doi.org/10.1038/s41467-020-20732-w
8. Nasuha Rohaizad et al. "1T-phase transition metal dichalcogenides (MoS_2, $MoSe_2$, WS_2, and WSe_2) with fast heterogeneous electron transfer: application on second-generation enzyme-based biosensor." *ACS Applied Materials & Interfaces* 9.46 (2017), pp. 40697–40706.
9. Hesameddin Ilatikhameneh et al. "Tunnel field-effect transistors in 2-D transition metal dichalcogenide materials." *IEEE Journal on Exploratory Solid-State Computational Devices and Circuits* 1 (2015), pp. 12–18.
10. Wei Zhe Teo et al. "Cytotoxicity of exfoliated transition-metal dichalcogenides (MoS_2, WS_2, and WSe_2) is lower than that of graphene and its analogues." *Chemistry–A European Journal* 20.31 (2014), pp. 9627–9632.
11. Ruitao Lv et al. "Transition metal dichalcogenides and beyond: synthesis, properties, and applications of single-and few-layer nanosheets." *Accounts of Chemical Research* 48.1 (2015), pp. 56–64.
12. Brajesh Rawat, MM Vinaya, and Roy Paily. "Transition metal dichalcogenide-based field-effect transistors for analog/mixed-signal applications." *IEEE Transactions on Electron Devices* 66.5 (2019), pp. 2424–2430.
13. Frank Schwierz. "Graphene transistors: status, prospects, and problems." *Proceedings of the IEEE* 101.7 (2013), pp. 1567–1584.
14. Frank Schwierz. "Graphene transistors." *Nature Nanotechnology* 5.7 (2010), pp. 487–496.
15. Shuiyuan Wang, Xiaoxian Liu, and Peng Zhou. "The road for 2D semiconductors in the silicon age." *Advanced Materials* 34.48 (2022), p. 2106886.
16. Jing Guo et al. "Metal–insulator–semiconductor electrostatics of carbon nanotubes." *Applied Physics Letters* 81.8 (2002), pp. 1486–1488.
17. Lian-Mao Peng, Zhiyong Zhang, and Sheng Wang. "Carbon nanotube electronics: recent advances." *Materials Today* 17.9 (2014), pp. 433–442.

18. Elena Gnani et al. "Quantum-mechanical analysis of the electrostatics in silicon-nanowire and carbon-nanotube FETs." *Solid-State Electronics* 50.4 (2006), pp. 709–715.
19. Muhammad A. Wahab et al. "Electrostatic dimension of aligned-array carbon nanotube field-effect transistors." *Acs Nano* 7.2 (2013), pp. 1299–1308.
20. MP Anantram and F Leonard. "Physics of carbon nanotube electronic devices." *Reports on Progress in Physics* 69.3 (2006), p. 507.
21. Zhihui Cheng et al. "How to report and benchmark emerging field-effect transistors." *Nature Electronics* 5.7 (2022), pp. 416–423.
22. Sheng-Kai Su et al. "Layered semiconducting 2D materials for future transistor applications." *Small Structures* 2.5 (2021), p. 2000103.
23. CJ Dorow et al. "Gate length scaling beyond Si: mono-layer 2D channel FETs robust to short channel effects." *2022 International Electron Devices Meeting (IEDM)*. IEEE. 2022, pp. 5–7.
24. Sobia Nisar et al. "2D materials in advanced electronic biosensors for point-of-care devices." *Advanced Science* 11.31 (2024), p. 2401386.
25. Shivam Nitin Kajale et al. "2D material based field effect transistors and nano-electromechanical systems for sensing applications." *IScience* 24.12 (2021), p. 103513.
26. Laixiang Qin et al. "Steep slope field effect transistors based on 2D materials." *Advanced Electronic Materials* 10.8 (2024), p. 2300625.
27. Sayan Kanungo et al. "2D materials-based nanoscale tunneling field effect transistors: current developments and future prospects." *npj 2D Materials and Applications* 6.1 (2022), p. 83.
28. R Marani and AG Perri. "Critical analysis of CNTFET-based electronic circuits design." *ECS Journal of Solid State Science and Technology* 12.5 (2023), p. 051005.
29. Pradyot Patel and Gagnesh Kumar. "Optimization of CNTFET and effect of channel length scaling on CNTFET and MOSFET threshold voltage." *2024 5th International Conference for Emerging Technology (INCET)*. IEEE. 2024, pp. 1–8.
30. Xuesong Yao et al. "Carbon nanotube field-effect transistor-based chemical and biological sensors." *Sensors* 21.3 (2021), p. 995.
31. Furqan Zahoor et al. "Carbon nanotube field effect transistors: an overview of device structure, modeling, fabrication and applications." *Physica Scripta* 98.8 (2023), p. 082003.
32. Mazin A. Zamzami et al. "Fabrication and characterization of field ef-fect transistor based on single walled carbon nanotubes." *Journal of King Saud University – Science* 34.6 (2022), p. 102137. DOI:10.1016/j.jksus.2022.102137
33. Kevin P O'Brien et al. "Advancing 2D monolayer CMOS through con-tact, channel and interface engineering." *2021 IEEE International Electron Devices Meeting (IEDM)*. IEEE. 2021, pp. 1–7.
34. Yue Zheng et al. "Ohmic contact engineering for two-dimensional materials." *Cell Reports Physical Science* 2.1 (2021), 100298.
35. Yaoqiang Zhou et al. "Contact-engineered reconfigurable two-dimensional Schottky junction field-effect transistor with low leakage currents." *Nature Communications* 14.1 (2023), p. 4270.
36. Min Sup Choi et al. "Recent progress in 1D contacts for 2D-material-based devices." *Advanced Materials* 34.39 (2022), p. 2202408.

37. Gihyeon Kwon et al. "Interaction-and defect-free van der Waals contacts between metals and two-dimensional semiconductors." *Nature Electronics* 5.4 (2022), pp. 241–247.
38. Jorge-Daniel Aguirre-Morales et al. "A large-signal monolayer graphene field-effect transistor compact model for RF-circuit applications." *IEEE Transactions on Electron Devices* 64.10 (2017), pp. 4302–4309.
39. Jorge Daniel Aguirre-Morales et al. "Physics-based electrical compact model for monolayer Graphene FETs." *2016 46th European Solid-State Device Research Conference (ESSDERC)*. IEEE. 2016, pp. 240–243.
40. Jorge-Daniel Aguirre-Morales et al. "An accurate physics-based compact model for dual-gate bilayer graphene FETs." *IEEE Transactions on Electron Devices* 62.12 (2015), pp. 4333–4339.
41. J. D. Aguirre-Morales et al. "A new physics-based compact model for bilayer graphene field-effect transistors." *2015 45th European Solid State Device Research Conference (ESSDERC)*. 2015, pp. 180–183. DOI:10.1109/ESSDERC.2015.7324743
42. Phaedon Avouris et al. "Carbon nanotube electronics." *Proceedings of the IEEE* 91.11 (2004), pp. 1772–1784.
43. Riichiro Saito et al. *Physical Properties of Carbon Nanotubes*. Vol. 203. London: Imperial College Press, 1998.
44. Ian O'Connor et al. "CNTFET modeling and reconfigurable logic-circuit design." *IEEE Transactions on Circuits and Systems I: Regular Papers* 54.11 (2007), pp. 2365–2379.
45. Zhihong Chen et al. "An integrated logic circuit assembled on a single carbon nanotube." *Science* 311.5768 (2006), pp. 1735–1735.
46. Arijit Raychowdhury and Kaushik Roy. "Carbon-nanotube-based voltage-mode multiple-valued logic design." *IEEE Transactions on Nanotechnology* 4.2 (2005), pp. 168–179.
47. Roman Sordan et al. "Exclusive-OR gate with a single carbon nanotube." *Applied Physics Letters* 88.5 (2006),p. 053119.
48. Ali Hajimiri and Thomas H Lee. "A general theory of phase noise in electrical oscillators." *IEEE Journal of Solid-State Circuits* 33.2 (2002), pp. 179–194.
49. S Rumyantsev et al. "Electrical and noise characteristics of graphene field-effect transistors: ambient effects, noise sources and physical mechanisms." *Journal of Physics: Condensed Matter* 22.39 (2010), p. 395302.
50. Qinghui Shao et al. "Flicker noise in bilayer graphene transistors." *IEEE Electron Device Letters* 30.3 (2009), pp. 288–290.
51. Zhihong Chen et al. "Graphene nano-ribbon electronics." *Physica E: Low-dimensional Systems and Nanostructures* 40.2 (2007), pp. 228–232.
52. Atindra Nath Pal, Ageeth A Bol, and Arindam Ghosh. "Large low-frequency resistance noise in chemical vapor deposited graphene." *Applied Physics Letters* 97.13 (2010).
53. Brice Grandchamp et al. "Characterization and modeling of graphene transistor low-frequency noise." *IEEE Transactions on Electron Devices* 59.2 (2011), pp. 516–519.
54. Chhandak Mukherjee et al. "Characterization and modeling of low-frequency noise in CVD-grown graphene FETs." *2015 45th European Solid State Device Research Conference (ESSDERC)*. IEEE. 2015, pp. 176–179.

55. Yu-Ming Lin and Phaedon Avouris. "Strong suppression of electrical noise in bilayer graphene nanodevices." *Nano Letters* 8.8 (2008), pp. 2119–2125.
56. Tianhua Yu et al. "Bilayer graphene system: transport and reliability." *ECS Transactions* 28.5 (2010), p. 39.
57. WJ Liu et al. "Positive bias-induced V_{th} instability in graphene field effect transistors." *IEEE Electron Device Letters* 33.3 (2012), pp. 339–341.
58. Jung-Kyu Lee et al. "Reliability of bottom-gate graphene field-effect transistors prepared by using inductively coupled plasma-chemical vapor deposition." *Applied Physics Letters* 98.19 (2011), p. 193504.
59. Chhandak Mukherjee et al. "Versatile compact model for graphene FET targeting reliability-aware circuit design." *IEEE Transactions on Electron Devices* 62.3 (2015), pp. 757–763. DOI:10.1109/TED.2015.2395134
60. B Huard et al. "Evidence of the role of contacts on the observed electron-hole asymmetry in graphene." *Physical Review B—Condensed Matter and Materials Physics* 78.12 (2008), p. 121402.
61. Bin Liu et al. "Bias temperature instability (BTI) characteristics of graphene field-effect transistors." *Proceedings of 2011 International Symposium on VLSI Technology, Systems and Applications*. IEEE. 2011, pp. 1–2.
62. Sikandar Aftab, Hosameldin Helmy Hegazy, and Fahmid Kabir. "Emerging trends in 2D flexible electronics." *Advanced Materials Technologies* 8.11 (2023), p. 2201897.
63. Ajit Kumar Katiyar et al. "2D materials in flexible electronics: recent advances and future prospectives." *Chemical Reviews* 124.2 (2023), pp. 318–419.
64. Fan Xia et al. "Carbon nanotube-based flexible ferroelectric synaptic transistors for neuromorphic computing." *ACS Applied Materials & Interfaces* 14.26 (2022), pp. 30124–30132.
65. Guiming Cao et al. "2D material based synaptic devices for neuromorphic computing." *Advanced Functional Materials* 31.4 (2021), p. 2005443.
66. Arnab Pal et al. "An ultra energy-efficient hardware platform for neuromorphic computing enabled by 2D-TMD tunnel-FETs." *Nature Communications* 15.1 (2024), p. 3392.
67. Parvin Fathi-Hafshejani et al. "Two-dimensional-material-based field-effect transistor biosensor for detecting COVID-19 virus (SARS-CoV-2)." *ACS Nano* 15.7 (2021), pp. 11461–11469.
68. Abdul Kaium Mia, M Meyyappan, and PK Giri. "Two-dimensional transition metal dichalcogenide based biosensors: from fundamentals to healthcare applications." *Biosensors* 13.2 (2023), p. 169.
69. David C Ferrier and Kevin C Honeychurch. "Carbon nanotube (CNT)-based biosensors." *Biosensors* 11.12 (2021), p. 486.
70. Wei Cao et al. "A mode-balanced reconfigurable logic gate built in a van der Waals strata." *npj 2D Materials and Applications* 5.1 (2021), p. 20.
71. Junru An et al. "Perspectives of 2D materials for optoelectronic integration." *Advanced Functional Materials* 32.14 (2022), p. 2110119.
72. Anhan Liu et al. "The roadmap of 2D materials and devices toward chips." *Nano-Micro Letters* 16.1 (2024), p. 119.
73. Saima Batool et al. "Electrical contacts with 2D materials: current developments and future prospects." *Small* 19.12 (2023), p. 2206550.

74. Tom Schram et al. "Challenges of wafer-scale integration of 2D semiconductors for high-performance transistor circuits." *Advanced Materials* 34.48 (2022), p. 2109796.
75. Sresta Valasa et al. "A critical review on performance, reliability, and fabrication challenges in nanosheet FET for future analog/digital IC applications." *Micro and Nanostructures* 170 (2022), p. 207374.
76. Saptarshi Das et al. "Transistors based on two-dimensional materials for future integrated circuits." *Nature Electronics* 4.11 (2021), pp. 786–799.
77. Changhao Dai, Yunqi Liu, and Dacheng Wei. "Two-dimensional field-effect transistor sensors: the road toward commercialization." *Chemical Reviews* 122.11 (2022), pp. 10319–10392.
78. Mukeshchand Thakur et al. "High durability and stability of 2D nanofluidic devices for long-term single-molecule sensing." *npj 2D Materials and Applications* 7.1 (2023), p. 11.

Chapter 4

Design and performance analysis of thin-film solar cell

Anany Saxena, Pradeep Kumar, Sunny Anand, Naveen Kumar, Prabhat Singh, Sandeep Kumar, and Navjeet Bagga

4.1 INTRODUCTION

The economic feasibility is evident in the less material usage and higher energy conversion efficiency. The requirement of high efficiency goal is fulfilled by the wafer technologies but, thin film is capable of meeting minimum material usage also [1]. The minimum material usage and the high efficiency, both goals are needed to be fulfilled simultaneously for facilitating the low-cost production of electricity and the high market penetration [2]. The amorphous-Si (α-Si), cadmium telluride (CdTe), and Copper Indium Gallium Selenide (CIGS) technologies are the three most widely commercialized thin-film solar cells. The common property that is found among these materials is the direct band gap, which incentivizes the usage of very thin material [3]. In addition to the above-mentioned property, the materials also possess a lower temperature coefficient. However, the complete dependence on CdTe or CIGS technologies will demand more than 75 times the semiconductor materials, like gallium or tellurium, for use in the photovoltaics industry, which is higher in comparison to the total production for rest of all uses [4]. Moreover, it is evident that in terms of material availability, silicon-based technologies are more advantageous since silicon is available in abundance when compared to gallium and tellurium [5]. The further market growth of technologies of thin-film solar cell may be facilitated by using of more robust materials and cell architectures so as to enhance reliability and minimize the dependence on rare elements. One of the most lucrative features of $\alpha - Si$ is its direct band gap, which fosters absorption of a significant amount of solar energy [6]. However, the short orders in amorphous material and the dangling bonds are responsible for the short minority carrier diffusion lengths and the abnormal electrical behavior. But, the capability of $\alpha - Si : H$ material to reduce dangling bond densities by a large order of magnitude is responsible for an increase in the minority carrier length. However, the hydrogenation of amorphous-Si material introduces the Staebler-Wronski light degradation effect [7]. Hence, in order to contain the effect to a certain magnitude, a multijunction or tandem structure incorporating an interface of $\alpha - Si : H$ with $\mu c - Si : H$ is utilized in this work. In addition to that,

DOI: 10.1201/9781003635123-4

the absorption rate could be increased by the usage of multijunction technologies since a fraction of the energy that may be lost in single-junction solar cells can be captured and converted [8]. The multijunction technologies also account for less light-induced degradation [9]. However, Staebler-Wronski light degradation effect has demarcated the market penetration of amorphous-Si at present to consumer electronics accounting for only 1% share in the global photovoltaics industry [9, 10]. Thus, in order to scale up the acceptance of amorphous-Si in the photovoltaics industry, a textured tandem thin-film solar cell architecture has been proposed in this chapter. Moreover, the optical absorption spectrum of $\alpha - Si : H$ is highly absorptive starting at 2 eV and transparent up to 1.7 eV [11]. Additional benefits of $\alpha - Si : H$ include shorter payback time and low manufacturing cost. In this chapter, a surface texturing technique in combination with antireflective coating is used for the purpose of better trapping of light thereby striving toward greater absorption of wavelengths in the visible and the near-infrared regions. The silicon oxide layer is used as an antireflective coating owing to its lesser absorption coefficient and easier deposition method over zinc oxide (ZnO). The ZnO serves a dual role as both a transparent conductive oxide and as an anode [12]. The indium tin oxide (ITO) is utilized as a common electrode at the junction of $\alpha - Si : H / \mu c - Si : H$ in order to prevent current leakages [13]. Further, the height of triangular textures is varied from 100 to 1000 nm, and its effect on the essential parameters governing the performance of textured tandem thin-film solar cell structure is analyzed. Additionally, a comparison between non-textured and textured tandem thin-film solar cells is also made.

4.2 DEVICE STRUCTURE

The textured tandem thin-film solar cell structure is described as possessing a SiO_2 layer of 80 nm at the top. The SiO_2 or antireflective layer is utilized to minimize the reflection of incident light from the cell surface. The technique of graded refractive index is adopted for better absorption of wavelengths by the cell. The SiO_2 layer is followed by a 920 nm layer of transparent conducting oxide, ZnO. The function of a transparent conducting oxide is to ensure minimum absorption by its material while passing the wavelengths onto the absorbing regions of the structure. In addition to that, ZnO is used as an anode. The ZnO layer is succeeded by the top cell of $\alpha - Si : H$ and the bottom cell of $\mu c - Si : H$. The $\alpha - Si : H$ material is utilized as top cell owing to its higher bandgap than $\mu c - Si : H$. The p–i–n structure is incorporated in both cells. The dimensions of the top cell are a 20 nm p-type doped layer, a 200 nm intrinsic layer, and a 20 nm n-type doped layer. Similarly, the dimensions of the bottom cell consist of a 20 nm p-type doped layer, a 1.6 μm intrinsic layer, and a 20 nm n-type doped layer. The thickness ratio of 8 is considered with regard to thickness of intrinsic layer of top cell to thickness

Table 4.1 The comparison between non-textured and textured thin-film solar cell structures

Regions	*Non-textured cell (nm)*	*Textured cell (nm)*
SiO_2	500	80
Anode	500 (ZnO)	920 (ZnO)
$\alpha - Si:H$	225	240
$\mu c - Si:H$	2225	1640
Cathode	200 (Al)	110 (Ag)
Common	10 (ITO)	10 (ITO)

of intrinsic layer of bottom cell in order to optimize current matching in the textured structure. The intrinsic layers act as the absorbing regions of the structure [14]. A common electrode of ITO of 10 nm is injected at the interface of the two cells with the intention of preventing the current leakages during current flow from anode to cathode. A layer of silver of 110 nm is used as back contact owing to its remarkable electrical properties in terms of conductivity and reflectivity [15]. Moreover, it also acts as the cathode. Table 4.1 depicts the difference in the device structures of textured and non-textured tandem thin-film solar cell architectures. The textures utilized in this work are periodic triangular textures owing to its capability that maxima or minima in absorption take place for lesser height extents and significantly specific periods, thereby leaving the space for further improvements through minimal adjustments in period and height.

The height of triangular textures is varied for the purpose of analyzing its effect on the cell's performance. The heights used for the accomplishment of this purpose are 100, 250, 500, 750, and 1000 nm. The height variation with regard to textured structure has influenced the dimensions of ZnO region only. Figure 4.1(a) showcases the device structures of non-textured tandem thin-film solar cells and textured tandem thin-film solar cells with varying heights of (b) 100 nm, (c) 250 nm, (d) 500 nm, (e) 750 nm, and (f) 1000 nm, respectively.

4.3 RESULTS AND DISCUSSIONS

The important parameters that are used to characterize the performance of solar cells are J_{SC}, V_{OC}, and FF, while the conversion efficiency (η) can be determined from these parameters [16]. Upon successful simulation of the proposed architecture, the obtained results indicate the merits of applying the surface texturing technique on tandem thin-film solar cells. The values of critical parameters obtained at different triangular texture heights are plotted and analyzed to decide the most suitable height for triangular textures.

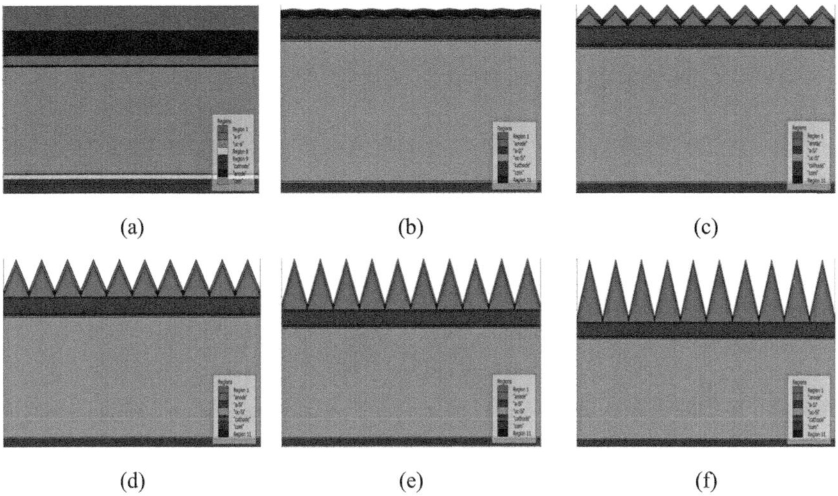

Figure 4.1 (a) The device structures of non-textured tandem thin-film solar cells and textured tandem thin-film solar cells with varying heights of (b) 100 nm, (c) 250 nm, (d) 500 nm, (e) 750 nm, and (f) 1000 nm, respectively.

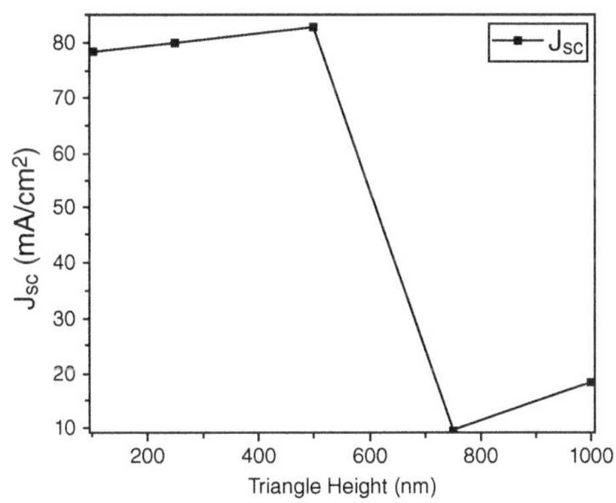

Figure 4.2 The short circuit current density (J_{SC}) curve with regard to different heights of triangular textures.

Figure 4.2 highlights the short circuit current density (J_{SC}) curve for different heights of triangular textures. The short circuit current density is dependent on the absorption rate of the solar cell, as the number of photons absorbed by the cell is accountable for better current generation. In other words, J_{SC} is a measure of the total current produced in the solar cell. From the curve, it is evident that J_{SC} is much higher than those of non-textured tandem thin-film solar cells for heights of 100, 250, 500, and 1000 nm, with an exception for the height of 750 nm, at which the value of J_{SC} is less than that of the non-textured tandem thin-film solar cell.

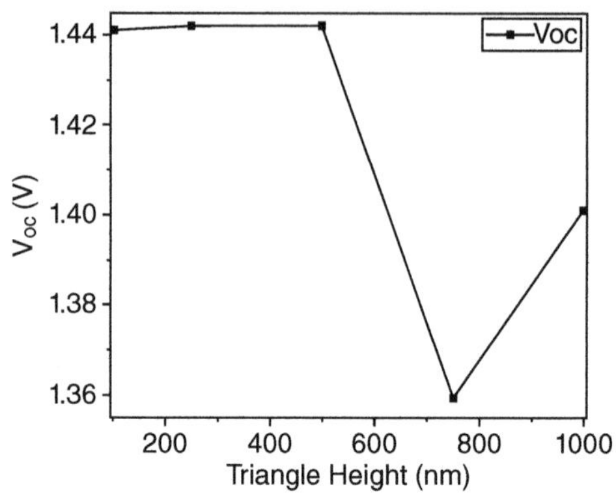

Figure 4.3 The open-circuit voltage (V_{OC}) curve with regard to different heights of triangular textures.

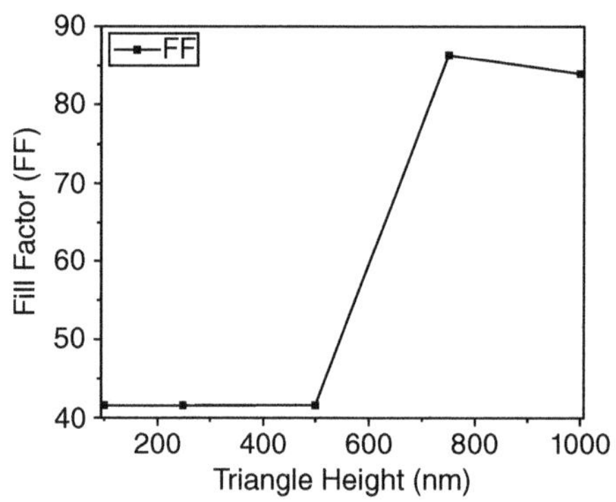

Figure 4.4 The fill factor (FF) curve with regard to different heights of triangular textures.

The forward bias at which the photocurrent density is equivalent to the dark current density is categorized as open-circuit voltage (V_{OC}) [17]. Figure 4.3 represents the open-circuit voltage (V_{OC}) curve with regard to different heights of triangular textures. The curve obtained indicates an increase in the V_{OC} for all heights of triangular textures taken for consideration.

Figure 4.4 shows the fill factor (FF) curve with regard to different heights of triangular textures. FF is defined as the ratio of the maximum power produced by a solar cell to the product of V_{OC} and J_{SC} [18]. From the curve, it can be comprehended that the value of FF for heights of 100, 250, and 500 nm is lower than that of the non-textured tandem thin-film solar cell, whereas for heights of 750 and 1000 nm, the FF is higher.

Figure 4.5 highlights the efficiency curve with regard to different heights of triangular textures. The ratio of the maximum power generated to the incident power is termed as the conversion efficiency and from the curve it is

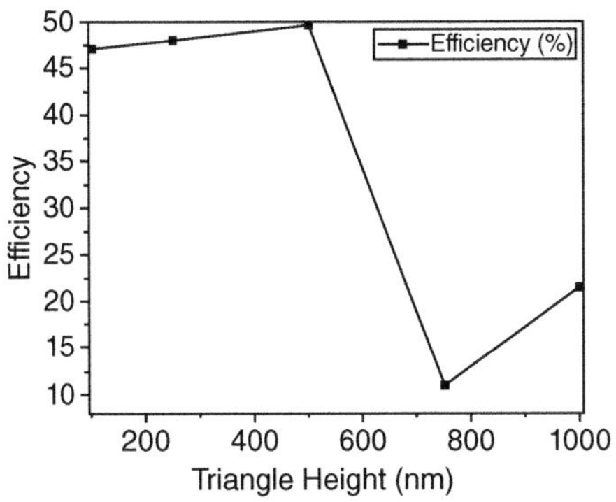

Figure 4.5 The efficiency curve with regard to different heights of triangular textures.

Table 4.2 The values of critical parameters at different heights of triangular textures

Triangular texture height (nm)	J_{SC} *(mA/cm²)*	V_{OC} *(V)*	*FF (%)*	η *(%)*
1000	18.33	1.40	85.45	21.95
750	9.396	1.36	86.32	11.03
500	82.86	1.44	41.59	49.69
250	80	1.44	41.55	47.94
100	78.52	1.44	41.56	47.04
Non-textured	10.21	1.36	71.95	9.97

depicted that the efficiency is increased for all heights of triangular textures taken under consideration [19]. Table 4.2 presents the values of J_{SC}, V_{OC}, FF, and η observed at heights of 100, 250, 500, 750, and 1000 nm, respectively, along with the values for non-textured tandem thin-film solar cell. From the table, it is observed that texturing the surface of tandem thin-film solar cell tend to have an optimistic influence on the values of J_{SC}, V_{OC}, FF, and η.

Figure 4.6 represents the cathode current versus anode voltage curves for different heights of triangular textures. The curves obtained indicate better cathode current for heights of 100, 250, 500, and 1000 nm. In contrast, at a height of 750 nm, the observed value of cathode current is the lowest among other heights and is comparable to that of a non-textured tandem thin-film solar cell.

The power versus anode voltage curves for different heights of triangular textures is highlighted in Figure 4.7. From the curves, it is visible that the curve for height of 750 nm follows the lowest trajectory when compared to the other heights. Therefore, based on the above-mentioned results, it can be

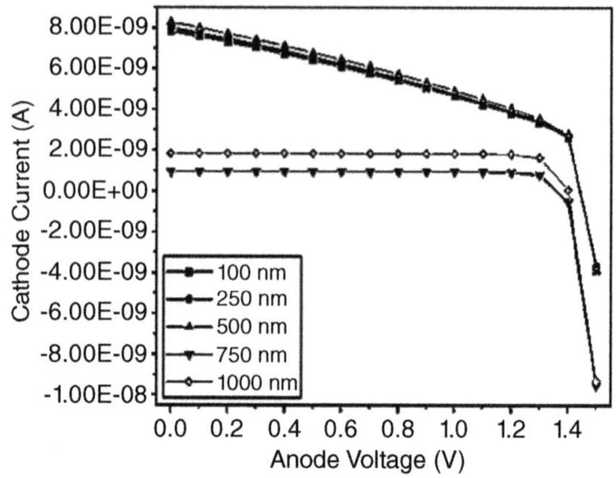

Figure 4.6 The cathode current versus anode voltage curves for different heights of triangular textures.

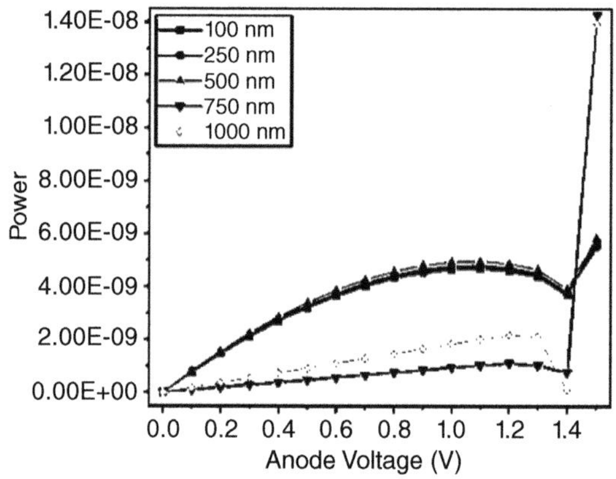

Figure 4.7 The power versus anode voltage curves with regard to different heights of triangular textures.

concluded that the application of the surface texturing technique on the tandem thin-film solar cell has contributed to an enhancement in the values of critical parameters like J_{SC}, V_{OC}, FF, and η. In addition, the usage of anti-reflective coating in combination with transparent conducting oxide and silver back contact has contributed to an increased absorption rate, thereby improving the cell performance and the energy conversion efficiency [15]. The results also highlight the suitability of 1000 nm height for triangular textures owing to its better performance compared with the other heights. There is no trade-off witnessed with respect to 1000 nm height when compared with non-textured tandem thin-film solar cells.

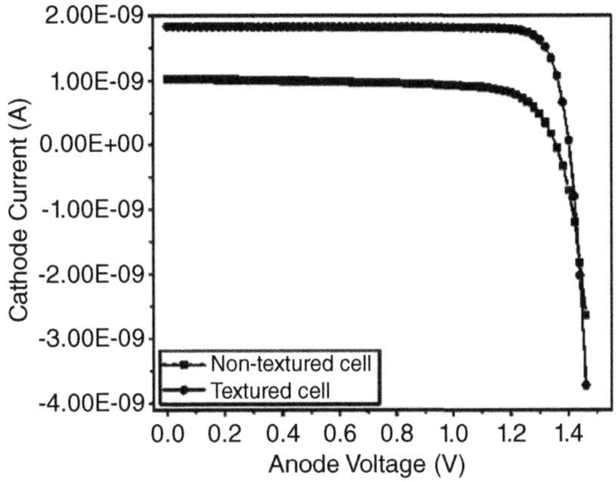

Figure 4.8 The cathode current versus anode voltage curve for non-textured and textured tandem thin-film solar cell architectures.

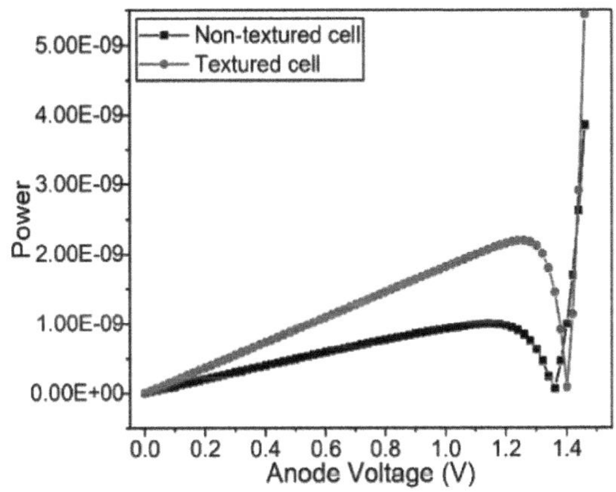

Figure 4.9 The power versus anode voltage curve for non-textured and textured tandem thin-film solar cell architectures.

Figure 4.8 represents the cathode current versus anode voltage curve for non-textured and textured tandem thin-film solar cell architectures. From the curve, it is evident that textured cells offer better cathode current owing to better optical properties. The power versus anode voltage curve for non-textured and the textured tandem thin-film solar cell architectures is showcased in Figure 4.9. The curve indicates better power performance of textured cell than its counterpart.

Table 4.3 shows the comparison between the textured thin-film solar cell structure and the other device structures. From the table, it is evident that the proposed textured thin-film solar cell structure produces better results.

Table 4.3 The comparison table highlighting the merits of a textured tandem thin-film solar cell

Devices	J_{SC} *(mA/cm²)*	V_{OC} *(V)*	*FF (%)*	η *(%)*
Textured cell	18.33	1.40	85.45	21.95
Non-textured cell	10.21	1.36	71.95	9.97
Reference [15]	22.60	0.85	68.21	13.96
Reference [16]	36	0.729	76.9	20.2
Reference [17]	3.76	0.51	38.4	0.73
Reference [18]	22.8	0.529	64.8	7.83

4.4 CONCLUSION

In this chapter, it is concluded that the application of a surface texturing technique in combination with an antireflecting coating has delivered better results than a non-textured structure owing to a better light-trapping mechanism and lesser reflectance. The conversion efficiency of 21.95% is comparable to CdTe and CIGS thin-film technologies. The market penetration of amorphous-Si technology got demarcated to consumer electronics but, the results achieved will incentivize its acceptance in the global photovoltaics industry.

DATA AVAILABILITY

All results presented in the figures and tables of the chapter have been obtained from simulation tools. All relevant data has been included in the chapter.

REFERENCES

1. Jian Wei Ho, Johnson Wong, Percis Teena, Christopher Subhodayam, Kwan Bum Choi, Divya, Anantha Narayanan, Samuel Raj, and Armin G. Aberle; "Device parameter extraction for loss analysis of silicon solar cells based on intelligent model fitting," Presented at *12th International Photovoltaic Power Generation and Smart Energy Conference & Exhibition (SNEC 2018)*, 2018.
2. N. Boukortt and B. Hadri; "Bifacial n-PERC solar cell characterization," *Indian Journal of Physics* (2019), DOI 10.1007/s12648-018-1274-5
3. N. Boukortt, S. Patane, and B. Hadri; "Development of high-efficiency PERC solar cells using Atlas Silvaco," *Indian Journal of Physics* (2019), DOI 10.1007/s12633-018-9838-8

4. Nour El Islam Boukortt and Baghdad Hadri; "Simulation of electrical characteristics of PERC solar cells," *Journal of Electronic Materials* (2018), DOI 10.1007/s11664-018-6456-6
5. ATLAS Device Simulation Software, Silvaco International, Santa Clara, CA, 2014.
6. J. J. Chaudhari and U. S. Joshi; "Fabrication of high-quality kesterite Cu2ZnSnS4 thin flms deposited by an optimized sol–gel sulphurization technique for solar cells," *Journal of Materials Science: Materials in Electronics* (2020), DOI 10.1007/s10854-020-04000-7
7. Subhash Chander; "Advancement in CdIn2Se4/CdTe based photoelectrochemical solar cells," *Advances in Energy Materials*, Shadia Jamil Ikhmayies (ed.), Advances in Material Research and Technology, Springer, 2020, DOI 10.1007/978-3-030-50108-2_2
8. Taesoo D. Lee and Abasifreke U. Ebong; "A review of thin film solar cell technologies and challenges," *Renewable and Sustainable Energy Reviews* (2017), DOI 10.1016/j.rser.2016.12.028
9. Min Su Kim, Ju Heon Lee, and Moon Kyu Kwak; "Review: surface texturing methods for solar cell efficiency enhancement," *International Journal of Precision Engineering and Manufacturing* (2020), DOI 10.1007/s12541-020-00337-5
10. Michael Berginski, Jürgen Hüpkes, Aad Gordijn, Wilfried Reetz, Timo Wätjen, Bernd Rech, and Matthias Wuttig; "Experimental studies and limitations of the light trapping and optical losses in microcrystalline silicon solar cells," *Solar Energy Materials and Solar Cells* 92(9), 1037–1042 (2008).
11. Y. Huang, W. Wang, W. Pan, W. Chen, Z. Wang, X. Tan, and W. Yan; "Comparative investigation on designs of light absorption enhancement of ultrathin crystalline silicon for photovoltaic applications," *Journal of Photonics Energy* 6(4), 047001 (2016).
12. A. Bozzola, P. Kowalczewski, and L. C. Andreani; "Towards high efficiency thin-film crystalline silicon solar cells: the roles of light trapping and non-radiative recombinations," *Journal of Applied Physics*, 115(9), 094501 (2014).
13. K. X. Wang, Z. Yu, V. Liu, Y. Cui, and S. Fan; "Absorption enhancement in ultrathin crystalline silicon solar cells with antireflection and light-trapping nanocone gratings," *Nano Letters* 12(3), 1616–1619 (2012).
14. W. Zhu, F. Xiao, I. D. Rukhlenko, J. Geng, X. Liang, M. Premaratne, and R. Jin; "Wideband visible-light absorption in an ultrathin silicon nanostructure," *Optic Express* 25(5), 5781–5786 (2017).
15. Waqas Farooq, Aimal Daud Khan, Adnan Daud Khan, Abdul Rauf, Sultan Daud Khan, Haider Ali, Javed Iqbal, Rehan Ullah Khan, and Muhammad Noman; "Thin-film tandem organic solar cells with improved efficiency," *IEEE Access* (2020), DOI 10.1109/ACCESS.2020.2988325
16. Anishkumar Soman, Ugochukwu Nsofor, Ujjwal Das, Tingyi Gu, and Steve Hegedus; "a-Si:H/c-Si interface hydrogenation for implied Voc=755 mV in Silicon heterojunction solar cell," *2019 IEEE 46th Photovoltaic Specialists Conference (PVSC)*, 2019, DOI 10.1109/pvsc40753.2019.8980808
17. Prabhat Singh, Ashish Raman, and Naveen Kumar; "Spectroscopic and simulation analysis of facile PEDOT: PSS layer deposition-silicon for perovskite solar cell," *Silicon* 12(8), 1769–1777 (2020).

18. Sreedevi Gedi, Hyunmin Jung, Salh Alhammadi, Vasudeva Reddy, Minnam Reddy, Yeongju Seo, Doohyung Moon, and Woo Kyoung Kim; "Studies on Co-sputtered Al–Zn–Sn–O films: a new TCO for thin film solar cells," *2019 IEEE 46th Photovoltaic Specialists Conference (PVSC)*, DOI 10.1109/pvsc40753.2019.8980962
19. Jaya Madan Shivani, Rahul Pandey, and Rajnish Sharma; "Designing of CZTSSe based SnS thin film solar cell for improved conversion efficiency: a simulation study with SCAPS," *2019 IEEE 46th Photovoltaic Specialists Conference (PVSC)*, DOI 10.1109/pvsc40753.2019.8980459

Chapter 5

Nano energy harvesting devices

Komatreddy Damodar Reddy, Naini Pavan Kumar, and Siva Chidambaram

5.1 INTRODUCTION TO NANO ENERGY HARVESTING

Since the beginning of the 21st century, emerging nations have experienced a surge in industrialization, a significant rise in population, and an increasing demand for enhanced living standards [1–3]. Energy demands have surged significantly due to technical developments in the production of diverse electronic goods and the swift population growth. At present, the majority of global electricity generation depends on fossil fuels (67%), followed by hydropower (16%) and nuclear energy (11%) [4–6]. Meanwhile, renewable sources such as solar energy (>2%) and wind (>4%) contribute a smaller yet steadily increasing share [7]. Previously, fossil fuels were primarily utilized to satisfy energy requirements; however, their usage has driven the planet into a climate danger zone. Moreover, the greenhouse effect and global warming are direct results of increased fossil fuel consumption [8, 9]. Also, recent years have seen significant advancements in Internet of things devices across diverse sectors, including industrial electronics, household appliances, wearable gadgets, and embedded technologies. Advanced flexible semiconductors have shown rapid progress that facilitates the transformation of conventional rigid devices into soft, lightweight, thin, long-lasting, and convenient devices equipped with sensing and responsive capabilities. Based on the present growth of IoT devices, it is anticipated that there will be over 50 billion IoT devices. Despite the minimal energy utilization of individual IoT devices, the aggregate power use of all such devices will attain a significantly substantial level, resulting in a growing demand for energy generation [10]. Most portable and embedded medical sensors rely heavily on efficient device functionality. Hence, researchers have begun concentrating on self-sustaining energy harvesting from plentiful ambient energy sources. A vast range of ambient energy sources exists in the environment, including mechanical, thermal, chemical, electromagnetic, and solar energy, each defined by a wide spectrum of energy densities [11]. To produce this energy, energy-harvester devices such as solar cells and piezoelectric, electromagnetic, triboelectric, and thermoelectric generators (TEGs) have been utilized [12–14].

DOI: 10.1201/9781003635123-5

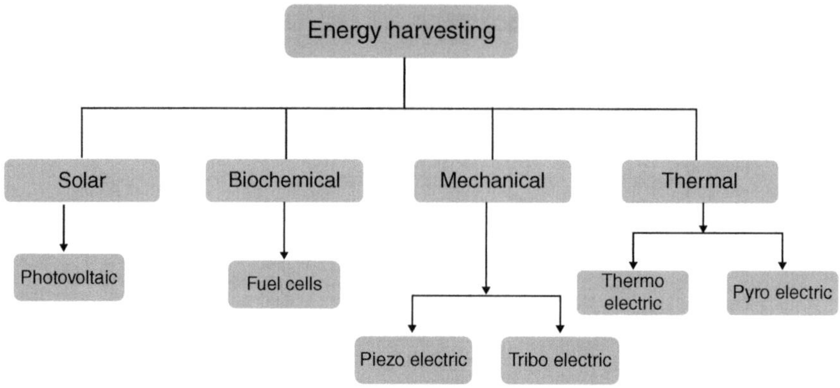

Figure 5.1 Different types of energy harvesting methods.

Nano energy harvesting devices are an innovative technology that employs nanoscale materials and structures to transform available ambient energy sources into useful electrical energy. These devices offer a viable alternative for energizing a variety of micro- and nanoelectromechanical systems (MEMS/NEMS), such as sensors, actuators, and wireless communication apparatuses [15, 16]. As the size and power requirements of electronic devices continue to decrease, energy harvesting from ambient sources becomes a feasible approach for powering autonomous systems. As shown in Figure 5.1, various renewable energy sources that are present in the environment for energy harvesting, such as solar, biochemical, mechanical, and thermal, have garnered the attention of the scientific society [17].

In this chapter, we mainly discuss energy harvesting from mechanical and thermal sources through piezoelectric, triboelectric, thermoelectric, and pyroelectric mechanisms.

5.2 PIEZOELECTRIC NANO ENERGY HARVESTING

5.2.1 Introduction to the piezoelectric effect

Various energy-harvesting materials and mechanisms have been investigated, largely inspired by the fundamental discovery of piezoelectricity. However, the progress in nanostructured piezoelectric materials is a more recent advancement in this domain. Ongoing research continues to refine the understanding of nanoscale effects on ferroelectric and piezoelectric properties, alongside the development of novel fabrication methods and emerging technological applications [18]. The term "piezoelectricity" originates from the Greek word "piezo," meaning pressure, and refers to the generation of electric charge under mechanical stress. This phenomenon was first identified in 1880 by Pierre and Jacques Curie, who observed that

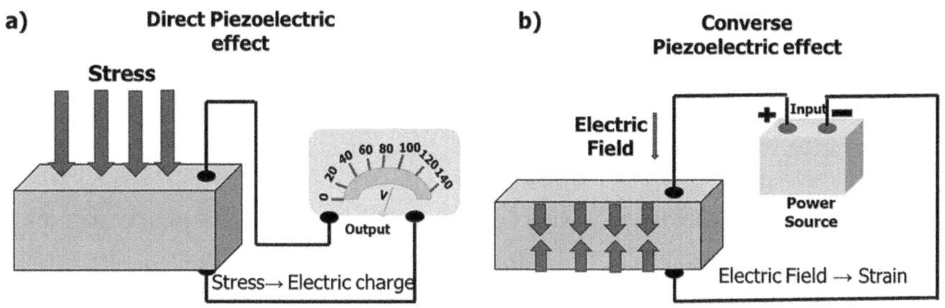

Figure 5.2 (a) Direct and (b) converse piezoelectric effect.

applying pressure to certain crystalline materials resulted in the accumulation of positive and negative charges on specific crystal surfaces [19]. The following year, in 1881, Hankel formally introduced the term "piezoelectricity" to describe this effect, which was analyzed concerning the symmetry properties of the crystals [19]. Certain crystalline substances, including quartz, barium titanate, topaz, Rochelle salt, and tourmaline, along with naturally occurring organic compounds such as cane sugar, exhibit piezoelectric properties. These materials generate an electric charge in response to applied mechanical stress, a phenomenon known as the direct piezoelectric effect. Conversely, when an external electric field is applied, these materials undergo mechanical deformation, demonstrating the converse piezoelectric effect, as shown in Figure 5.2. Both mechanisms are widely used in various applications: the direct effect is employed in sensors and energy harvesting systems, while the converse effect serves as the operating principle for actuators. The constitutive equations of direct and converse piezoelectric effects are as follows [1]:

$$\varepsilon = S\sigma + dE \tag{5.1}$$

$$D = d\sigma + \varepsilon E \tag{5.2}$$

where σ and e are the stress and the strain; D and E are the electric displacement and the electric field intensity; and s, d, and ε are the elastic compliance, the piezoelectric coefficient, and the permittivity, respectively [20].

The key feature of piezoelectric materials is the presence of noncoincident positive and negative charge centers (i.e., noncentrosymmetric materials) within the material. Various piezoelectric materials have been advanced for piezoelectric nano generators PENGs over the past decades, such as barium titanate ($BaTiO_3$), zinc oxide (ZnO), aluminum nitride (AlN), lead zirconate titanate (PZT), $Pb(Mg_{1/3}Nb_{2/3})O_3$–$PbTiO_3$ (PMN-PT), molybdenum sulfide (MoS_2), and poly(vinylidene fluoride) (PVDF) [21].

5.2.2 Piezoelectric nanogenerators

Mechanical energy is converted into electrical energy through technologies based on three fundamental scientific principles: the electrostatic effect, electromagnetic induction, and the piezoelectric effect [22]. In electromagnetic energy harvesters, ambient mechanical vibrations induce variations in the magnetic flux linked to a closed conducting loop, thereby generating electrical power. The fundamental working principle of this energy conversion method necessitates the presence of magnetic materials [23]. Although this approach has been widely employed in various power grids for harvesting ambient mechanical energy, achieving a power density of approximately 400 $\mu W/m^3$ with an efficiency of around 30%, it is associated with several limitations [24]. These include high installation costs, the requirement for a large operational area, dependence on specific magnetic materials, and a lack of portability.

In electrostatic generators, the electrostatic force arises between two charged capacitor plates with their respective motion, and the work done against this force by applied mechanical vibrations is transformed into electricity [24, 25]. Although very efficient, the requirements for pre-charging electrets and the expensive production process have rendered this method uncommon in practical portable applications.

Piezoelectric energy harvesting occurs when an active material transforms mechanical deformations, motion, fluid dynamics, impact forces, or vibrations into electrical energy [26]. These types of generators have garnered significant attention recently due to their low production cost, excellent compactness and flexibility in wearable form, and adequate energy conversion efficiency. The applications of nanostructured piezoelectrics are a recent advancement, and the comprehension of nanoscale size effects on ferro- and piezoelectricity is still evolving. Many developments are connected to novel tools and applications of ferroelectric materials. Nonetheless, the utilization of piezoelectric nanostructures for energy harvesting has proliferated significantly over the past decade, resulting in a vast array of documented systems [27].

In 2006, the idea of nanogenerators (NGs) was first introduced by Wang and Song, who demonstrated that the NG device can convert any irregular mechanical vibrations into electricity using ZnO [28]. Superior portability has made this technique useful for various real-life applications. Later, research on the development of NG devices has received intensive attention due to their several advantages, such as the ability to operate at low frequencies, enhanced piezoelectric performance, high sensitivity, and superior mechanical properties compared to micro-mechanical devices [29]. To enhance the output of PENGs, several studies focus on material selection, new device designs, and structural integrations. Figure 5.3 shows the increasing trend of publications on PENGs from 2005 to 2024 (Collected data from Scopus by searching with the keyword '*Piezoelectric nanogenerators*').

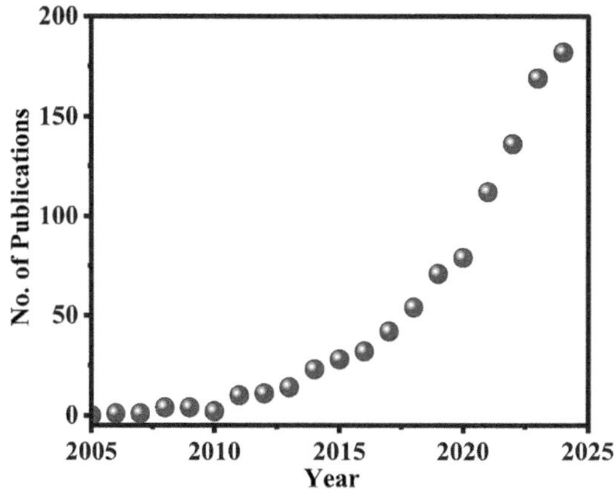

Figure 5.3 No. of publications on PENGs from 2005 to 2024 (As per the Scopus list).

5.2.3 Piezoelectric materials for energy harvesting applications

The choice of piezoelectric matrix materials significantly influences the performance of piezoelectric nanogenerators (PNGs). Variations in the piezoelectric material used can lead to substantial differences in the efficiency and output characteristics of these devices. For instance, Hu et al. demonstrated the structures and output performance graphs of PNGs fabricated using ZnO, PZT, $BaTiO_3$, and PVDF nanowires [30].

Zinc oxide (ZnO) was among the first semiconducting piezoelectric materials to be investigated. It is a highly functional material due to its advantageous optical, semiconducting, and piezoelectric characteristics, thereby making it a promising candidate for developing PNGs. Zinc oxide (ZnO) is a wide-band gap semiconductor with an energy gap of 3.37 eV. It is a naturally abundant metal oxide known for its low cost, chemical stability in air, and biocompatibility [31]. Under ambient conditions, ZnO predominantly crystallizes in the hexagonal wurtzite phase, in which four O^{2-} anions tetrahedrally coordinate each Zn^{2+} cation and vice versa. This tetrahedral coordination results in a non-centrosymmetric crystal structure. In the absence of external stress, the centers of mass of the positive and negative ions coincide at the center of the tetrahedral configuration [32]. Flexible PNGs utilizing patterned ZnO nanowire arrays synthesized via the hydrothermal method are designed for efficient energy harvesting applications [31]. The maximum output current was determined to be around 150 nA, which is adequate for operating select micro- and nanoelectronic devices [33]. Numerous review articles have discussed the advancements in research on ZnO-based PNGs [32, 34].

In another study, the fabrication of the piezoelectric nanocomposites involves the dispersion of PMN-PT nanorods within the piezoelectric PVDF polymer. The 20% PMN-PT nanorods-PVDF composite shows the highest output voltage of 110.3 V, with an output current of 46 nA. This configuration yielded a 13-fold increase in output voltage and a 4.5-fold increase in output current compared to the pure PVDF piezoelectric polymer. The outstanding performance of output voltage and current density can be ascribed to the elevated piezoelectric constant and the distinctive piezoelectric nanostructures. PMN-PT/PVDF nanocomposites exhibit significant potential for applications in harvesting electricity and self-sustaining sensing [35].

$BaTiO_3$, recognized for its lead-free composition and impressive piezoelectric coefficient, along with a high dielectric constant, has been extensively studied for applications in NGs, capacitors, sensors, and more [36, 37]. Considerable advancements have been achieved in the development of lead-free $BaTiO_3$-based piezoelectric materials in recent years. Chen et al. introduced an efficient, flexible PNG utilizing a piezoelectrically enhanced nanocomposite micropillar array composed of polyvinylidene fluoride-trifluoroethylene (P(VDF-TrFE))/barium titanate ($BaTiO_3$) aimed at energy harvesting and highly sensitive self-powered sensing [38]. Vertically aligned P(VDF-TrFE)/$BaTiO_3$ nanocomposite micropillar arrays with enhanced piezoelectric properties were synthesized using a reproducible and scalable nanoimprinting technique. The resulting piezoelectric device demonstrates a significant performance improvement, generating a voltage of 13.2 V and a current density of 0.33 $\mu A/cm^2$, representing a 7.3-fold increase compared to the unmodified P(VDF-TrFE) bulk film [38].

Yan et al. investigated the incorporation of silver nanoparticles as a secondary phase into porous barium calcium zirconate titanate to modulate its ferroelectric, dielectric, and piezoelectric characteristics. Their findings indicate that an optimal concentration of silver particles facilitates a reduction in ferroelectric domain size, thereby enhancing the piezoelectric coefficient and the energy harvesting figure of merit. Furthermore, by implementing an array-structured design, the piezoelectric composite energy harvester achieved a maximum output voltage of 76 V and a current density of 340 μA, corresponding to a peak power density of 0.97 mW/cm^2 [39].

5.3 TRIBOELECTRIC NANO ENERGY HARVESTING

5.3.1 Introduction to the TENGs

Triboelectric nanogenerators (TENGs) have emerged at the conjunction of tribology and interfacial charge transfer, establishing a novel framework for efficiently capturing mechanical energy from everyday environments [40]. The TENG, first introduced in 2012, harnesses the principles of contact electrification and electrostatic induction to efficiently transform mechanical energy into electrical power [41]. Triboelectrification occurs when two distinct materials make contact and transfer electrical charges. Variations in the separation or contact area between these materials induce an electric potential, driven by the displacement current resulting from the movement of static charges [13]. This phenomenon holds significant potential for applications in displacement sensing and the conversion of mechanical energy into electrical energy [42]. TENGs exhibit superior flexibility and energy density, making them particularly advantageous for miniaturized and planar-integrated systems compared with piezoelectric and electrostatic energy harvesters. A TENG represents an innovative technology with diverse applications, encompassing energy harvesting, self-sustained sensing, biomedical treatments, and electronic devices [43]. TENG is capable of capturing various forms of mechanical energy present in everyday life, including human motion, walking, vibrations, mechanical stimuli, rotational movement, wind, vehicular motion, flowing water, raindrops, tides, and ocean waves. They offer advantages such as affordability, lightweight construction, straightforward fabrication, environmental compatibility, and versatile design, while also possessing the capability to generate high voltages due to their extremely low capacitance [44]. As a result, it represents a pioneering method of energy harvesting. The open-circuit output voltage of TENGs, denoted as V_{TENG}, can be expressed as follows [45]:

$$V_{TENG} = \frac{Q_{sc}}{C(x)} \tag{5.3}$$

where Q_{sc} represents the amount of charge transferred during a short-circuit condition, and $C(x)$ represents the capacitance between the two electrodes, which varies with displacement x. For any TENG with displacement ranging from 0 to x_{max}, the maximum output voltage remains unaffected by the frequency of the mechanical input [46]. Thus, TENGs are well suited for harvesting low-frequency mechanical energy, such as that from ocean waves. Additionally, TENGs typically generate a relatively high output voltage (~100 V), which helps mitigate losses during rectification and regulation, making them effective for battery charging [47].

5.3.2 Triboelectric operational modes and materials

To enhance TENG output performance, significant research has been conducted on both system and circuit design. As shown in Figure 5.4, four distinct operational modes—contact–separation [48], sliding [49], single-electrode [50], and free-standing [51]—have been established. Additionally, various device architectures incorporating elastic spacers or arc-shaped substrates have been proposed. A comparative analysis of these operational modes is presented in Table 5.1.

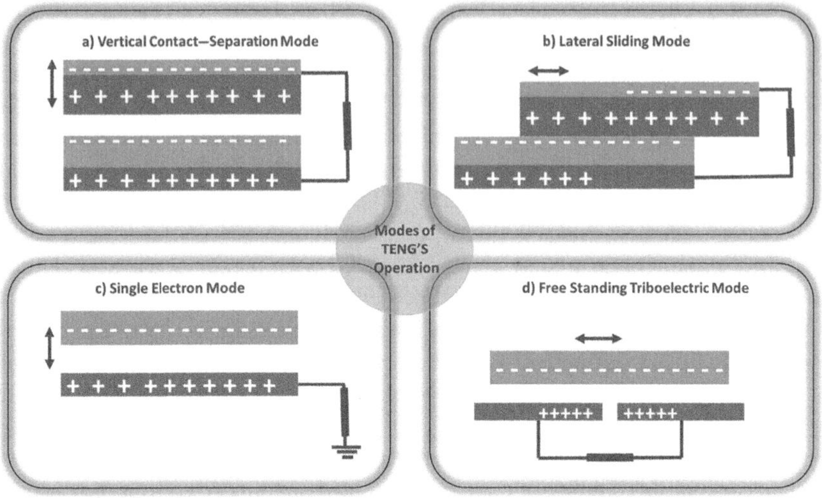

Figure 5.4 Working mechanism of triboelectric nanogenerators.

Table 5.1 Comparison of operational modes in triboelectric nanogenerators (TENGs)

S. no.	*Modes of operation*	*Pros*	*Cons*
1	Contact–separation mode	Stable and high output, but the air breakdown condition is attained with difficulty	The structure and fabrication process are more complex
2	Sliding mode	Electric charges are produced due to oscillatory motion	Issues of wear and tear
3	Single-electrode mode	Easy to design and fabricate and has one free layer	Negatively impacted by the electrostatic shield effect
4	Free-standing mode	Highest performance figure of merits in CS design and one freely moving layer	Complex structure and fabrication process

Source: Mathew et al. [59].

5.3.3 TENG materials

A wide range of materials serve as triboelectric layers, including polymers, metals, and inorganic compounds. Among them, dielectric polymers, including polytetrafluoroethylene [52], fluorinated ethylene propylene [53], polydimethyl siloxane [54], and Kapton [55], are the most frequently used. Recently, there has been growing interest in environmentally friendly materials, particularly those derived from cellulose [56]. The integration of cellulose-based materials into TENGs has led to notable enhancements in their performance [57]. The selection of materials plays a crucial role in the performance of TENGs, as their triboelectric properties are fundamental to device operation. When designing a TENG for specific applications, several key parameters must be taken into account, including power density, operational stability, environmental sustainability, and mechanical flexibility [58].

A highly efficient and environmentally friendly triboelectric nanogenerator (FG-TENG) utilizing cellulose-based tribolayers has been reported [56]. This device achieves an output power density exceeding 300 W/m^2, setting a new benchmark for TENGs composed of sustainable materials. The exceptional performance of the FG-TENG is attributed to the high positive charge density of regenerated cellulose. Moreover, the device demonstrates remarkable stability, maintaining consistent operation over 30,000 cycles under humidity conditions ranging from 30% to 84%.

Further, the TENG was constructed using Ar-treated Kapton with ZnO particles as the positively charged triboelectric layer and pristine Kapton as the negatively charged triboelectric layer, functioning in the vertical contact–separation mode. The Kapton-ZnO device produces an electrical output of 57 V, 1.21 μA current, and a 1.34 nC charge. It demonstrates stable performance, enabling the charging of conventional capacitors and the operation of electronic devices such as wristwatches and calculators [55].

Conventional metal electrodes often face challenges such as limited durability, which constrains the broader application of TENGs. In contrast, graphene, with its distinctive structure and superior electrical characteristics, presents a promising alternative for enhancing TENG efficiency and reliability [60]. In 2021, Yan et al. developed an intelligent human–machine interface (HMI) system with wireless control capabilities for managing personal electronic devices. This system incorporated a 9-digit arrayed touch panel utilizing a laser-induced graphene (LIG)-patterned TENG [61]. Subsequently, in 2022, Luo et al. introduced a self-powered vibration sensor based on LIG, demonstrating its applicability for deployment on curved surfaces and its ability to monitor the intensity of human activity. This sensor effectively differentiates between various motions, such as slipping, walking, jumping, and running [62]. In the same year, further advancements were made by Xia et al., who designed an integrated self-powered measurement and control system composed of LIG-based pressure sensors and a TENG [63]. Additionally, Hu et al. developed an ultra-sensitive hydrogel-based pressure–strain sensor, which was capable of recognizing distinct utterances when placed on the throat's surface [64]. This hydrogel sensor also functioned as a bioelectrode for detecting human electrophysiological signals.

5.4 THERMOELECTRIC NANO ENERGY HARVESTING

5.4.1 Introduction to thermoelectric nano generators (TENs)

Over two-thirds of the globally consumed energy is lost as waste heat, dissipating into the environment without being utilized. Harnessing this largely unexploited thermal energy for clean, emission-free power generation presents significant economic and environmental advantages [65]. Thermoelectric materials and devices, which enable the direct conversion of heat into electrical energy, have emerged as promising candidates for addressing global energy challenges [66]. The thermoelectric effect was first introduced by T. Seebeck in 1821, enabling the direct conversion of thermal energy into electrical energy due to a temperature difference between two electrodes of thermoelectric materials (Figure 5.5a) [67, 68].

Thermoelectric conversion technologies, especially TEGs, offer an innovative approach to sustainable energy harvesting [69–71]. TEGs, which transform heat into electrical energy through the Seebeck effect, have been developed. A key advantage of TEGs is their reliability, as they can function continuously without the need for light or mechanical movement, provided a heat source is available. TEGs can harness thermal energy from diverse sources, including the human body, automotive systems, and industrial processes, thereby expanding the practical applications of thermoelectric technology. Since the early 19th-century discoveries of thermoelectric effects by Seebeck and Peltier, extensive research has been conducted on various materials for thermoelectric power generation, ranging from metals and ceramics to advanced semiconductors.

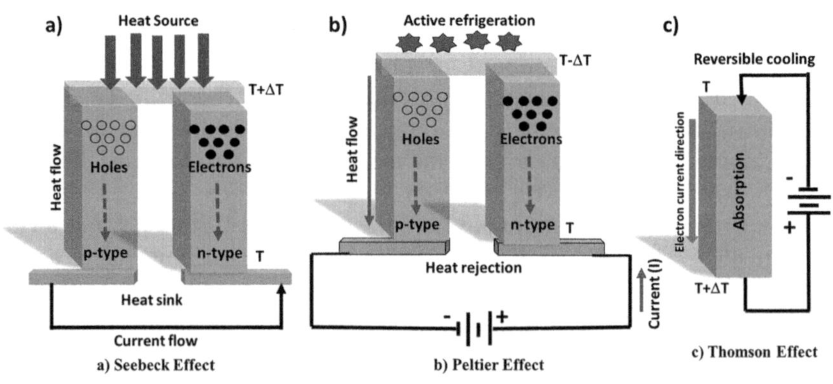

Figure 5.5 Schematic representation of (a) Seebeck effect, (b) Peltier effect, and (c) Thomson effect.

Thermoelectric performance is evaluated by a dimensionless figure-of-merit, ZT, defined as

$$ZT = \left(\frac{S^2\sigma T}{k}\right) \tag{5.4}$$

where S denotes the Seebck coefficient, T denotes the absolute temperature, σ denotes the electrical conductivity, k denotes the thermal conductivity, and $S^2\sigma$ indicates the power factor. Unlike electrical conductivity (σ), which increases proportionally with the charge carrier concentration (n), the Seebeck coefficient (S) generally exhibits an inverse relationship, decreasing as the charge carrier concentration increases. Thus, insulating materials with electronic structures characterized by large energy gaps are expected to exhibit Seebeck coefficients of 1 mV/K or higher. However, their extremely low electrical conductivity (below 10^{-12} S/cm) poses significant challenges for accurate thermoelectric voltage measurements [72]. For converting heat into electricity, the thermoelectric power factor $(S^2\sigma)$ serves as a key parameter, typically reaching its peak within an intermediate charge concentration range of semiconducting materials. This behavior arises because these materials generally exhibit thermal conductivity (k) below unity, primarily due to the lack of long-range crystalline order.

On the other hand, thermoelectric coolers (TECs), which utilize the Peltier effect to convert electrical energy into heat, are employed in various applications such as cooling electronic devices, medical instruments, and portable refrigeration systems [73, 74]. In 1834, J. C. A Peltier discovered that when an electric current traverses a circuit composed of two distinct conductors, both cooling and heating effects are detected at separate junctions (Figure 5.5b). The shift in temperatures at the connections is called the Peltier effect. The Peltier effect may be utilized for both heating and cooling, although its primary purpose is cooling. The temperature of the junctions varies depending on the direction of current flow within the circuit, resulting in either heating or cooling effects. TEGS and TECs collectively enhance the versatility of thermoelectric technology, making it an essential tool for both energy conversion and thermal regulation [75, 76]. The voltage induced across the thermoelectric materials is directly proportional to the temperature gradient, analogous to the Seebeck effect. In addition to the Peltier and Seebeck effects, as shown in Figure 5.5c, the other thermoelectric phenomenon is the Thomson effect, which pertains to reversible heating or cooling that occurs in a conductor when both an electric current and temperature gradient are present [75].

5.5 THERMOELECTRIC MATERIALS

In the pursuit of high-efficiency devices, extensive investigations have been conducted on both organic and inorganic flexible thermoelectric materials [77]. While organic materials offer numerous benefits, including favorable

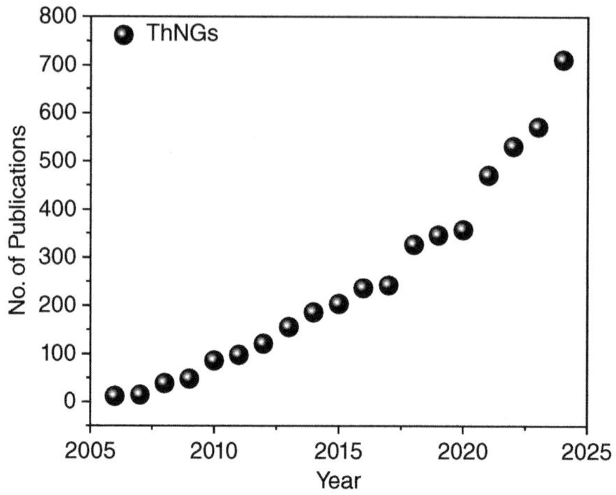

Figure 5.6 No. of papers published on TENs each year from 2006 to 2024 (as per Scopus list).

mechanical properties, affordability, and plentiful availability, they typically exhibit a significantly low Seebeck coefficient, resulting in reduced ZT values [78, 79]. Figure 5.6 shows the increasing trend of thermoelectric NG articles from 2006 to 2024 as per the Scopus list.

Bismuth telluride (Bi_2Te_3) is fundamentally recognized as a highly effective thermoelectric material with extreme conversion efficiency at ambient temperature. Recent research has garnered significant attention in energy harvesting, semiconductor cooling, sensor technology, and various domains of materials science due to its promising applications [80]. Bi_2Te_3 constitutes a class of narrow band gap semiconductor materials characterized by relatively elevated electrical conductivity, a significant Seebeck coefficient, and reduced thermal conductivity in the 300–500 K temperature range. To date, bismuth telluride, along with its alloys with antimony telluride or bismuth selenide, has emerged as the most advanced thermoelectric materials system suitable for applications at ambient temperatures [81]. Some commercial thermoelectric devices and modules for power generation and solid-state cooling in this temperature range have been successfully developed based on bismuth telluride and its alloy. Enhancing the electrical conductivity and Seebeck coefficient can improve thermoelectric performance. Nevertheless, governing these parameters is difficult due to their connection. Nanowires possess a large surface area, which helps reduce thermal conductivity. To realize high ZT values for BTS, many researchers have adopted different strategies, such as nanostructuring, point defects [82], ordered architectures, nanocompositing [83], and structure texturing [84], to reduce lattice thermal conductivity and to improve its TE performance.

The report by Bushra et al. demonstrates that incorporating 3D topological Bi_2Se_3 insulator nanoparticles into the n-type $Bi_2Te_2Se_{0.3}$ nanocomposites significantly enhances the power factor by 36% and lowers thermal conductivity by 16%. It has been shown that the addition of 0.3 vol% of Bi_2Se_3 to the $Bi_2Te_2Se_{0.3}$ achieves a high ZT of 1.28 at 375 K and an average ZT of 1.16 (300–525 K) [85]. A flexible thin film was fabricated using p-type $Bi_xSb_{2-x}Te_3$ via the thermal diffusion method, which shows a maximum ZT value of 1.11 near room temperature (RT). Furthermore, a flexible device fabricated with p-type $Bi_xSb_{2-x}Te_3$ and n-type Ag_2Se thin films exhibited a remarkable power density of nearly 1028 μW/cm^2 at a 25 K temperature difference. These advantageous performances of $Bi_xSb_{2-x}Te_3$ are because of its exceptional crystallinity, low modulus, and improved carrier concentration [86].

Mg_2X (X = Si, Sn, or Ge) samples are of significant interest due to their nontoxic nature, high elemental presence, and favorable thermoelectric properties in the medium-temperature range. Modification of these systems using suitable methods such as doping, heterojunctions, and structure texturing can provide improved thermoelectric properties. Hence, Yongbin et al. followed a heterogeneous structure strategy to enhance ZT and to reduce the thermal conductivity. The authors developed heterogeneous nanocomposites of $(Mg_2Sn_{0.85}Sb_{0.15})_{1-x}$–$(Mg_3Sb_2)_x$ and achieved a ZT value of 1.4 at 773 K for $x = 0.075$ sample. The attained figure of merit is 14% higher than the $x = 0$ sample (i.e., $Mg_2Sn_{0.85}Sb_{0.15}$) [87].

In recent times, sulfur-based materials have been recognized for thermoelectric applications, and considerable advancement has been made in optimizing thermal and electrical transport properties due to their equivalent structural and chemical properties to tellurides and selenides [88]. The sulfur-based TE material $Cu_5SnMgGeZnS_9$ shows a ZT value of 0.58 at a temperature difference of 773 K [89]. Yu et al. examined the impact of various dopants on improving the performance and thermal stability of $Cu_{2-x}S$. Among the evaluated samples, Pb-doped $Cu_{2-x}S$ demonstrates the most significant enhancement in stability concerning sulfur volatilization. Furthermore, the incorporation of Pb facilitates the adjustment of charge carrier concentration, leading to a notable enhancement of the power factor. At an optimal Pb concentration of 0.3%, a ZT value of 2.03 was achieved at 880 K, representing the highest value recorded for $Cu_{2--x}S$ [90]. The synthesis procedure and conditions significantly influence the microstructural, crystalline, and morphological studies, as well as their thermoelectric properties [91].

5.6 PYROELECTRIC NANO ENERGY HARVESTING

5.6.1 Introduction to PyENs

As previously discussed, thermoelectric materials facilitate thermal energy conversion into electrical energy. Alternatively, pyroelectric energy harvesting presents a compelling approach for utilizing heat energy by exploiting temperature fluctuations to generate electrical power [92]. Unlike thermoelectric materials, which generate electricity based on spatial temperature gradients, pyroelectric materials operate in response to temporal temperature variations [92, 93]. The pyroelectric effect is typically observed in ferroelectric materials that demonstrate temperature-dependent variations in spontaneous polarization. These materials generate a transient electrical potential in response to thermal variations [94]. Over the years, advancements in materials science, nanotechnology, and energy harvesting methodologies have significantly contributed to the refinement and expansion of this phenomenon [95, 96]. These materials are capable of converting small temperature fluctuations into electrical energy while exhibiting a high-frequency response [97]. For the same temperature fluctuation, the theoretical efficiency of pyroelectric energy harvesters can surpass that of TEGs based on the Seebeck effect [93]. Pyroelectric materials offer several benefits over thermoelectric materials, including high sensitivity, rapid response times, and simpler designs, since thermoelectric systems need large heat sinks to sustain the temperature gradient, etc. [98]. Pyroelectric materials have been extensively employed across diverse sensing technologies, including self-powered thermal sensors, gas and fire detection systems, pollution monitoring devices, radiometers, and laser detectors [99–103]. Additionally, they play a crucial role in thermal imaging and electrochemical processes, such as hydrogen generation, dye decomposition, and sterilization/disinfection [104–107]. Furthermore, these materials are integrated into thin-film calorimeters, NGs, and thermal energy harvesting systems, thereby contributing to advancements in energy conversion and sustainable technology [108].

Pyroelectricity was known more than 2400 years ago by the Greek philosopher Theophrastus; however, it has been used mainly for detector and sensor applications since the 1960s [109]. Over time, numerous pyroelectric materials have been identified and studied. Notable examples include $BaTiO_3$ (BTO), $Ba_{0.5}Sr_{0.5}TiO_3$ (BST), $Pb(Mg_{1/3}Nb_{2/3})O_3$–$PbTiO_3$(PMN-PT), $Pb_{0.52}Zr_{0.48}TiO_3$ (PZT), $SrBaNb_2O_6$(SBN), and polyvinylidene fluoride (PVDF) [110–115].

Pyroelectric materials garnered significant attention due to their superior thermodynamic efficiency [116, 117]. Achieving periodic temperature fluctuations remains a key technical challenge in utilizing these materials for energy harvesting. To address this issue, certain studies have employed cyclic thermal pumping to convert a static temperature gradient into a dynamic temperature variation [118, 119]. For practical implementation, the energy consumption associated with cyclic pumping must remain low relative to the harvested power. Furthermore, to enhance output current, pyroelectric materials should possess an increased surface area and optimized pyroelectric coefficients and be subjected to rapid temperature variations [91].

5.6.2 Pyroelectric effect

Pyroelectric materials belong to a specialized category of dielectric materials because of their polar symmetry. These materials exhibit spontaneous polarization (P_s) even in the absence of an external electric field. In ionically bonded crystalline structures, polarization primarily results from the inherent arrangement of ions within the lattice [120]. Conversely, in polymeric systems, polarization arises from the alignment of covalently bonded atoms along the molecular chains [121]. Similar to the converse piezoelectric effect, pyroelectric materials also demonstrate a reverse phenomenon known as the electrocaloric effect, wherein a temperature change occurs as a result of variations in polarization [122, 123].

For a homogeneous pyroelectric material in which the temperature T remains uniformly distributed at all times, the resulting pyroelectric current can be expressed as follows:

$$I = \frac{dQ}{dt} = S\lambda \frac{dT}{dt}$$

where Q is the induced charge, t is the time, and S is the electrode surface of the sample and the pyroelectric coefficient λ can be given as follows [124]:

$$\lambda = \frac{dP_s}{dT}$$

where P_s is the spontaneous polarization, and T is the temperature.

For pyroelectric layers positioned between two metallic electrodes in a flat-plate capacitor configuration and polarized along the axis normal to the plates, the spontaneous polarization (P_s) is oriented perpendicular to the capacitor surface, with its magnitude corresponding to the surface charge density.

Figure 5.7 shows the pyroelectric mechanism with a diagrammatic illustration of the dipole orientation in response to the temperature changes. Thermal fluctuations will cause the electric dipoles to fluctuate randomly from their respective alignment axes. As shown in Figure 5.7a, when the

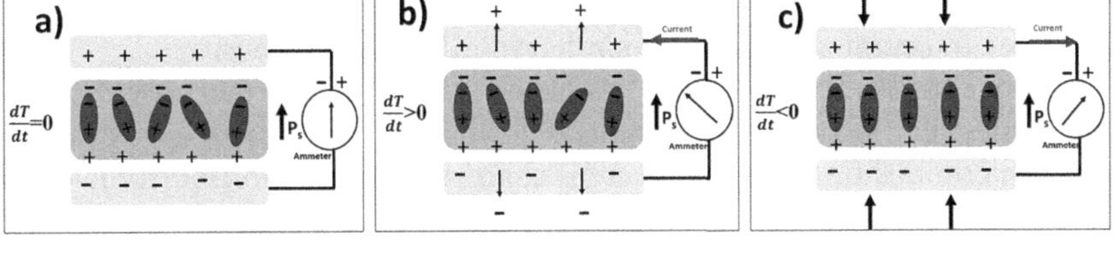

Figure 5.7 The working mechanism of the pyroelectric effect.

temperature remains constant $\left(\frac{dT}{dt}=0\right)$, the total average polarization strength of the electric dipoles also remains constant, resulting in no current. When the temperature of the pyroelectric material is raised from room temperature to a higher level $\left(\frac{dT}{dt}>0\right)$, the electric dipoles oscillate more vigorously around their alignment axes. These heightened oscillations reduce the average spontaneous polarization due to widening angles of oscillation. Consequently, as shown in Figure 5.7b, the number of induced charges on the electrodes is reduced, leading to a current in the external circuit. Conversely, cooling the material $\left(\frac{dT}{dt}<0\right)$, instead of heating it, enhances spontaneous polarization, as the electric dipoles oscillate with a narrower range of angles due to decreased thermal activity. As a result, the overall magnitude of polarization increases, and the induced charges at the electrodes rise, causing the electrons to flow in the opposite direction, as illustrated in Figure 5.7c [125, 126].

5.6.3 Pyroelectric materials

Over the past 20 years, extensive research has been conducted on pyroelectric materials, yielding significant experimental advancements in energy harvesting applications. Several review articles have been published, discussing various pyroelectric materials and the techniques utilized for thermal energy harvesting [92]. Lang et al. presented a historical perspective on pyroelectricity, while previous reviews have also examined pyroelectric materials and associated energy harvesting methods [127]. Pyroelectric materials can be classified based on their composition and structure, including single crystals, composites, ceramics, polymers, and inorganic thin films. The perovskite-structured ferroelectric ceramics have gained greater popularity compared to polymers for Pyroelectric nanogenerator (PyENG) applications. Ceramics exhibit remarkable robustness and maintain stability under high humidity or vacuum conditions, demonstrating superior thermal stability compared to polymers. Furthermore, these materials exhibit a significant pyroelectric coefficient along with minimal dielectric loss, making them ideal for energy-harvesting applications. The pyroelectric properties of these ceramic materials are influenced by their composition, enabling the engineering of many ceramics to optimize the maximum harvestable energy density.

So far, considerable attention has been focused on improving pyroelectric properties. For example, Ti-doped $Pb_{0.99}Nb_{0.02}[(Zr_{0.57}Sn_{0.43})]_{0.98}O_3$ ceramics showed compositional and structural changes near the morphotropic phase boundary (MPB), leading to enhanced pyroelectric properties. It was observed that as the Ti^{4+} content increased, the ferroelectricity of the system gradually enhanced due to the suppression of antiferroelectric behavior.

However, the sample with x = 0.06 ($Pb_{0.99}Nb_{0.02}[(Zr_{0.57}Sn_{0.43})_{0.94}]_{0.98}Ti_{0.06}O_3$) exhibited no pyroelectric response, which was attributed to its predominantly antiferroelectric nature. In contrast, with slight compositional variations while maintaining a similar microstructure, the ceramics with x = 0.07 and 0.075 demonstrated pyroelectric coefficients (p) of 683 and 630 μC/m^{2}K at room temperature, respectively. Notably, these samples exhibited significantly higher pyroelectric coefficients of 695.5 × 10^{-4} C/m^{2}K at the ferroelectric-antiferroelectric phase transition temperature (~36.7°C) and 2240.3 × 10^{-4} C/m^{2}K at ~58.4°C [128]. Also, $Pb[Mg_{1/3}Nb_{2/3}]_{0.71}Ti_{0.29}O_3$ modified with Sm shows enhanced pyroelectric properties near the MPB by exhibiting a pyroelectric coefficient of 4610 μC/m^{2}K [129].

However, because of the environmental and health hazards associated with lead-based materials, there has been a growing interest among researchers in lead-free alternatives such as $BaTiO_3$, $K_{0.5}Na_{0.5}NbO_3$, and $Na_{0.5}Bi_{0.5}TiO_3$. In this regard, Priya et al. prepared ferroelectric $BaTiO_3$–ZnO-based composites for PyENGs, emphasizing exceptional performance near ambient temperatures. Bi and Li were doped into $BaTiO_3$ ($Ba_{0.85}(Bi_{0.5}Li_{0.5})_{0.15}TiO_3$—BBLT) to increase their ferroelectric and pyroelectric characteristics. The prepared BBLT sample exhibits a pyroelectric coefficient of 240 μC/m^{2}K at room temperature and 3340 μC/m^{2}K at T_c (~45°C). Additionally, BBLT is employed to fabricate the composite utilizing semiconductor ZnO, owing to its superior thermal conductivity (~30 W/mK) compared to other semiconductors [130]. The composite made of 10% ZnO shows the highest pyroelectric coefficient of 7250 μC/m^{2}K at 45°C. The incorporation of ZnO in the grain boundary areas of BBLT could strengthen the thermal conduction network, thereby enhancing the pyroelectric response through improved heat transmission within the composite. [131, 132].

The other lead-free ferroelectric NBT-0.06BT ceramic doped with La exhibited an improved pyroelectric coefficient, which is necessary to achieve high efficiency of energy harvesting. The ferroelectric NBT-0.06 shows 214 μC/m^2 °C at RT, which increased to ~2390 μC/m^2 °C at 90 °C (T_d). The La-doped NBT-0.06BT sample augments the pyroelectric coefficient to 742 μC/m^2 °C at RT and 10,540 μC/m^2 °C at T_d, ascribed to the existence of MPB [133]. Many reports utilize lead-free pyroelectric materials for energy harvesting applications.

Further, polymer-based pyroelectric materials were used to prepare flexible and wearable devices. Pyroelectric NGs based on polymers mainly utilize PVDF and P(VDF-TrFE). Although PVDF exhibits a low pyroelectric coefficient, it possesses significant biocompatibility and mechanical strength, making it suitable for wearable devices. A wearable pyroelectric NG device utilizing PVDF films was developed for harvesting energy from human respiration [134]. This device also functions as a self-generating multifunctional sensor for monitoring human breathing and ambient temperature.

Table 5.2 Pyroelectric coefficient values of some existing pyroelectric materials

		Pyroelectric coefficient ($\mu C/m^2K$)	
1	(Li^++Ta^{5+}) co-doped $Bi_{0.5}Na_{0.5}TiO_3$	107,911 (at T_d)	[137]
2	PMN-PMS-PZT	5524 at RT	[138]
3	$0.94(Bi_{0.5}Na_{0.5})TiO_3$–$0.06Ba(Ti_{0.75}Zr_{0.25})O_3$	2720	[139]
4	$Pb_{1-x}La_x(Zr_{0.4}Ti_{0.6})O_3$	98,000	[140]
5	$0.94Na_{0.5}Bi_{0.5}TiO_3$–$0.06Ba_{1+x}TiO_3$	74,070	[141]
6	$K_{0.48}Na_{0.52}NbO_3$–$Bi_{0.5}Ag_{0.5}ZrO_3$–0.2%Fe_2O_3	509	[142]
7	$Na_{0.995}Ag_{0.005}NbO_3$–0.1%Mn	255	[143]
8	$Sr_{0.5}Ba_{0.5}Nb_2O_6$	550	[144]
9	$BaTi_{0.95}Sn_{0.05}O_3$	300	[145]
10	PMN-PMS-PZT-0.2AN	4146	[146]

Combining pyroelectric ceramic particles with a polymer matrix results in a composite sheet that is both flexible and exhibits enhanced pyroelectric activity compared to pure polymers. Since polymers are renowned for their strength, elasticity, and ease of processing, the polymer matrix gives excellent mechanical resistance and flexibility to the composite sheet, making it a suitable structure for a variety of applications [135]. Taking advantage of these, flexible composite sheets composed of potassium sodium niobate and polyvinylidene fluoride (KNN-PVDF) have been developed to create a highly efficient pyroelectric energy harvester. The incorporation of KNN in the PVDF matrix leads to an improvement in the polarization, exhibiting a higher polarization at 30% KNN, attributed to the presence of the polar β-phase of PVDF in conjunction with the KNN. The composite 30% KNN-PVDF shows a higher pyroelectric coefficient of 147×10^{-5} C/m^2K and a figure of merit of 49.47×10^{-11} m^3/J [125]. Furthermore, a study involving P(VDF-TrFE) demonstrates that dissolving the polymer in various solvents yields higher pyroelectric coefficient values when higher dipole moment solvents are utilized. This indicates that the selection of suitable solvents with elevated dipole moments is essential for optimizing energy efficiency [136]. Table 5.2 presents the pyroelectric coefficient of existing materials reported in the previous reports.

5.7 CONCLUSION

Since the dawn of human civilization, its advancement has been heavily reliant on energy resources. However, contemporary society is facing significant challenges due to the escalating global energy crisis. In addition to the limited availability of conventional energy resources, the energy sources present several other challenges, including high production costs and significant losses during transmission and distribution. Hence, various alternative and renewable energy sources, like wind, solar, geothermal, mechanical, and biomass energy, have been increasingly developed in recent years. This chapter discussed the harvesting of energy through piezo-, tribo-, thermo-, and pyroelectric mechanisms. The basic mechanisms and some existing energy-harvesting devices were also discussed.

DATA AVAILABILITY

Data will be made available on request.

REFERENCES

1. N. Sezer, M. Koç, A comprehensive review on the state-of-the-art of piezoelectric energy harvesting, *Nano Energy* 80 (2021) 105567. https://doi.org/10.1016/J.NANOEN.2020.105567
2. M.H. Shubbak, Advances in solar photovoltaics: Technology review and patent trends, *Renewable and Sustainable Energy Reviews* 115 (2019) 109383. https://doi.org/10.1016/J.RSER.2019.109383
3. D.R. Komatireddy, S. Andem, P.P. Gotipamul, V.V. Rajankumar, S. Chidambaram, Harnessing the potential of transition metal tungstates (MWO4, M = Ni, Co, Cu, and Zn) for high-performance asymmetric supercapacitors, *Journal of Energy Storage* 100 (2024) 113557. https://doi.org/10.1016/J.EST.2024.113557
4. A. Kalair, N. Abas, M.S. Saleem, A.R. Kalair, N. Khan, Role of energy storage systems in energy transition from fossil fuels to renewables, *Energy Storage* 3 (2021). https://doi.org/10.1002/est2.135
5. B. Ozcan, S. Kılıç Depren, M.T. Kartal, Impact of nuclear energy and hydro electricity consumption in achieving environmental quality: Evidence from load capacity factor by quantile based non-linear approaches, *Gondwana Research* 129 (2024) 412–424. https://doi.org/10.1016/J.GR.2023.05.014
6. I. Pioro, R.B. Duffey, P.L. Kirillov, R. Pioro, A. Zvorykin, R. MacHrafi, Current status and future developments in nuclear-power industry of the world, *Journal of Nuclear Engineering and Radiation Science* 5 (2019). https://doi.org/10.1115/1.4042194
7. T.Z. Ang, M. Salem, M. Kamarol, H.S. Das, M.A. Nazari, N. Prabaharan, A comprehensive study of renewable energy sources: Classifications, challenges and suggestions, *Energy Strategy Reviews* 43 (2022) 100939. https://doi.org/10.1016/J.ESR.2022.100939
8. U.A. Bhatti, M.A. Bhatti, H. Tang, M.S. Syam, E.M. Awwad, M. Sharaf, Y.Y. Ghadi, Global production patterns: Understanding the relationship between greenhouse gas emissions, agriculture greening and climate variability, *Environmental Research* 245 (2024) 118049. https://doi.org/10.1016/J.ENVRES.2023.118049

9. A.B. Lumb, Smoking and air pollution, in: A.B. Lumb (Ed.), *Nunn's Applied Respiratory Physiology*, Elsevier, 2016: pp. 281–290. https://doi.org/10.1016/B978-0-7020-6294-0.00019-8
10. M. Safaei, H.A. Sodano, S.R. Anton, A review of energy harvesting using piezoelectric materials: State-of-the-art a decade later (2008–2018), *Smart Materials and Structures* 28 (2019). https://doi.org/10.1088/1361-665X/ab36e4
11. C. Yu, G. Li, Electrical energy harvesting by connected form stable phase change material composites, *Energy Conversion and Management* 299 (2024). https://doi.org/10.1016/j.enconman.2023.117851
12. A.S. Al-Ezzi, M.N.M. Ansari, Photovoltaic solar cells: A review, *Applied System Innovation* 5 (2022). https://doi.org/10.3390/asi5040067
13. S. Du, P. Basset, H. Guo, D. Galayko, A. Karami, Power management technologies for triboelectric nanogenerators, *MRS Bulletin* (2025). https://doi.org/10.1557/s43577-025-00860-8
14. K. Damodar Reddy, N. Pavan Kumar, D. Xavier, P. Sowjanya, V. Subramanian, C. Siva, Magnetoelectric behaviour of Pb-free Na0.5Bi0.5TiO_3-Ni0.5Co0.5Fe2O4 particulate novel ceramic composites, *Ceramics International* 50 (2024) 10402–10415. https://doi.org/10.1016/j.ceramint.2023.12.352
15. K. Tao, J. Wu, A.G.P. Kottapalli, S.W. Lye, J. Miao, MEMS/NEMS-enabled vibrational energy harvesting for self-powered and wearable electronics, in: L. Ting, Z. Liu (Ed.), *Outlook and Challenges of Nano Devices, Sensors, and MEMS*, Springer International Publishing, 2017: pp. 271–297. https://doi.org/10.1007/978-3-319-50824-5_9
16. S.G. Kim, S. Priya, I. Kanno, Piezoelectric MEMS for energy harvesting, *MRS Bulletin* 37 (2012) 1039–1050. https://doi.org/10.1557/mrs.2012.275
17. Y. Hu, Z.L. Wang, Recent progress in piezoelectric nanogenerators as a sustainable power source in self-powered systems and active sensors, *Nano Energy* 14 (2014) 3–14. https://doi.org/10.1016/j.nanoen.2014.11.038
18. M. Zhu, Z. Yi, B. Yang, C. Lee, Making use of nanoenergy from human – Nanogenerator and self-powered sensor enabled sustainable wireless IoT sensory systems, *Nano Today* 36 (2021) 101016. https://doi.org/10.1016/J.NANTOD.2020.101016
19. A. Arnau, D. Soares, Fundamentals of piezoelectricity, in: A.A. Vives (Ed.), *Piezoelectric Transducers and Applications*, Springer, 2008: pp. 1–38. https://doi.org/10.1007/978-3-540-77508-9_1
20. F. Narita, M. Fox, A review on piezoelectric, magnetostrictive, and magnetoelectric materials and device technologies for energy harvesting applications, *Advanced Engineering Materials* 20 (2018). https://doi.org/10.1002/adem.201700743
21. B. Saravanakumar, K. Thiyagarajan, S.K. Balasingam, T. Siva, R.K. Pai, A. Ramadoss, Recent advances and innovations in the piezoelectrochemical process for energy and the environment: A review, *Journal of Energy Storage* 83 (2024) 110576. https://doi.org/10.1016/J.EST.2024.110576
22. S. Yun, Y. Zhang, Q. Xu, J. Liu, Y. Qin, Recent advance in new-generation integrated devices for energy harvesting and storage, *Nano Energy* 60 (2019) 600–619. https://doi.org/10.1016/j.nanoen.2019.03.074
23. S. Bhandari, G. Vajpayee, L.L. da Silva, M. Hinterstein, G. Franchin, P. Colombo, A review on additive manufacturing of piezoelectric ceramics: From feedstock development to properties of sintered parts, *Materials Science*

and Engineering: R: Reports 162 (2025) 100877. https://doi.org/10.1016/j.mser.2024.100877
24. M.R. Sarker, M.H.M. Saad, J.L. Olazagoitia, J. Vinolas, Review of power converter impact of electromagnetic energy harvesting circuits and devices for autonomous sensor applications, *Electronics (Basel)* 10 (2021). https://doi.org/10.3390/electronics10091108
25. S.P. Beeby, M.J. Tudor, N.M. White, Energy harvesting vibration sources for microsystems applications, *Measurement Science and Technology* 17 (2006) R175. https://doi.org/10.1088/0957-0233/17/12/R01
26. Z. Turar, M. Sembay, A. Mubarak, A. Belgibayeva, L. Kong, G. Kalimuldina, Advances in porous structure design for enhanced piezoelectric and triboelectric nanogenerators: A comprehensive review, *Global Challenges* 9 (2025) 2400224. https://doi.org/10.1002/gch2.202400224
27. J. Briscoe, S. Dunn, Piezoelectric nanogenerators – A review of nanostructured piezoelectric energy harvesters, *Nano Energy* 14 (2015) 15–29. https://doi.org/10.1016/J.NANOEN.2014.11.059
28. Z.L. Wang, J. Song, Piezoelectric nanogenerators based on zinc oxide nanowire arrays, *Science* (1979) 312 (2006) 242–246. https://doi.org/10.1126/science.1124005
29. B. Kumar, S.-W. Kim, Recent advances in power generation through piezoelectric nanogenerators, *Journal of Materials Chemistry* 21 (2011) 18946–18958. https://doi.org/10.1039/C1JM13066H
30. D. Hu, M. Yao, Y. Fan, C. Ma, M. Fan, M. Liu, Strategies to achieve high performance piezoelectric nanogenerators, *Nano Energy* 55 (2019) 288–304. https://doi.org/10.1016/J.NANOEN.2018.10.053
31. Z. Zhang, Y. Chen, J. Guo, ZnO nanorods patterned-textile using a novel hydrothermal method for sandwich structured-piezoelectric nanogenerator for human energy harvesting, *Physica E: Low-dimensional Systems and Nanostructures* 105 (2019) 212–218. https://doi.org/10.1016/J.PHYSE.2018.09.007
32. A.T. Le, M. Ahmadipour, S.Y. Pung, A review on ZnO-based piezoelectric nanogenerators: Synthesis, characterization techniques, performance enhancement and applications, *Journal of Alloys and Compounds* 844 (2020) 156172. https://doi.org/10.1016/J.JALLCOM.2020.156172
33. D. Yang, Y. Qiu, Q. Jiang, Z. Guo, W. Song, J. Xu, Y. Zong, Q. Feng, X. Sun, Patterned growth of ZnO nanowires on flexible substrates for enhanced performance of flexible piezoelectric nanogenerators, *Applied Physics Letters* 110 (2017) 63901. https://doi.org/10.1063/1.4975477
34. R.K. Pandey, J. Dutta, S. Brahma, B. Rao, C.-P. Liu, Review on ZnO-based piezotronics and piezoelectric nanogenerators: aspects of piezopotential and screening effect, *Journal of Physics: Materials* 4 (2021) 44011. https://doi.org/10.1088/2515-7639/ac130a
35. C. Li, W. Luo, X. Liu, D. Xu, K. He, PMN-PT/PVDF nanocomposite for high output nanogenerator applications, *Nanomaterials* 6 (2016). https://doi.org/10.3390/nano6040067
36. H. Zheng, J. Wang, S.E. Lofland, Z. Ma, L. Mohaddes-Ardabili, T. Zhao, L. Salamanca-Riba, S.R. Shinde, S.B. Ogale, F. Bai, D. Viehland, Y. Jia, D.G. Schlom, M. Wuttig, A. Roytburd, R. Ramesh, Multiferroic $BaTiO_3$-CoFe2O4 nanostructures, *Science* (1979) 303 (2004) 661–663. https://doi.org/10.1126/science.1094207

37. S. Korkmaz, I.A. Kariper, $BaTiO_3$-based nanogenerators: fundamentals and current status, *Journal of Electroceramics* 48 (2022) 8–34. https://doi.org/10.1007/s10832-021-00266-3
38. X. Chen, X. Li, J. Shao, N. An, H. Tian, C. Wang, T. Han, L. Wang, B. Lu, High-performance piezoelectric nanogenerators with imprinted P(VDF-TrFE)/$BaTiO_3$ nanocomposite micropillars for self-powered flexible sensors, *Small* 13 (2017) 1604245. https://doi.org/10.1002/smll.201604245
39. M. Yan, Y. Zhang, S. Liu, J. Liang, X. Gu, Q. Xu, J. Zhang, X. Yuan, K. Zhou, D. Zhang, Boosting energy harvesting performance in piezoelectric composites with aligned porosity via a dual structure design strategy, *ACS Applied Materials & Interfaces* 17 (2025) 6530–6540. https://doi.org/10.1021/acsami.4c17248
40. M. Li, H.-W. Lu, S.-W. Wang, R.-P. Li, J.-Y. Chen, W.-S. Chuang, F.-S. Yang, Y.-F. Lin, C.-Y. Chen, Y.-C. Lai, Filling the gap between topological insulator nanomaterials and triboelectric nanogenerators, *Nature Communications* 13 (2022) 938. https://doi.org/10.1038/s41467-022-28575-3
41. F.R. Fan, Z.Q. Tian, Z. Lin Wang, Flexible triboelectric generator, *Nano Energy* 1 (2012) 328–334. https://doi.org/10.1016/J.NANOEN.2012.01.004
42. W. Harmon, D. Bamgboje, H. Guo, T. Hu, Z.L. Wang, Self-driven power management system for triboelectric nanogenerators, *Nano Energy* 71 (2020) 104642. https://doi.org/10.1016/J.NANOEN.2020.104642
43. T. Cheng, J. Shao, Z.L. Wang, Triboelectric nanogenerators, *Nature Reviews Methods Primers* 3 (2023) 38. https://doi.org/10.1038/s43586-023-00230-1
44. Q. Zhou, K. Lee, K.N. Kim, J.G. Park, J. Pan, J. Bae, J.M. Baik, T. Kim, High humidity- and contamination-resistant triboelectric nanogenerator with superhydrophobic interface, *Nano Energy* 57 (2019) 903–910. https://doi.org/10.1016/J.NANOEN.2018.12.091
45. U. Khan, S.-W. Kim, Triboelectric nanogenerators for blue energy harvesting, *ACS Nano* 10 (2016) 6429–6432. https://doi.org/10.1021/acsnano.6b04213
46. Y. Zi, H. Guo, Z. Wen, M.-H. Yeh, C. Hu, Z.L. Wang, Harvesting low-frequency (<5 Hz) irregular mechanical energy: A possible killer application of triboelectric nanogenerator, *ACS Nano* 10 (2016) 4797–4805. https://doi.org/10.1021/acsnano.6b01569
47. Z. Wen, H. Guo, Y. Zi, M.-H. Yeh, X. Wang, J. Deng, J. Wang, S. Li, C. Hu, L. Zhu, Z.L. Wang, Harvesting broad frequency band blue energy by a triboelectric–electromagnetic hybrid nanogenerator, *ACS Nano* 10 (2016) 6526–6534. https://doi.org/10.1021/acsnano.6b03293
48. S. Niu, S. Wang, L. Lin, Y. Liu, Y.S. Zhou, Y. Hu, Z.L. Wang, Theoretical study of contact-mode triboelectric nanogenerators as an effective power source, *Energy & Environmental Science* 6 (2013) 3576–3583. https://doi.org/10.1039/C3EE42571A
49. S. Niu, Y. Liu, S. Wang, L. Lin, Y.S. Zhou, Y. Hu, Z.L. Wang, Theory of sliding-mode triboelectric nanogenerators, *Advanced Materials* 25 (2013) 6184–6193. https://doi.org/10.1002/adma.201302808
50. S.W. Chen, X. Cao, N. Wang, L. Ma, H.R. Zhu, M. Willander, Y. Jie, Z.L. Wang, An ultrathin flexible single-electrode triboelectric-nanogenerator for mechanical energy harvesting and instantaneous force sensing, *Advanced Energy Materials* 7 (2017) 1601255. https://doi.org/10.1002/aenm.201601255

51. Z. Zhao, X. Pu, C. Du, L. Li, C. Jiang, W. Hu, Z.L. Wang, Freestanding flag-type triboelectric nanogenerator for harvesting high-altitude wind energy from arbitrary directions, *ACS Nano* 10 (2016) 1780–1787. https://doi.org/10.1021/acsnano.5b07157
52. P. Zhao, N. Soin, K. Prashanthi, J. Chen, S. Dong, E. Zhou, Z. Zhu, A.A. Narasimulu, C.D. Montemagno, L. Yu, J. Luo, Emulsion electrospinning of polytetrafluoroethylene (PTFE) nanofibrous membranes for high-performance triboelectric nanogenerators, *ACS Applied Materials & Interfaces* 10 (2018) 5880–5891. https://doi.org/10.1021/acsami.7b18442
53. H. Wattanasarn, T. Ngennam, T. Sumpao, C. Thanachayanont, T. Seetawan, The charging performance in contact electrification of fluorinated ethylene propylene surfaces by electrode bridge, *Sensors and Actuators, A: Physical* 346 (2022) 113881. https://doi.org/10.1016/J.SNA.2022.113881
54. G.-Z. Li, G.-G. Wang, D.-M. Ye, X.-W. Zhang, Z.-Q. Lin, H.-L. Zhou, F. Li, B.-L. Wang, J.-C. Han, High-performance transparent and flexible triboelectric nanogenerators based on PDMS-PTFE composite films, *Advanced Electronic Materials* 5 (2019) 1800846. https://doi.org/10.1002/aelm.201800846
55. M. Sahu, S. Šafranko, S. Hajra, A.M. Padhan, P. Živković, S. Jokić, H.J. Kim, Development of triboelectric nanogenerator and mechanical energy harvesting using argon ion-implanted kapton, zinc oxide and kapton, *Materials Letters* 301 (2021) 130290. https://doi.org/10.1016/J.MATLET.2021.130290
56. R. Zhang, C. Dahlström, H. Zou, J. Jonzon, M. Hummelgård, J. Örtegren, N. Blomquist, Y. Yang, H. Andersson, M. Olsen, M. Norgren, H. Olin, Z.L. Wang, Cellulose-based fully green triboelectric nanogenerators with output power density of 300 W m^{-2}, *Advanced Materials* 32 (2020) 2002824. https://doi.org/10.1002/adma.202002824
57. R. Zhang, H. Olin, Material choices for triboelectric nanogenerators: A critical review, *EcoMat* 2 (2020) e12062. https://doi.org/10.1002/eom2.12062
58. Y. Zhou, J.-H. Zhang, F. Wang, J. Hua, W. Cheng, Y. Shi, L. Pan, Recent advances in flexible self-powered sensors in piezoelectric, triboelectric, and pyroelectric fields, nanoenergy *Advances* 4 (2024) 235–257. https://doi.org/10.3390/nanoenergyadv4030015
59. A.A. Mathew, A. Chandrasekhar, S. Vivekanandan, A review on real-time implantable and wearable health monitoring sensors based on triboelectric nanogenerator approach, *Nano Energy* 80 (2021) 105566. https://doi.org/10.1016/J.NANOEN.2020.105566
60. B. Xie, Y. Guo, Y. Chen, H. Zhang, J. Xiao, M. Hou, H. Liu, L. Ma, X. Chen, C. Wong, Advances in graphene-based electrode for triboelectric nanogenerator, *Nano-Micro Letters* 17 (2024) 17. https://doi.org/10.1007/s40820-024-01530-1
61. Z. Yan, L. Wang, Y. Xia, R. Qiu, W. Liu, M. Wu, Y. Zhu, S. Zhu, C. Jia, M. Zhu, R. Cao, Z. Li, X. Wang, Flexible high-resolution triboelectric sensor array based on patterned laser-induced graphene for self-powered real-time tactile sensing, *Advanced Functional Materials* 31 (2021) 2100709. https://doi.org/10.1002/adfm.202100709
62. H. Luo, Y. Lu, Y. Xu, G. Yang, S. Cui, D. Han, Q. Zhou, X. Ouyang, H. Yang, T. Cheng, K. Xu, A fully soft, self-powered vibration sensor by laser direct writing, *Nano Energy* 103 (2022) 107803. https://doi.org/10.1016/J.NANOEN.2022.107803

63. S.Y. Xia, Y. Long, Z. Huang, Y. Zi, L.Q. Tao, C.H. Li, H. Sun, J. Li, Laser-induced graphene (LIG)-based pressure sensor and triboelectric nanogenerator towards high-performance self-powered measurement-control combined system, *Nano Energy* 96 (2022) 107099. https://doi.org/10.1016/J.NANOEN.2022.107099
64. K. Hu, Z. Zhao, Y. Wang, L. Yu, K. Liu, H. Wu, L. Huang, L. Chen, Y. Ni, A tough organohydrogel-based multiresponsive sensor for a triboelectric nanogenerator and supercapacitor toward wearable intelligent devices, *Journal of Materials Chemistry A* 10 (2022) 12092–12103. https://doi.org/10.1039/D2TA01503J
65. M. Martín-González, O. Caballero-Calero, P. Díaz-Chao, Nanoengineering thermoelectrics for 21st century: Energy harvesting and other trends in the field, *Renewable and Sustainable Energy Reviews* 24 (2013) 288–305. https://doi.org/10.1016/J.RSER.2013.03.008
66. E. Karvannan, V. Vijay, T.S. Nivin, M. Navaneethan, J. Archana, Tailoring charge carrier dynamics for improved thermoelectric properties in nickel-incorporated Bi_2S_3, *Materials Chemistry and Physics* 335 (2025) 130490. https://doi.org/10.1016/J.MATCHEMPHYS.2025.130490
67. A.W. Van Herwaarden, P.M. Sarro, Thermal sensors based on the seebeck effect, *Sensors and Actuators* 10 (1986) 321–346. https://doi.org/10.1016/0250-6874(86)80053-1
68. E. Karvannan, V. Vijay, T.S. Nivin, J. Archana, M. Navaneethan, A. Karthigeyan, Enhanced thermoelectric performance of iso-valent Al substituted Bi2S3 via carrier tuning and multiscale phonon scattering, *Materials Chemistry and Physics* 312 (2024) 128506. https://doi.org/10.1016/J.MATCHEMPHYS.2023.128506
69. D. Champier, Thermoelectric generators: A review of applications, *Energy Conversion and Management* 140 (2017) 167–181. https://doi.org/10.1016/J.ENCONMAN.2017.02.070
70. J. He, K. Li, L. Jia, Y. Zhu, H. Zhang, J. Linghu, Advances in the applications of thermoelectric generators, *Applied Thermal Engineering* 236 (2024) 121813. https://doi.org/10.1016/J.APPLTHERMALENG.2023.121813
71. M.A. Zoui, S. Bentouba, J.G. Stocholm, M. Bourouis, A review on thermoelectric generators: Progress and applications, *Energies (Basel)* 13 (2020). https://doi.org/10.3390/EN13143606
72. C. Wood, Materials for thermoelectric energy conversion, *Reports on Progress in Physics* 51 (1988) 459. https://doi.org/10.1088/0034-4885/51/4/001
73. V.A. Drebushchak, The Peltier effect, *Journal of Thermal Analysis and Calorimetry* 91 (2008) 311–315. https://doi.org/10.1007/s10973-007-8336-9
74. W.-Y. Chen, X.-L. Shi, J. Zou, Z.-G. Chen, Thermoelectric coolers: Progress, challenges, and opportunities, *Small Methods* 6 (2022) 2101235. https://doi.org/10.1002/smtd.202101235
75. D. Michael Rowe, *CRC Handbook of Thermoelectrics*, CRC Press, 1995: p. 701.
76. S.M. Pourkiaei, M.H. Ahmadi, M. Sadeghzadeh, S. Moosavi, F. Pourfayaz, L. Chen, M.A. Pour Yazdi, R. Kumar, Thermoelectric cooler and thermoelectric generator devices: A review of present and potential applications, modeling and materials, *Energy* 186 (2019). https://doi.org/10.1016/j.energy.2019.07.179
77. M. Bharti, A. Singh, S. Samanta, D.K. Aswal, Conductive polymers for thermoelectric power generation, *Progress in Materials Science* 93 (2018) 270–310. https://doi.org/10.1016/J.PMATSCI.2017.09.004

78. L. Zhang, X.L. Shi, Y.L. Yang, Z.G. Chen, Flexible thermoelectric materials and devices: From materials to applications, *Materials Today* 46 (2021) 62–108. https://doi.org/10.1016/J.MATTOD.2021.02.016
79. J. Liang, T. Wang, P. Qiu, S. Yang, C. Ming, H. Chen, Q. Song, K. Zhao, T.-R. Wei, D. Ren, Y.-Y. Sun, X. Shi, J. He, L. Chen, Flexible thermoelectrics: from silver chalcogenides to full-inorganic devices, *Energy & Environmental Science* 12 (2019) 2983–2990. https://doi.org/10.1039/C9EE01777A
80. H. Mamur, M.R.A. Bhuiyan, F. Korkmaz, M. Nil, A review on bismuth telluride (Bi2Te3) nanostructure for thermoelectric applications, *Renewable and Sustainable Energy Reviews* 82 (2018) 4159–4169. https://doi.org/10.1016/J.RSER.2017.10.112
81. J. Pei, B. Cai, H.-L. Zhuang, J.-F. Li, Bi2Te3-based applied thermoelectric materials: research advances and new challenges, *National Science Review* 7 (2020) 1856–1858. https://doi.org/10.1093/nsr/nwaa259
82. L. Hu, T. Zhu, X. Liu, X. Zhao, Point defect engineering of high-performance bismuth-telluride-based thermoelectric materials, *Advanced Functional Materials* 24 (2014) 5211–5218. https://doi.org/10.1002/adfm.201400474
83. B. Jabar, X. Qin, D. Li, J. Zhang, A. Mansoor, H. Xin, C. Song, L. Huang, Achieving high thermoelectric performance through constructing coherent interfaces and building interface potential barriers in n-type Bi2Te3/Bi2Te2.7Se0.3 nanocomposites, *Journal of Materials Chemistry A* 7 (2019) 19120–19129. https://doi.org/10.1039/C9TA05798F
84. B. Zhu, Z.Y. Huang, X.Y. Wang, Y. Yu, L. Yang, N. Gao, Z.G. Chen, F.Q. Zu, Attaining ultrahigh thermoelectric performance of direction-solidified bulk n-type Bi2Te2.4Se0.6 via its liquid state treatment, *Nano Energy* 42 (2017) 8–16. https://doi.org/10.1016/J.NANOEN.2017.10.034
85. B. Jabar, X. Qin, A. Mansoor, H. Ming, L. Huang, M.H. Danish, J. Zhang, D. Li, C. Zhu, H. Xin, C. Song, Enhanced power factor and thermoelectric performance for n-type Bi2Te2.7Se0.3 based composites incorporated with 3D topological insulator nanoinclusions, *Nano Energy* 80 (2021) 105512. https://doi.org/10.1016/J.NANOEN.2020.105512
86. Z.-H. Zheng, Y.-M. Zhong, Y.-L. Li, M. Nisar, A. Mansoor, F. Li, S. Chen, G.-X. Liang, P. Fan, D. Xu, M. Wei, Y.-X. Chen, Ultrahigh thermoelectric properties of p-type $BiSb_2$–Te_3 thin films with exceptional flexibility for wearable energy harvesting, *Carbon Energy* 6 (2024) e541. https://doi.org/10.1002/cey2.541
87. Y. Zhu, R. Shu, F. Jiang, A. Elsukova, P.O.Å. Persson, P. Eklund, W. Liu, Enhanced thermoelectric properties of Mg_2Sn–Mg_3Sb_2 nanocomposites by tailoring matrix/inclusion interface toward energy harvesting applications, *ACS Applied Nano Materials* 6 (2023) 6133–6140. https://doi.org/10.1021/acsanm.3c00439
88. F.H. Sun, H. Li, J. Tan, L. Zhao, X. Wang, H. Hu, C. Wang, T. Mori, Review of current ZT > 1 thermoelectric sulfides, *Journal of Materiomics* 10 (2024) 218–233. https://doi.org/10.1016/j.jmat.2023.05.011
89. R.-Z. Zhang, F. Gucci, H. Zhu, K. Chen, M.J. Reece, Data-driven design of eco-friendly thermoelectric high-entropy sulfides, *Inorganic Chemistry* 57 (2018) 13027–13033. https://doi.org/10.1021/acs.inorgchem.8b02379
90. Y. Zhang, C. Xing, Y. Liu, M.C. Spadaro, X. Wang, M. Li, K. Xiao, T. Zhang, P. Guardia, K.H. Lim, A.O. Moghaddam, J. Llorca, J. Arbiol, M. Ibáñez, A. Cabot, Doping-mediated stabilization of copper vacancies to promote

thermoelectric properties of $Cu_{2-x}S$, *Nano Energy* 85 (2021) 105991. https://doi.org/10.1016/J.NANOEN.2021.105991
91. A. Sultana, M.M. Alam, T.R. Middya, D. Mandal, A pyroelectric generator as a self-powered temperature sensor for sustainable thermal energy harvesting from waste heat and human body heat, *Applied Energy* 221 (2018) 299–307. https://doi.org/10.1016/J.APENERGY.2018.04.003
92. D. Zhang, H. Wu, C.R. Bowen, Y. Yang, Recent advances in pyroelectric materials and applications, *Small* 17 (2021) 2103960. https://doi.org/10.1002/smll.202103960
93. S. Korkmaz, A. Kariper, Pyroelectric nanogenerators (PyNGs) in converting thermal energy into electrical energy: Fundamentals and current status, *Nano Energy* 84 (2021) 105888. https://doi.org/10.1016/J.NANOEN.2021.105888
94. Z. Liu, T. Lu, X. Dong, G. Wang, Y. Liu, Ferroelectric ceramics for pyroelectric detection applications: A review, *IEEE Transactions on Ultrasonics, Ferroelectrics, and Frequency Control* 68 (2021) 242–252. https://doi.org/10.1109/TUFFC.2020.3025168
95. I. Kriaa, A. Maalej, Pyroelectric energy harvesting, storage properties and electrocaloric effect of BNBT-ST ferroelectric ceramics, *Journal of Energy Storage* 102 (2024) 114044. https://doi.org/10.1016/J.EST.2024.114044
96. A. Thakre, A. Kumar, H.C. Song, D.Y. Jeong, J. Ryu, Pyroelectric energy conversion and its applications—flexible energy harvesters and sensors, *Sensors (Switzerland)* 19 (2019). https://doi.org/10.3390/s19092170
97. M.-H. You, X.-X. Wang, X. Yan, J. Zhang, W.-Z. Song, M. Yu, Z.-Y. Fan, S. Ramakrishna, Y.-Z. Long, A self-powered flexible hybrid piezoelectric–pyroelectric nanogenerator based on non-woven nanofiber membranes, *Journal of Materials Chemistry A* 6 (2018) 3500–3509. https://doi.org/10.1039/C7TA10175A
98. G. Sebald, D. Guyomar, A. Agbossou, On thermoelectric and pyroelectric energy harvesting, *Smart Materials and Structures* 18 (2009) 125006. https://doi.org/10.1088/0964-1726/18/12/125006
99. X. Wang, Y. Dai, R. Liu, X. He, S. Li, Z.L. Wang, Light-triggered pyroelectric nanogenerator based on a pn-junction for self-powered near-infrared photosensing, *ACS Nano* 11 (2017) 8339–8345. https://doi.org/10.1021/acsnano.7b03560
100. M. Zirkl, G. Scheipl, B. Stadlober, A. Haase, L. Kuna, J. Magnien, G. Jakopic, J.R. Krenn, A. Sawatdee, P. Bodö, P. Andersson, Fully printed, flexible, large area organic optothermal sensors for human-machine-interfaces, *Procedia Engineering* 5 (2010) 725–729. https://doi.org/10.1016/J.PROENG.2010.09.211
101. J. Cobian-Iñiguez, M. Gollner, S. Saha, J. Avalos, E. Ameri, Improved fire safety in the wildland-urban interface through smart technologies, in: H. Xinyan and W.C. Tam (Ed.), *Intelligent Building Fire Safety and Smart Firefighting*, Springer Nature, 2024: pp. 165–198. https://doi.org/10.1007/978-3-031-48161-1_8
102. Z. Bielecki, J. Mikolajczyk, J. Wojtas, A review of thermal detectors of THz radiation operated at room temperature, *Sensors* 24 (2024). https://doi.org/10.3390/s24216784
103. N. Argirusis, A. Achilleos, N. Alizadeh, C. Argirusis, G. Sourkouni, I.R. Sensors, Related materials, and applications, *Sensors* 25 (2025). https://doi.org/10.3390/s25030673

104. A. Gaur, N. Saurabh, S. Patel, Pyrocatalysis-driven dye degradation by Zn Doped NBT-BT: A comparative analysis between powder and bulk, *Materials Chemistry and Physics* 315 (2024) 128940. https://doi.org/10.1016/J.MATCHEMPHYS.2024.128940
105. S. Touili, M. Amjoud, D. Mezzane, Z. Kutnjak, I.A. Luk'Yanchuk, M. Jouiad, M. El Marssi, An overview of low-carbon hydrogen production via water splitting driven by piezoelectric and pyroelectric catalysis, *International Journal of Hydrogen Energy* 78 (2024) 218–235. https://doi.org/10.1016/J.IJHYDENE.2024.06.209
106. W. Zhang, J. Xiong, X. Lin, Y. Liu, T. Gan, H. Hu, Z. Huang, Y. Zhang, Advanced pyro-/piezoelectric catalytic disinfection of water via porous spontaneously polarized ceramic with strong local electric field induced by micro-nano bubbles, *Chemical Engineering Journal* 499 (2024) 155958. https://doi.org/10.1016/J.CEJ.2024.155958
107. E. Roake, Q. Liu, Q. Wang, E. Alsheim, B.L. Patenall, M. Hopkins, V. Jarkov, T. Jenkins, C.R. Bowen, Pyrocatalytic disinfection of water using lead-free ferroelectric ceramics, in: *2024 IEEE Ultrasonics, Ferroelectrics, and Frequency Control Joint Symposium (UFFC-JS)*, 2024: pp. 1–11. https://doi.org/10.1109/UFFC-JS60046.2024.10793794
108. B. Wu, H. Ouro-Koura, S.H. Lu, H. Li, X. Wang, J. Xiao, Z.D. Deng, Functional materials for powering and implementing next-generation miniature sensors, *Materials Today* 69 (2023) 333–354. https://doi.org/10.1016/J.MATTOD.2023.09.001
109. S.B. Lang, Pyroelectricity: From ancient curiosity to modern imaging tool, *Physics Today* 58 (2005) 31–36. https://doi.org/10.1063/1.2062916
110. M. Aggarwal, G. Sharma, A.K. Singh, S. Kumar, Investigation of structural, pyroelectric and energy storage properties of lead-free Ba0.85Sr0.15Zr0.09Sn0.01Ti0.9O3 ceramics, in: *2024 IEEE Ultrasonics, Ferroelectrics, and Frequency Control Joint Symposium (UFFC-JS)*, 2024: pp. 1–4. https://doi.org/10.1109/UFFC-JS60046.2024.10793813
111. M. Es-Souni, Potentials and LIMITS of PMN-PT and PIN-PMN-PT single crystals for pyroelectric energy harvesting, *Crystals (Basel)* 14 (2024). https://doi.org/10.3390/cryst14030236
112. D. Zhang, J. Luo, Efficient pyroelectric poling of electrooptic polymers using commercial PZT discs, *ACS Applied Electronic Materials* 6 (2024) 2457–2466. https://doi.org/10.1021/acsaelm.4c00110
113. A.M. Pugachev, A.A. Sokolov, I.O. Akhundov, N. V. Protasevich, Two-dimensional pyroelectric material with heat-transfer modulation, *Journal of Surface Investigation: X-ray, Synchrotron and Neutron Techniques* 18 (2024) 536–541. https://doi.org/10.1134/S102745102470006X
114. B. Zeng, J. Yang, Z. Ni, Y. Fan, Z. Hang, C. Feng, Improved pyroelectric performances of functionally graded graphene nanoplatelet reinforced polyvinylidene fluoride composites: Experiment and modelling, *Composites. Part A, Applied Science and Manufacturing* 176 (2024) 107883. https://doi.org/10.1016/J.COMPOSITESA.2023.107883
115. N. Saurabh, S. Patel, Analysis of pyroelectric properties in quenched NBT-BT composition, *Materials Today Communications* 41 (2024) 110784. https://doi.org/10.1016/J.MTCOMM.2024.110784

116. M.H. Park, M. Hoffmann, C.S. Hwang, Pyroelectric and electrocaloric effects and their applications, in: U. Schroeder, C.S. Hwang, H. Funakubo (Ed.), *Ferroelectricity in Doped Hafnium Oxide: Materials, Properties and Devices*, Elsevier, 2019: pp. 217–244. https://doi.org/10.1016/B978-0-08-102430-0.00013-9
117. N. Saurabh, S. Patel, Pyroelectric performance of zinc oxide-added 0.94Na0.5Bi0.5TiO3–0.06BaTiO_3 composite, *Journal of Materials Science: Materials in Electronics* 35 (2024) 1384. https://doi.org/10.1007/s10854-024-13108-z
118. Q. Cao, P. Zou, F. Yang, J. Zhang, J. Wu, X. Wang, H. Bi, The pumping performance, mechanism and challenges of low voltage and non-magnetic ion pump based on pyroelectric principle, *Vacuum* 228 (2024) 113543. https://doi.org/10.1016/J.VACUUM.2024.113543
119. S. Hur, S. Kim, H.S. Kim, A. Kumar, C. Kwon, J. Shin, H. Kang, T.H. Sung, J. Ryu, J.M. Baik, H.C. Song, Low-grade waste heat recovery scenarios: Pyroelectric, thermomagnetic, and thermogalvanic thermal energy harvesting, *Nano Energy* 114 (2023) 108596. https://doi.org/10.1016/J.NANOEN.2023.108596
120. V. Gupta, Z. Mallick, A. Choudhury, D. Mandal, On-demand MXene-coupled pyroelectricity for advanced breathing sensors and IR data receivers, *Langmuir* 40 (2024) 8897–8910. https://doi.org/10.1021/acs.langmuir.4c00074
121. H. Nguyen, A. Navid, L. Pilon, Pyroelectric energy converter using co-polymer P(VDF-TrFE) and Olsen cycle for waste heat energy harvesting, *Applied Thermal Engineering* 30 (2010) 2127–2137. https://doi.org/10.1016/J.APPLTHERMALENG.2010.05.022
122. Z. Ma, F. Han, H. Wang, Y. Wang, L. Liu, X. Chen, W. Dong, Y. Li, Y. Bai, D. Wang, L. Zheng, Q. Zhang, B. Peng, Unleashing the power of giant negative electrocaloric effect through heterojunctions, *Nano Energy* 134 (2025) 110578. https://doi.org/10.1016/J.NANOEN.2024.110578
123. Y. Meng, S. Tang, Z. Wang, X. Niu, H. Zhang, D. Wang, Y. Bai, B. Peng, S.-G. Lu, Q. Ke, L. Liu, Significantly enhanced electrocaloric effect by composition modulation in lead-free $BaTiO_3$-based ceramics, *Journal of Materiomics* 11 (2025) 100903. https://doi.org/10.1016/J.JMAT.2024.05.011
124. R. Mondal, M.A.M. Hasan, J.M. Baik, Y. Yang, Advanced pyroelectric materials for energy harvesting and sensing applications, *Materials Today* 66 (2023) 273–301. https://doi.org/10.1016/J.MATTOD.2023.03.023
125. B. Sharma, A. Sharma, R. Gupta, A. Chowdhuri, M. Verma, M. Tomar, Flexible sheets of lead free KNN-PVDF composite: A sustainable pyroelectric energy harvester, *Ceramics International* 50 (2024) 46982–46992. https://doi.org/10.1016/J.CERAMINT.2024.09.049
126. Z.L. Wang, G. Zhu, Y. Yang, S. Wang, C. Pan, Progress in nanogenerators for portable electronics, *Materials Today* 15 (2012) 532–543. https://doi.org/10.1016/S1369-7021(13)70011-7
127. S.B. Lang, S. Muensit, Review of some lesser-known applications of piezoelectric and pyroelectric polymers, *Applied Physics A* 85 (2006) 125–134. https://doi.org/10.1007/s00339-006-3688-8
128. L. Li, Y. Dai, X. Tang, S. Wang, J. Wang, Y. Wang, High pyroelectric response in Pb0.99Nb0.02[(Zr0.57Sn0.43)1−Ti]0.98O3 ceramics, *Journal of the American Ceramic Society* 106 (2023) 4202–4210. https://doi.org/10.1111/jace.19056

129. C. Singh, G. Vats, A. Kumar, Improved piezoelectric, ferroelectric, pyroelectric and electrocaloric properties of Sm-substituted PMN-PT, *Journal of Alloys and Compounds* 965 (2023) 171355. https://doi.org/10.1016/J.JALLCOM.2023.171355
130. S. Sulaiman, S. Izman, M.B. Uday, M.F. Omar, Review on grain size effects on thermal conductivity in ZnO thermoelectric materials, *RSC Advances* 12 (2022) 5428–5438. https://doi.org/10.1039/D1RA06133J
131. K.S. Priya, S. Pal, M. Mohan, P. Murugavel, High performance near-room-temperature pyroelectric energy harvesting characteristics of ferroelectric–semiconductor composites, *ACS Applied Electronic Materials* 5 (2023) 3790–3797. https://doi.org/10.1021/acsaelm.3c00511
132. S. Pandya, J. Wilbur, J. Kim, R. Gao, A. Dasgupta, C. Dames, L.W. Martin, Pyroelectric energy conversion with large energy and power density in relaxor ferroelectric thin films, *Nature Materials* 17 (2018) 432–438. https://doi.org/10.1038/s41563-018-0059-8
133. A.M. Balakt, C.P. Shaw, Q. Zhang, Enhancement of pyroelectric properties of lead-free 0.94Na0.5Bi0.5TiO3-0.06$BaTiO_3$ ceramics by La doping, *Journal of the European Ceramic Society* 37 (2017) 1459–1466. https://doi.org/10.1016/J.JEURCERAMSOC.2016.12.021
134. H. Xue, Q. Yang, D. Wang, W. Luo, W. Wang, M. Lin, D. Liang, Q. Luo, A wearable pyroelectric nanogenerator and self-powered breathing sensor, *Nano Energy* 38 (2017) 147–154. https://doi.org/10.1016/J.NANOEN.2017.05.056
135. M. Habib, I. Lantgios, K. Hornbostel, A review of ceramic, polymer and composite piezoelectric materials, *Journal of Physics D: Applied Physics* 55 (2022) 423002. https://doi.org/10.1088/1361-6463/ac8687
136. J. Kim, J.H. Lee, H. Ryu, J.-H. Lee, U. Khan, H. Kim, S.S. Kwak, S.-W. Kim, High-performance piezoelectric, pyroelectric, and triboelectric nanogenerators based on P(VDF-TrFE) with controlled crystallinity and dipole alignment, *Advanced Functional Materials* 27 (2017) 1700702. https://doi.org/10.1002/adfm.201700702
137. L. Wen, X. Wu, J. Yin, Y. Zhang, D. Yang, J. Wu, Realization of the Giant Pyroelectric Response via Modulated Polar Structures, *Small* 20 (2024) 2307326. https://doi.org/10.1002/smll.202307326
138. Q. Wang, Q. Sun, Z. Xu, H. Luo, S. Li, T. Wu, Improving waste heat harvesting using pyroelectric ceramics, *Surfaces and Interfaces* 51 (2024) 104657. https://doi.org/10.1016/J.SURFIN.2024.104657
139. M. Shen, W. Li, M.Y. Li, H. Liu, J. Xu, S. Qiu, G. Zhang, Z. Lu, H. Li, S. Jiang, High room-temperature pyroelectric property in lead-free BNT-BZT ferroelectric ceramics for thermal energy harvesting, *Journal of the European Ceramic Society* 39 (2019) 1810–1818. https://doi.org/10.1016/J.JEURCERAMSOC.2019.01.012
140. V. Mahlawat, M. Tomar, R. Gupta, Pyroelectric thermal energy harvester based on sol-gel-derived PLZT films, *Physica Status Solidi (a)* 221 (2024) 2400041. https://doi.org/10.1002/pssa.202400041
141. A.M. Balakt, C.P. Shaw, Q. Zhang, The effects of Ba2+ content on depolarization temperature and pyroelectric properties of lead-free 0.94Na0.5Bi0.5TiO_3–0.06Ba1+xTiO_3 ceramics, *Journal of Materials Science: Materials in Electronics* 27 (2016) 12947–12954. https://doi.org/10.1007/s10854-016-5433-1

142. Y. Huang, H. Xue, T. Zheng, J. Wu, Enhanced pyroelectric performance in potassium sodium niobate-based ceramics via multisymmetries coexistence design, *Journal of the American Ceramic Society* 106 (2023) 1113–1126. https://doi.org/10.1111/jace.18815
143. Z. Lv, S. Guo, Z. Liu, C. Yao, G. Wang, Simultaneously achieving superior pyroelectric performance and ultrahigh depolarization temperature in NaNbO3-based ceramics, *Applied Physics Letters* 125 (2024) 092904. https://doi.org/10.1063/5.0222066
144. T. Lukasiewicz, M.A. Swirkowicz, J. Dec, W. Hofman, W. Szyrski, Strontium–barium niobate single crystals, growth and ferroelectric properties, *Journal of Crystal Growth* 310 (2008) 1464–1469. https://doi.org/10.1016/J.JCRYSGRO.2007.11.233
145. K. Srikanth, S. Patel, R. Vaish, Pyroelectric performance of $BaTi_{1-x}Sn_xO_3$ ceramics, *International Journal of Applied Ceramic Technology* 15 (2018) 546–553. https://doi.org/10.1111/ijac.12814
146. Q. Wang, C.R. Bowen, W. Lei, H. Zhang, B. Xie, S. Qiu, M.-Y. Li, S. Jiang, Improved heat transfer for pyroelectric energy harvesting applications using a thermal conductive network of aluminum nitride in PMN–PMS–PZT ceramics, *Journal of Materials Chemistry A* 6 (2018) 5040–5051. https://doi.org/10.1039/C8TA00235E

Chapter 6

Storage devices

Memory and its types

M. Saravanan, Eswaran Parthasarathy, Shiromani Balmukund Rahi, S. Sreejith, and Samuel Tensingh

6.1 INTRODUCTION

Nano energy systems need memory technology for energy-efficient devices. Nano energy applications require high-performance memory devices that can store data during power outages. This makes the system more efficient, reduces energy consumption, and enable data retention during power outages. Energy demand is forcing changes in design and manufacture of memory devices. This applies to smart device nanoscale energy harvesting and the Internet of Things. These memory systems must be energy-efficient, reliable, and fast. Nano energy system memory can quickly store and analyze data. As energy storage and collection methods shrink, memory devices must be small, energy-efficient, reliable, and fast. Nonvolatile memory (NVM) is useful in low-power environments due to its ability to store data without power. Digital systems, especially nano energy systems with energetic limits, require memory technology. System performance, energy efficiency, and dependability can be affected by the memory choice. Applications of volatile and NVM differ. Unlike volatile memory, nonvolatile memory may retain information even when electricity is not present. This classification affects nano energy device architecture, power consumption, and efficacy.

6.2 TYPES OF SEMICONDUCTOR MEMORY

A wide range of memory devices are available for data storage. Typically, memory types are divided into two categories: nonvolatile memories and volatile memories. An outline of the main memory systems is shown in Figure 6.1. NVM stores information indefinitely, whereas volatile memory rapidly loses its information when the power is switched off. The memory hierarchy of modern Von Neumann computers uses a variety of memory technologies to provide the fastest performance at the lowest cost.

DOI: 10.1201/9781003635123-6

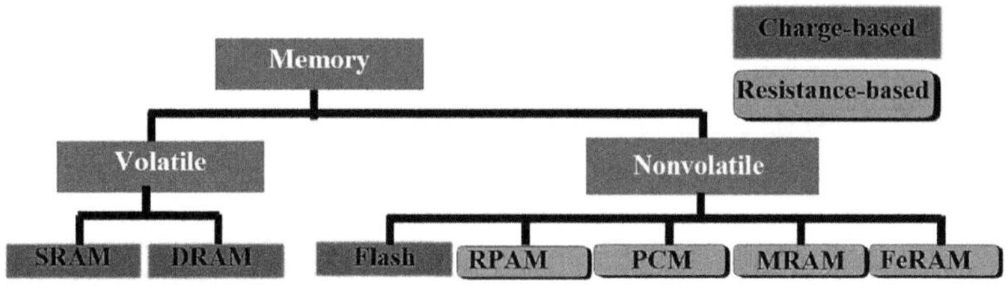

Figure 6.1 Memory classification.

Memory in digital systems can be broadly classified based on the following:

1. Volatility (Volatile vs. Nonvolatile)
2. Access Method (Random, Sequential, and Direct)
3. Technology (Semiconductor, Magnetic, Optical, etc.)
4. Hierarchy (Registers → Cache → Primary Memory → Secondary Storage)

Volatile memory devices and NVM devices exist. Memory's ability to retain data during power outages distinguishes them. Dynamic and static random access memory (RAM) are volatile memories. They store data using energy and are speedier, but power outages can erase their data. Read only memory (ROM), Flash, and FeRAM (ferroelectric RAM) preserve data after a power loss. Thus, data storage must last. NVM technologies are essential for nano energy systems because they store key system data without power.

6.2.1 RAM (random access memory)

A RAM system's block diagram usually includes a few essential parts for data storage and retrieval. Its main component is the memory array, which contains the data and is made up of rows and columns of cells. As shown in Figure 6.2, a row decoder and a column decoder are coupled to the address

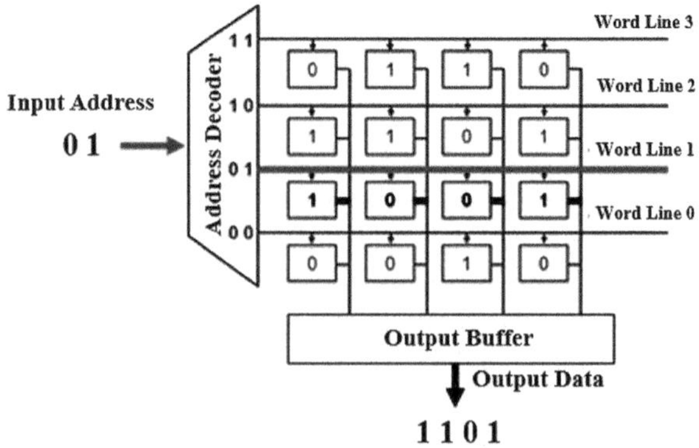

Figure 6.2 Block diagram of a random access memory (RAM) system.

lines. They use the binary input address to identify the exact region in the memory array for read or write operations. The data input/output lines make it easier for data to enter and exit the memory cells. The process is coordinated, and its correct operation is guaranteed by the read/write control software in accordance with the control signals [1, 2]. The tiny voltages that are extracted from the cells are detected and amplified by an optional sensing amplifier. This improves the stability and quickness of reading. The RAM system allows for quick and easy access to data at any position in memory by interpreting address inputs and controlling data flow with particular control signals. It enables fast, direct access to data at any memory location by decoding address inputs and controlling data flow using precise control signals.

6.2.2 SRAM (static random access memory)

The 6T SRAM (static RAM) cells are made up of six transistors, as shown in Figure 6.3. One bit of data (0 or 1) is stored by two cross-coupled inverters, made by four transistors, while the other two connect the cell to the bit lines (BLs) for reading or writing. During a write operation, the access transistors are activated by turning on the word line (WL). The data to be recorded is transmitted via the BLs, altering the internal states of the inverters. The WL is reactivated during a read operation, and the access transistors transmit the stored value to the BLs. The sense amplifiers pick up the stored bit, which is used to charge and discharge the BLs. When powered, 6T SRAM stores data without refreshing, making it fast and energy-efficient for cache memory applications.

An available sense amplifier picks up and amplifies very low voltages from the cells. This enhances reading speed and stability. The RAM system comprehends address inputs and regulates data flow using designated control signals. This enables rapid and direct access to data in any memory segment. The data is retained as long as power is provided, eliminating the necessity for refresh, rendering 6T SRAM rapid and energy-efficient for cache memory applications. A discretionary sensing amplifier identifies and enhances the subtle voltages derived from the cells. This improves reading

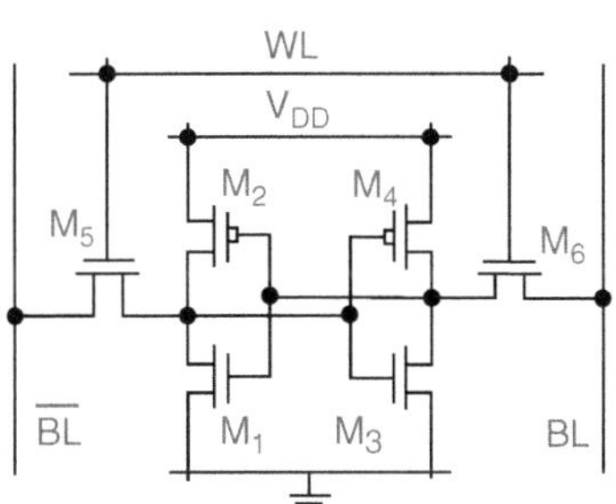

Figure 6.3 6T SRAM.

speed and stability [3, 4]. The RAM system decodes address inputs and manages data transfer through designated control signals, facilitating swift and direct access to data at any memory location. It interprets address inputs and employs accurate control signals to regulate data flow, facilitating rapid and direct access to data at any memory location.

6.2.3 DRAM (dynamic random access memory)

The 1T1C DRAM cell illustrated in Figure 6.4 consists of a single transistor (1T) and a single capacitor (1C). This cell stores one bit of current data. When you read or write, the access transistor, which links the capacitor to the BL, receives instructions from the WL. The capacitor retains a charge, representing "1" (charged) or "0" (discharged) in logic. The BL determines the amount of charge to be applied to the capacitor based on the data transmitted. The WL activates the transistor during a write operation. At read time, a sense amplifier detects the BL voltage, which is affected by the capacitor and WL charges, and reactivates the transistor.

Upon receipt of the data, the original value must be reinstated. Failure to do so will result in the deletion of the stored charge. DRAM consumes more power than SRAM due to charge leakage and requires periodic refreshing to maintain data integrity. Therefore, it is less portable and more suitable for large-capacity memory amplifiers. Conversely, accessing the data depletes the stored charge, necessitating the immediate restoration of the previous value. Charge leakage necessitates the periodic refreshing of DRAM to preserve data integrity. This indicates that although it is smaller and more suitable for large-capacity memory, it consumes more power than SRAM. The stored charge is erased if the original value is not restored (refreshed) immediately after the data is received. DRAM uses less power than SRAM but is more portable and suitable for large-capacity memory amplifiers because of the periodic refresh cycles required to protect the data from charge leakage. Reading the data destroys the stored charge, so the original value needs to be restored (refreshed) right away. Furthermore, because of charge leakage,

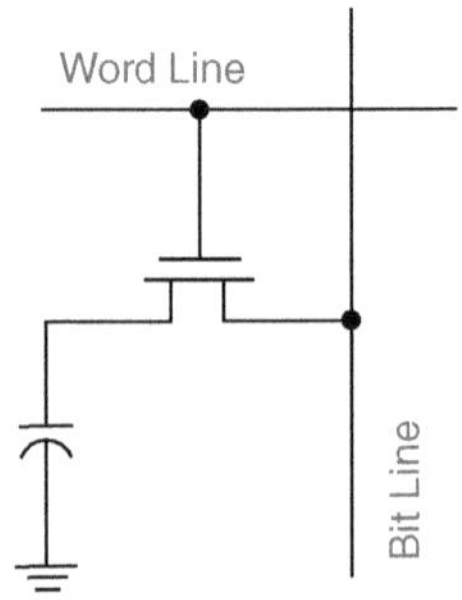

Figure 6.4 DRAM.

periodic refresh cycles are necessary to preserve the data, which makes DRAM less power-efficient than SRAM, but more portable and appropriate for large-capacity memory [5].

6.3 ROM (READ ONLY MEMORY)

See Table 6.1.

NVM, like ROM, retains its contents after a power loss. ROM is often programmed during chip manufacturing. A matrix of immutable numbers makes the data unchangeable after creation. ROM types include programmable ROM (PROM), erasable programmable ROM (EPROM), electrically erasable programmable ROM (EEPROM), and Mask ROM. Unchangeable classic ROM is suitable for essential system functions due to its stability and dependability.

Reprogramming and erasing are made feasible via EPROM and EEPROM. An EEPROM of this type stores data in a grid of floating-gate transistor-based memory cells. Nanocrystals are utilized in novel Flash memory technologies to enhance speed and improve data storage efficiency, even at reduced dimensions.

6.3.1 Flash Memory

Flash memory stores data using floating-gate transistors (Figure 6.5). These transistors can store data after a power outage. MOSFET structure, control gate, and floating gate are common memory cell features. During programming, hot electron injection (Fowler–Nordheim tunneling) injects electrons into the floating gate, thereby changing the transistor's threshold voltage to "0" or "1." In a read operation, the transistor current is tracked while the control gate is given some voltage. Trapped electrons in the floating gate boost the threshold voltage, making the transistor nonconductive. In alternate logic, it activates. Flash memory stores and deletes data in chunks, making it useful for memory cards, USB devices, and SSDs. While reading,

Table 6.1 Read only memory classification

Classification	*Attributes*	*Alterability*
ROM	Hard-wired, permanent	No
PROM	Programmable once	One-time programmable
EPROM	Using UV light erasable	Reprogrammable
EEPROM	Using an erasable electrical signal	Byte-wise erasable
Flash	Electrically erasable in blocks	Widely used in SSDs and USBs

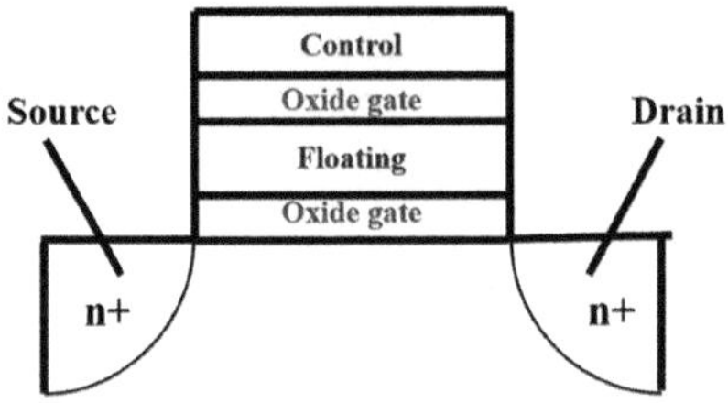

Figure 6.5 Flash memory.

the transistor current is checked. Electrons in the floating gate raise the threshold voltage, turning off the transistor. Otherwise, alternate logic is activated. Flash memory is perfect for USB drives, SSDs, and memory cards due to its large data storage capacity and simultaneous block erasure.

See Table 6.2.

Table 6.2 ROM fabrication technique and its challenges

Memory type	*Fabrication technique*	*Challenges*
ROM	Photolithography and mask programming	Fixed data and no flexibility for changes
PROM	Photolithography and masking	One-time programmability
EPROM	Photolithography, floating-gate, and UV erasure	UV erasure complexity
EEPROM	Floating-gate transistor and electrical erasure	Limited endurance
Flash Memory	Charge trapping and masked programming	Scaling and wear-out issues

6.4 MODERN NONVOLATILE MEMORY TECHNOLOGIES

Type	*Key feature*	*Working principle*	*Examples*
MRAM	Magnetic storage	Magnetic tunnel junctions.	Everspin and Samsung
STT-MRAM	Spin-polarized current	Spin-transfer torque.	Global Foundries.
ReRAM (RRAM)	Resistive switching	Metal–insulator–metal (MIM)	Crossbar Inc.
PCM (phase-change memory)	Heat-induced crystallization	Amorphous/ crystalline phases.	Intel Optane

(*Continued*)

Type	*Key feature*	*Working principle*	*Examples*
FeRAM	Ferroelectric capacitor	Polarization reversal	Fujitsu
CBRAM	Conductive-bridge formation	Ionic motion	Adesto
ECRAM	Electrochemical modulation	Ion intercalation	AI accelerators
Memristors	Resistance memory	Hysteresis loop behavior	HP Labs

6.5 MAGNETIC MEMORIES: MRAM AND STT-MRAM

MRAM has evolved across generations:

- **The first generation,** "Toggle MRAM," debuted in the early 2000s. It moved magnetic bits using magnetic fields but struggled with scalability and writing energy.
- **Second generation (TA-MRAM)**: Thermal assistance during writes improves speed and scalability.
- **Third generation (STT-MRAM)**: Spin-polarized currents change magnetic states, enabling faster, more scalable, and less energy-intensive writing.
- Improve performance, energy economy, and durability for future generations (SOT-MRAM and VC-MRAM).

6.5.1 Construction of MRAM cell

MRAM cells often have magnetic tunnel junctions (MTJs). A set reference magnetic layer with a specified magnetization makes up this junction. A non-uniform magnetic layer exhibiting spatial variations in magnetic field intensity. The tunnel barriers are often made of magnesium oxide [6]. The resistance of the MTJ depends on the configuration of the magnetizations, as these layers are stacked sequentially:

- **Parallel alignment**: Minimal resistance (logic 0)
- **Nonparallel alignment**: Elevated resistance (logic 1)

Magnetic memory, specifically MRAM and STT-MRAM, is a potential NVM technology. Data is stored using magnetic states rather than electric charge, making it fast, energy-efficient, and durable. MRAM is being intensively researched for embedded systems, automotive, aerospace, and new neuromorphic platforms [7].

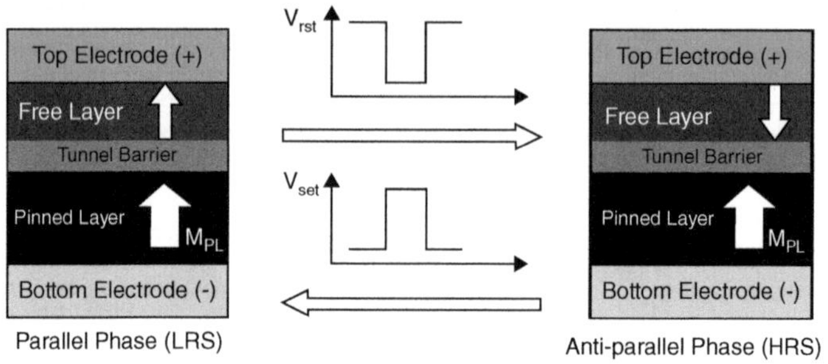

Figure 6.6 Basic structure of MRAM (MTJ).

MRAM: MRAM stores information using magnetic components, which keep it intact even in the case of a power outage. It is necessary to modify the magnetic alignment of these components to retrieve the data.

STT-MRAM: This type of memory modifies its magnetic state using the spin-transfer torque effect [8]. Compared to conventional MRAM, it exhibits greater density and lower power consumption.

An MTJ has two ferromagnetic layers divided by a very thin insulating tunnel barrier, which is usually made of magnesium oxide, as shown in Figure 6.6. The stationary magnetic layer is referred to as the reference layer. The magnetic orientation of the free layer can be modified. The resistance is minimal when the two layers are aligned parallelly, signifying synchronization; conversely, when the layers are aligned in an antiparallel, the resistance is substantial, showing a lack of synchronization. This fluctuation in resistance is identified during a read operation. Spin-transfer torque (STT) and spin-orbit torque (SOT) are employed to alter the magnetization of the free layer during a write operation. MRAM is applicable for both embedded memory and high-performance computing due to its rapid speed, data retention capability during inactivity, and extended longevity [9].

The working of MRAM is based on the tunnel magnetoresistance (TMR) effect. The write operation is performed by changing the orientation of the free layer, using either magnetic fields (Toggle MRAM) or spin-polarized currents (STT-MRAM). The read operation measures the resistance across the MTJ:

- If the resistance is low, the bit is 0.
- If the resistance is high, the bit is 1.

6.5.2 Physical architecture of STT-MRAM

The fundamental element of STT-MRAM is the MTJ, as shown in Figure 6.7. The device features a magnesium oxide (MgO) barrier that inhibits electrical conductivity, a free magnetic layer with adjustable orientation,

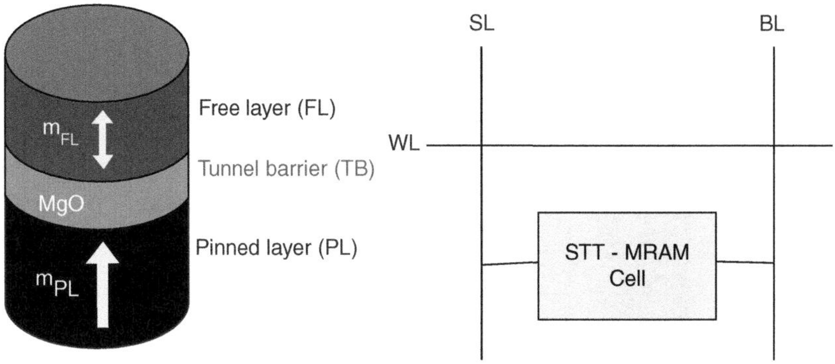

Figure 6.7 (a) MTJ device organization. (b) STT-MRAM cell.

and a fixed magnetic layer (reference layer). The resistance across the MTJ is negligible when the two magnetic layers are aligned parallelly [10]. The resistance is significant when aligned in an antiparallel configuration. The types of resistance classified as "0" and "1" are identical in nature.

During a write operation, the MTJ experiences an electric current. The magnetization of the free layer is oriented either parallel (low resistance, logic "0") or antiparallel (high resistance, logic "1") to the reference layer by spin-polarized electrons, which are affected by the current's direction. Figure 6.8 indicates an n x m array of STT-MRAM cells. A low voltage is applied to the MTJ during a read operation, and the resistance is measured to ascertain the stored bit without modifying the data. STT-MRAM data remains intact during power outages, exhibiting both speed and durability.

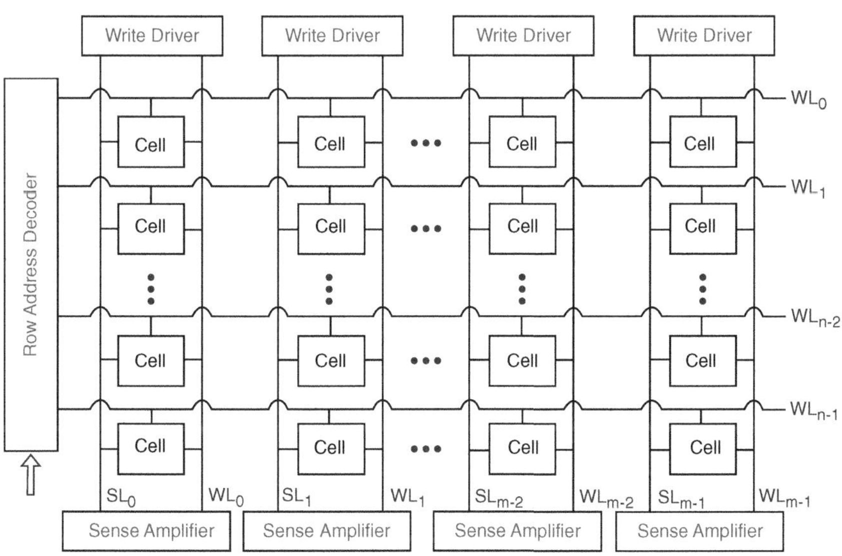

Figure 6.8 Array of STT-MRAM cells.

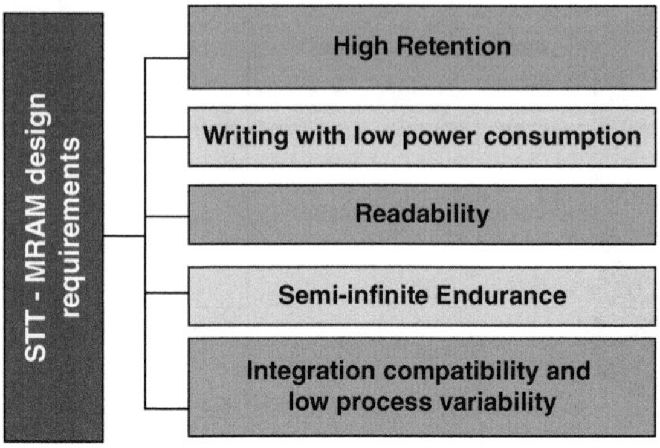

Figure 6.9 STT-MRAM cell design requirements.

In embedded and cache memory applications, it may function as a substitute for SRAM or DRAM. Figure 6.9 illustrates STT-MRAM cell design requirements.

See Table 6.3.

Table 6.3 Key features of MRAM

Parameter	*Details*
Read mechanism	Tunneling magnetoresistance (TMR) detection via resistance difference
Write mechanism	Spin-transfer torque (STT)/spin-orbit torque (SOT)
Endurance	$>10^{15}$ write cycles (very high)
Retention	≥ 10 years at 85°C
Write speed	~10–50 ns (varies with type and technology)
Read speed	~2–20 ns (comparable to SRAM)
CMOS compatibility	Fully compatible with standard CMOS processes
Operating voltage	0.2–3.3 V (depends on the process node)
Target areas	Embedded systems, cache, IoT, aerospace, automotive, and data centers
Types	Toggle MRAM, STT-MRAM, SOT-MRAM, and p-MRAM
Vendors	Everspin, Samsung, TSMC, GlobalFoundries, and Avalanche Tech

6.5.3 Properties of materials for STT-MRAM

See Table 6.4.

Table 6.4 TT-MRAM material properties

Component	*Material used*	*Required properties*
Free layer	CoFeB (cobalt iron boron)	Soft ferromagnetic, high spin polarization, low coercivity, and easy to switch magnetization
Reference layer	CoFeB with a synthetic antiferromagnet (SAF)	Fixed magnetization, high thermal stability, and spin polarization stability
Tunnel barrier	MgO (magnesium oxide)	Thin (~1 nm), high tunneling magnetoresistance (TMR), and good insulation
Pinned layer	CoFe or CoFeB	High magnetic anisotropy and stable fixed magnetic direction
Seed layer	Ta (tantalum), Ru	Promotes crystallographic orientation and smooth interfaces
Capping layer	Ta, Ru, or W (tungsten)	Prevents oxidation and ensures thermal and chemical stability
Electrodes	Cu, Al, or W	High conductivity, good adhesion, and compatibility with CMOS processing
Substrate	Silicon (Si)	CMOS-compatible and supports integration with backend of line (BEOL)

6.5.4 STT switching mechanism

In STT-MRAM, a spin-polarized current transfers spin angular momentum to the magnetic moments of the free layer as it traverses the MTJ. STT elucidates this motion. It produces a torque that alters the orientation of the free layer, resulting in a new data state. Large-scale STT consumes less energy than field-based switching. In STT-MRAM, a spin-polarized current in the MTJ conveys spin angular momentum from electrons to the magnetic moments of the free layer. STT elucidates this motion. By generating torque, it can alter the orientation of the open layer to record fresh data. STT consumes less energy than field-based switching and is scalable (Table 6.5).

Table 6.5 MRAM versus STT-MRAM technologies

Parameter	*MRAM*	*STT-MRAM*
Write speed	10 ns	10–50 ns
Endurance	10^6	10^{15}
Power usage	Moderate	Low
Nonvolatility	Yes	Yes
Scalability	Moderate	High

6.5.5 Manufacturing procedure for MRAM and STT-MRAM

- Delicate layers of ferromagnetic and oxide materials (such as CoFeB and MgO) must be meticulously arranged to fabricate the MTJ—the principal component.
- Magnetron sputtering is a method used for producing magnetic layers.
- Electron beam lithography is employed to fabricate patterns for junctions smaller than 50 nm.
- Ion beams are employed for etching while preserving the integrity of the tunnel barrier.
- To enhance TMR and fabricate the MgO barrier.

6.5.6 Challenges in fabrication

- Reliable management of magnetic unevenness in very small devices.
- Making sure that there is a low RA (resistance-area) result and a high TMR.
- Making that MgO does not get damaged during etching processes.
- Making that the device remains stable at back-end-of-line (BEOL) integration temperatures.
- Keeping write error rates and variations at small nodes.

6.5.7 Applications of MRAM and STT-MRAM

- Automotive ECUs (Electronic Control Units).
- IoT devices and smart sensors.
- Persistent memory in AI accelerators.
- Replacement for embedded SRAM and NOR Flash.

6.5.8 Future directions and innovations

- Voltage-Controlled MRAM (VC-MRAM) to reduce switching energy.
- SOT-MRAM (Spin-Orbit Torque MRAM) for faster, more efficient switching.
- Integration with FinFET and GAAFET platforms for sub-7 nm nodes.
- 3D MRAM structures and stacked magnetic layers for density scaling.

6.6 RESISTIVE MEMORIES (RRAM/CBRAM)

This memory stores data through fluctuations in resistance. Conductive-bridge RAM (CBRAM), or Conductive-Bridge RAM, is a metal–insulator–metal (MIM) device characterized by a changeable insulator resistance contingent upon voltage. CBRAM utilizes ionic conduction between electrodes to store data. Production challenges encompass the development of stable memory device materials, the assurance of endurance, and the regulation of switching voltage. Resistive RAM (RRAM) is one of the most promising technologies in neuromorphic computing and memory.

6.6.1 Evolution of RRAM

Despite the original observation of the resistive switching effect in the 1960s, comprehensive research into RRAM devices only became feasible in the 2000s. Enhancing device reliability and scalability, together with understanding switching mechanisms like filament creation, constitute some of the most significant achievements. Entities such as Crossbar, HP, and Panasonic have demonstrated significant interest in RRAM technology.

6.6.2 Construction of a RRAM cell

The MIM structure of a typical RRAM cell consists of the top electrode (TE), the Resistive Switching Layer (which may be an oxide or another dielectric), and the Bottom Electrode (BE). Prevalent switching materials comprise TiO_2, HfO_2, and NiO. A selection device could be integrated into the design to regulate access and avert unauthorized paths in crossbar arrays [11]. Figure 6.10 illustrates the basic structure of the RRAM device and its working principle.

6.6.3 Working principle of RRAM

Low Resistance State (LRS) signifies logic "1," while High Resistance State (HRS) signifies logic "0." RRAM operates by modifying the resistance of the switching layer in response to an applied voltage.

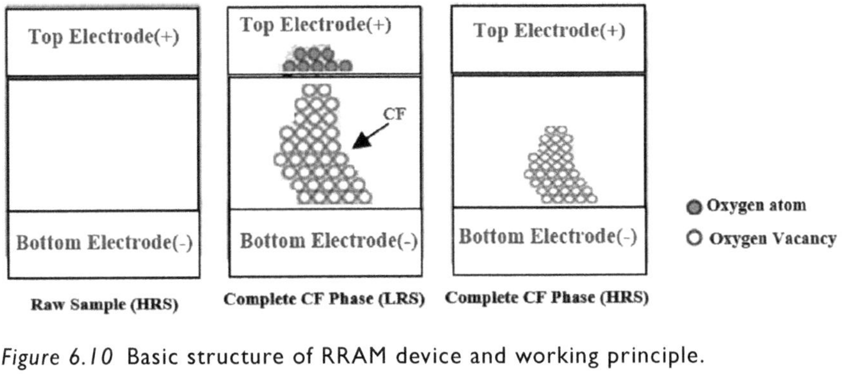

Figure 6.10 Basic structure of RRAM device and working principle.

6.6.3.1 *Filamentary switching*

The formation and dissolution of conductive filaments (CFs) composed of metal ions or oxygen vacancies. This switching method alters resistance by modifying the contact between the electrode and the insulator [12]. A positive voltage induces CFs, transitioning the device to LRS. A negative voltage disrupts the wires, reverting the gadget to HRS. RRAM operates by modulating the resistance of its switching layer between an HRS and an LRS under the application of voltage. This change is nonvolatile and forms the basis for storing binary data (1 and 0).

There are two key processes:

- SET Process (Write "1"):
  - A positive voltage is applied to the TE.
  - This causes the formation of CFs in the switching layer by the migration of oxygen vacancies or metal ions.
  - The cell switches to LRS, representing a logic "1".
- RESET Process (Write "0"):
  - A reverse or higher voltage is applied.
  - The CF is ruptured or dissolved due to recombination or Joule heating.
  - The device returns to HRS, representing a logic "0" (Table 6.6).

6.6.4 Fabrication challenges of RRAM

- Resistive memories, including RRAM and CBRAM, present significant potential as the forthcoming generation of NVM owing to their simplicity of fabrication, rapid switching capabilities, and scalability [13]. Nevertheless, certain manufacturing concerns must be addressed prior to their market availability.
- **Diversity in the electroforming process:** The electroforming phase exhibits significant variability, leading to the formation of conductive strands within the dielectric. This process can permanently modify the

Table 6.6 RRAM technical specifications

Parameter	*Typical values*
Switching speed	1–100 ns
Endurance	10^4–10^{12} cycles (material dependent)
Retention	>10 years at 85°C
Energy per bit	<1 pJ (ultra-low-power designs)
On/off resistance ratio	~10^2–10^6
Material systems	HfO_2, TiO_2, Ta_2O_5, Al_2O_3, and NiO
Integration capability	Compatible with CMOS backend

device's attributes, resulting in varied functionality across a memory array, contingent upon the initial distribution of flaws.

- **Filament Management and Operational Consistency**: Continuous generation and disruption of conductive pathways is essential in filament-based resistive switching. Variations in filament size, number, and position diminish the stability of resistance states and switching thresholds. Consequently, the equipment is less reliable and has a reduced lifespan.
- **Material interface instability**: The interface between the electrode and the switching layer must maintain stability despite several cycles. Ag, Cu, and TiN are examples of electrode materials that may lose efficacy if they fail to adhere or scatter properly. The modulation of RRAM is substantially affected by the type and positioning of oxygen vacancies.
- **Low-temperature compatibility**: RRAM and CBRAM must be fabricated at temperatures below 400°C to be compatible with CMOS BEOL processes. Identifying materials and processing methods that align with these constraints is a significant challenge.
- **Sustaining and seeking its disadvantages**: Filaments are often manufactured with increased durability for enhanced retention, which may necessitate higher reset currents or lead to diminished endurance. Device design is becoming complex due to the trade-off between managing numerous program/erase operations and ensuring optimal data retention.
- **Addressing current overshoot**: SET utilization may lead to excessive filament growth from current overshoot, potentially resulting in short circuits or device malfunction. Precision current control circuits are necessary to regulate filament formation and limit the compliance current.
- **Uniformity across extensive arrays**: A primary problem is achieving consistent resistive switching across millions of memory cells. Device shape differences, material quality, and process consistency affect array-level speed, yield, and manufacturing capabilities.
- **Read/write disturb effects**: Parasitic currents in crossbar arrays might inadvertently modify inactive cells (sneak routes). This may result in interruptions in reading and writing. We must create solutions to alleviate these effects, such as supplementary resistive switches or selection devices.
- **Advancing to sub-10 nm technological nodes**: Challenges such as heightened leakage currents, elevated forming voltage, and diminished retention occur when RRAM and CBRAM are miniaturized. Regulating switches at the atomic scale gets increasingly challenging at these dimensions.
- **Environmental stability and packaging**: Moisture can dissolve or oxidize metallic filaments, posing significant challenges for CBRAM. Specialized packaging and encapsulating processes are essential to ensure long-term reliability across diverse environmental conditions.

Progress in materials research, process engineering, and circuit design is essential to address these manufacturing challenges. Novel techniques to address these problems include the development of robust selection elements, the use of 2D materials, and atomic-layer deposition (ALD) for conformal switching layers. Collaboration between academic institutions, foundries, and system developers is essential for developing manufacturing-scalable RRAM and CBRAM technologies.

6.6.5 RRAM device architectures

See Table 6.7.

Table 6.7 RRAM device architectures

Type	*Description*
1T-1R	One transistor—one RRAM (low sneak path risk)
1D-1R	One diode—one RRAM (used in crossbar arrays)
2R	Complementary RRAM structure for bipolarity

6.6.6 Prospective research areas

- **Neuromorphic applications**: Analog resistance states enable synaptic emulation.
- **3D stacking**: Vertical integration for ultra-high-density memory.
- **Self-selective RRAM**: Eliminates the need for external selectors.
- **Hybrid memory**: Integration with SRAM/Flash for hierarchical memory systems.

6.7 PHASE-CHANGE MEMORY (PCM)

Phase-change memory (PCM) is based on the ability of certain materials, such as Ge2Sb2Te5, to change between amorphous and solid states. Each state's binary value is unique. A transient electrical discharge produces heat in the phase-transition material [14]. As a result, it crystallizes or becomes amorphous, making it a data storage medium. Nanowires are being studied as possible PCM materials due to their high surface-to-volume ratio and excellent electrical and thermal characteristics. Maintaining the stability of the cells for high-speed, high-density applications and managing the hardening process are the most challenging parts of the memory cell production process.

6.7.1 Evolution of PCM

The idea of phase-change materials has been around since the 1960s. However, PCM could not be sold in stores until the 2000s. Intel and STMicroelectronics introduced PCM products in the late 2000s. Performance, endurance, and retention have all improved over time as a result of developments in materials engineering, cell design, and integration methods. The main focus of current research is on 3D PCM arrays and how they might be applied to artificial intelligence and neuromorphic systems.

6.7.2 Construction of a PCM cell

In Figure 6.11, the Top Electrode (TE), Phase-Change (PC) layer, and Bottom Electrode (BE) of a basic PCM cell are separated by a chalcogenide material, such as $Ge_2Sb_2Te_5$ or GST. The structure includes a selection device, which may be a transistor or a diode, to access the memory cell. An electric current heats the GST material, converting it from an amorphous to a crystalline state [15]. This provides the cell with directives.

6.7.3 Working principle of PCM

The chalcogenide material in PCM can transition between a crystalline state (low resistance, logic "1") and an amorphous state (high resistance, logic "0"). Data is inscribed by electrical pulses that thermally alter the material. RESET (Amorphize): A quick, high-intensity pulse rapidly liquefies the material and subsequently cools it swiftly to achieve an amorphous state. SET (Crystallize): Atoms may align in a crystalline configuration due to a medium-strength, prolonged impulse [16]. Read operations evaluate

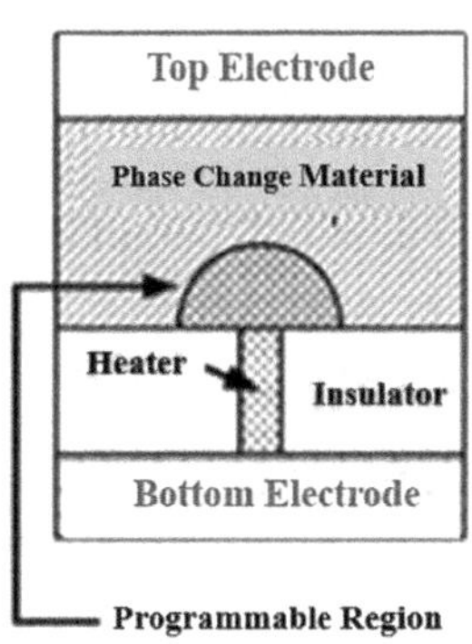

Figure 6.11 Schematic of a conventional PCM cell.

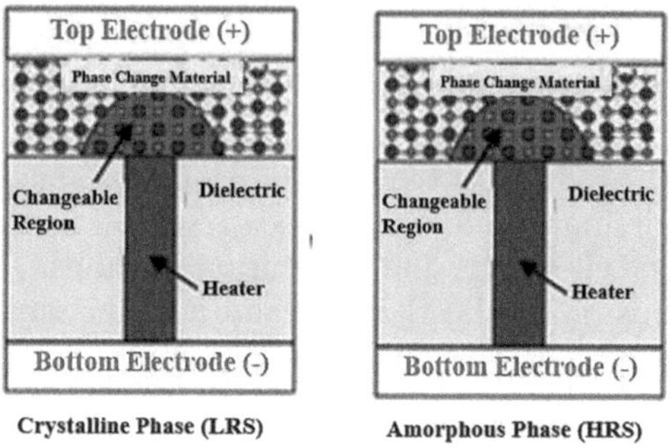

Figure 6.12 Mushroom-shaped structure of PCM device.

resistance with a low current without modifying the state. Figure 6.12 illustrates the mushroom-shaped structure of the PCM device.

(a) **SET operation (write "1")**: Crystallization
- A moderate-voltage **pulse** (long duration, ~100 ns, lower amplitude) is applied across the electrodes.
- The temperature rises above the **crystallization temperature** (~200°C–300°C) but remains **below the melting point.**
- The GST transforms from an amorphous to a **crystalline phase.**
- **Result**: LRS (logic "1").

(b) **RESET operation (write "0")**: Amorphization
- A high-voltage, short pulse (e.g., 10–50 ns) is applied.
- Temperature in the small contact region exceeds GST's melting point (~600°C).
- When the pulse is removed rapidly, the molten material quenches (cools rapidly), resulting in the formation of an amorphous phase.
- **Result**: HRS (logic "0").

(c) Read operation
- A low voltage (~0.1–0.3 V) is applied that does not disturb the phase.
- Current is measured: High for crystalline and low for amorphous.
- Nondestructive read (Table 6.8).

6.7.4 Phase-change material parameters versus PCM device performance

Among the most important variables influencing the performance of a PCM technology are the melting point, electrical resistivity, thermal conductivity, and switching speed of the phase-change material [16]. Higher crystallization temperature enhances data storage but slows SET, while a lower freezing

Table 6.8 Mushroom-shaped PCM device: structural description

Region	*Material*	*Function*
Bottom electrode:	TiN and W	Supplies current and heat from the bottom
Phase-change layer	GST ($Ge_2Sb_2Te_5$)	Stores the data by switching between phases
Heater contact area:	Narrow contact under GST	Creates localized heating—forms the "mushroom"
Top electrode:	TiN, W	Completes the circuit and applies programming voltage
Dielectric sidewalls	SiO_2	Isolates the active region

point lowers RESET energy. The disparity in resistivity between crystalline and amorphous states underpins read margin and sensing reliability. Reduced thermal conductivity minimizes electricity usage by retaining heat in the congested area. Long-term data storage and high-speed switching require a stable amorphous phase and rapid crystallization. Additionally, a slight volume variation during phase changes enhances endurance by lowering mechanical stress. Improved material properties, such as those of GST ($Ge_2Sb_2Te_9$), directly impact the speed, power efficiency, durability, and expansion of PCM devices (Table 6.9).

Table 6.9 Phase-change material parameters versus PCM device performance

Parameter	*Typical range/ example*	*Influence on PCM device performance*
Crystallization temperature (T_c)	~150°C–350°C	Affects SET speed and data retention; higher T_c improves retention but slows down crystallization
Melting temperature (T_m)	~600°C–650°C (e.g., GST)	Determines RESET pulse amplitude and power; lower T_m reduces RESET current.
Thermal conductivity	0.2–0.5 W/m·K	Low thermal conductivity improves thermal confinement, thereby reducing power consumption
Electrical resistivity (amorphous)	~10^5–10^8 Ω·cm	High resistivity provides a clear resistance contrast for HRS
Electrical resistivity (crystalline)	~10^2–10^3 Ω·cm	Low resistivity ensures fast and stable read operation in LRS
TCR (temp. coeff. of resistance)	Negative	Affects read-margin stability under thermal variation
Switching energy	<10 pJ/bit	Lower energy enables low-power memory operation

(Continued)

Table 6.9 (Continued)

Parameter	*Typical range/ example*	*Influence on PCM device performance*
Phase stability	Years (for amorphous)	Impacts data retention; a stable amorphous state is essential for long-term storage
Cycling endurance	~10^6–10^9 cycles	Determines device lifetime; influenced by material stress and atomic migration
Volume change (ΔV)	5%–10% (between phases)	Causes mechanical stress, influencing endurance and scaling
Optical contrast	High (used in optical storage)	Not essential in electrical PCM, but important for optical-read applications
Latent heat of fusion	~100 J/g	Higher latent heat requires more RESET energy, which affects the thermal design
Band gap (E_g)	~0.5–0.7 eV	Affects leakage current and data retention at high temperatures
Density difference (Δρ)	~5%–10%	Can induce stress-related failure over multiple switching cycles

6.7.5 Key materials: example ($Ge_2Sb_2Te_5$—GST)

See Table 6.10.

Table 6.10 Technical specifications of Ge_2Sb_2Te

Property	*Value*
Crystallization temperature	~150°C–200°C
Melting temperature	~600°C–620°C
Resistivity (amorphous)	~10^6 Ω·cm
Resistivity (crystalline)	~10^3 Ω·cm
SET speed	~50–100 ns
RESET speed	~10–50 ns
Endurance	~10^8 cycles
Data retention	≥10 years at 85°C

The cell resistance in nonvolatile memories like RRAM and PCM can be adjusted using a closed-loop method, known as the Program and Verify Algorithm with Pulse Slope Adjustment, as shown in Figure 6.13. At first, programming pulse factors like amplitude, width, and slope are left at their default values. A program pulse is sent to the memory cell, and its resistance is subsequently measured as part of the proof phase. The resistance had to

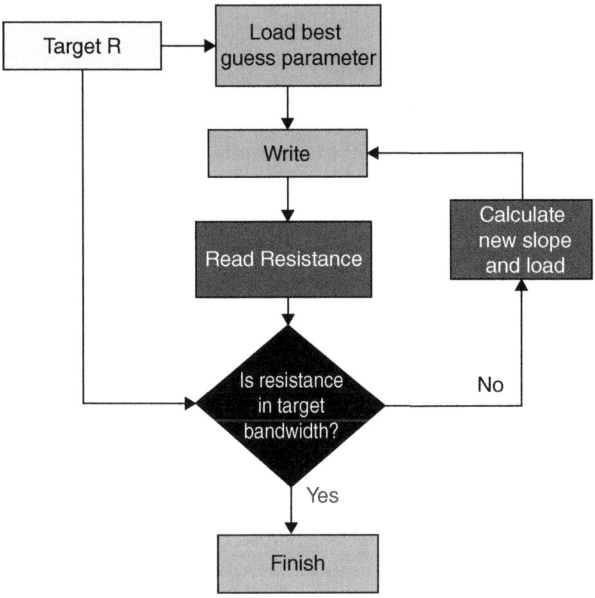

Figure 6.13 Program and verify the algorithm with the adjustment of the pulse slope.

stay within the specified range for the setup to be considered successful. Otherwise, the algorithm retransmits the pulse after modifying its slope to be flatter or steeper, contingent upon whether the resistance is too high or too low. In this cycle, the pulse is modified and repeatedly verified until the desired resistance or a maximum number of iterations is reached. By doing this, device stress is decreased and write accuracy is increased while ensuring that the data is consistently and accurately preserved (Table 6.11).

Table 6.11 Fabrication challenges

Thermal crosstalk	Heat from programming can affect adjacent cells.
Reset current	High current required to melt the material increases power consumption.
Material stability	Phase-change materials can degrade after repeated cycling.
Contact engineering	Need for confined structures (e.g., mushroom-type) to localize heating.
3D integration	Complex thermal budget and patterning in vertical PCM stacks.
Crystallization kinetics	Requires tight process control for speed and endurance.
Scaling limitations	Controlling phase stability at the nanoscale remains challenging.

6.7.6 Applications

- Storage-Class Memory (SCM)
- Main memory replacement for high-speed computing
- In-memory computing (e.g., matrix-vector multiplication)
- Neuromorphic systems
- Automotive and aerospace systems (high-temperature stability)

6.8 FERROELECTRIC RAM (FeRAM/FRAM)

Ferroelectric RAM, commonly referred to as FRAM or FeRAM, is an NVM that integrates the optimal features of ROM and RAM technologies. The machine has an extended lifespan, functions rapidly, and utilizes less electricity. FeRAM retains its polarization state in the absence of power and utilizes a ferroelectric layer for data storage. This renders it appropriate for medical electronics, embedded systems, and automotive applications.

FeRAM, or FRAM, is a type of NVM that utilizes the unique characteristics of ferroelectric materials for data storage [17]. Standard DRAM retains data as an electrical charge on a capacitor, but FeRAM stores it through the polarization states of a ferroelectric capacitor. This allows for quick operation, low-power consumption, and durability over billions of read/write cycles, rendering it appropriate for embedded devices and nano energy systems.

6.8.1 Invention and evolution

In the 1950s, the idea of using ferroelectric materials for memory was put forward. Ramtron International achieved numerous breakthroughs when it brought FeRAM to the market in the 1980s and 1990s. Due to its strong ferroelectric qualities, lead zirconate titanate (PZT) is the most widely utilized material in FeRAM. New and more complex applications have been made possible by the combination of CMOS and FeRAM technologies. Basic FRAM structure is shown in Figure 6.14.

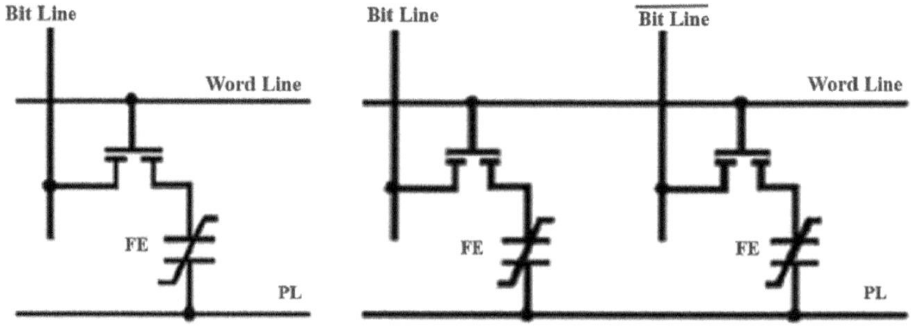

Figure 6.14 (a) 1T1C (b) 2T2C.

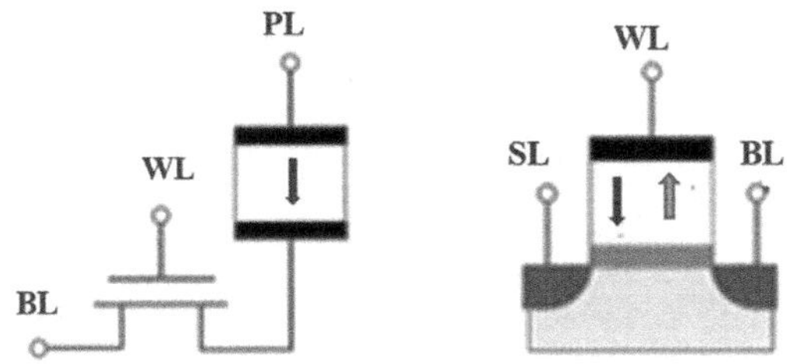

Figure 6.15 Switching mechanism of (a) 1T1C (b) 2T2C.

6.8.2 1T1C FRAM cell operation (one transistor–one capacitor)

The 1T1C FRAM cell comprises a single ferroelectric capacitor and an access transistor. The capacitor retains information in the polarization state of the ferroelectric material. The switching mechanism of 1T1C and 2T2C is shown in Figure 6.15.

(a) **Write operation:** Upon activation of the WL during a write operation, the access transistor is engaged, and a voltage is applied to the ferroelectric capacitor. The orientation of the ferroelectric layer changes in accordance with the voltage direction, indicating either a logic "1" or "0".
(b) **Read operation:** The charge on the capacitor is ascertained by detecting a voltage. The sense amplifier produces a displacement current indicative of a bit when the polarity changes during the read process. This action alters the stored state, necessitating an update (rewriting) each time it is accessed.

6.8.3 2T2C FRAM cell operation (two transistors–two capacitors)

Each bit of the 2T2C FRAM cell comprises two ferroelectric capacitors and two access transistors. This indicates that one capacitor retains a logic "1" while the other retains a logic "0." Each bit preserves complementary information [18]. This design enhances noise resilience and reliability.

(a) **Write operation:** Complementary voltage signals are transmitted across both capacitors via active WLs. One capacitor's "high" state represents logic "1," while the other's "low" state signifies logic "0."
(b) **Read operation:** The read operation involves the simultaneous reading of both capacitors. A differential sense amplifier evaluates the charges from two capacitors, enhancing reading accuracy. As this is also a destructive read, the accurate state is overwritten upon detection.

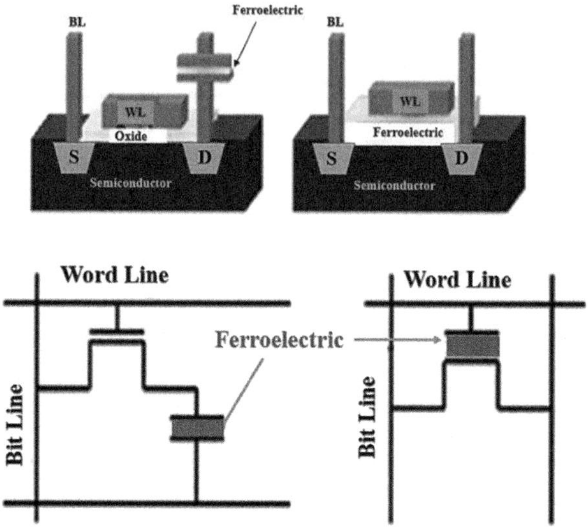

Figure 6.16 (a) 1T1C FRAM (alternate to DRAM) (b) 1T FeFET (alternate to NVRAM).

Figure 6.16 illustrates two novel configurations for memory cells: the 1T1C FRAM cell, an alternative to DRAM, has one access transistor and one ferroelectric capacitor. Storing data in the polarization state of ferroelectric material facilitates rapid read/write operations with little power usage [19]. Nonetheless, it resembles DRAM in that its read operation is destructive, necessitating a refresh following each read. (a) The 1T FeFET cell incorporates a ferroelectric layer into the gate stack of a MOSFET, serving as a replacement for NVRAM. Data is retained in the ferroelectric layer, with the polarization direction dictating the regulation of the transistor's threshold voltage [18]. FeFET serves as an excellent alternative to Flash in high-speed, low-power applications due to its nondestructive readout, nonvolatility, and compatibility with CMOS, in contrast to FRAM (Table 6.12).

Table 6.12 Comparison of FeRAM with other nonvolatile memory technologies

Property	*FeRAM*	*Flash*	*MRAM*	*RRAM*
Write speed	10–100 ns	μs–ms	10–50 ns	10–100 ns
Endurance	$>10^{12}$ cycles	10^4–10^5	10^8–10^{11}	10^6–10^9
Power consumption	Very low	Moderate	Low	Low
CMOS compatibility	High ($HfZrO_2$)	High	Moderate	High
Scalability	Moderate to high	Very high	Moderate	High

6.8.4 Fabrication process and challenges

- The fabrication of FeRAM involves integrating ferroelectric materials, such as PZT, into the semiconductor process. Key fabrication steps include the following:
  - Deposition of ferroelectric films using sputtering or sol–gel methods.
  - High-temperature annealing to achieve the ferroelectric phase.
  - Patterning and etching of ferroelectric layers.
  - Integration with CMOS transistors.

6.8.5 Challenges

- High processing temperatures incompatible with BEOL CMOS.
- Material compatibility and uniformity.
- Scaling below 28 nm is difficult due to material limitations.
- Retention degradation due to repeated cycling and fatigue.

6.8.6 Applications

FeRAM continues to evolve through research into new materials like hafnium oxide-based ferroelectrics that are more CMOS-compatible. Applications include the following:

- Automotive ECUs
- Secure smart cards and RFID tags
- Medical implants
- Industrial control systems

6.8.7 Future directions

- Integration with advanced logic platforms (e.g., GAAFETs).
- Development of 3D FeRAM structures for ultra-dense memory.
- Use in neuromorphic systems with multi-bit storage capability.
- Ferroelectric tunneling junctions for ultra low energy switching.

6.9 ELECTROCHEMICAL RANDOM ACCESS MEMORY (ECRAM)

6.9.1 Invention

Electrochemical memory was initially utilized in early electrochemical devices. In the 2010s, the development of the present electrochemical RAM (ECRAM) was primarily driven by the need for effective neuromorphic memory [20]. ECRAM technology was developed by experts from IBM, Stanford University, and MIT, among others. These were among the initial devices to utilize materials such as WO_0 (tungsten trioxide), HfO_2 (hafnium oxide), and two-dimensional materials like MoS_2, demonstrating analog resistive switching by electrochemical doping [21].

6.9.2 Structure

A novel memory system, termed ECRAM, integrates electrochemistry with solid-state electronics to produce memory devices that are scalable, energy-efficient, and nonvolatile. It operates by adding or expelling ions, thereby modifying the electrical conductivity of a channel material. ECRAM is extremely desirable for neuromorphic and in-memory computing systems that need the utilization of analog memory functionalities. Its basic structure [22] is shown in Figure 6.17 (Table 6.13).

6.9.3 Working principle

ECRAM devices have three terminals, analogous to a transistor. The primary components include the source, the drain, the channel (often composed of an oxide or layered material), the electrolyte, and the gate. An electric charge from the gate electrode attracts or repels ions from the liquid in and out of the channel. The ion mobility alters the electrical conductivity of the channel by modifying the ion's oxidation state [23]. The alteration in resistance persists after power is disengaged due to the retention of ions within the channel material.

Key features of ECRAM operation include the following:

- Ion doping of channel material
- Analog modulation of resistance
- Nonvolatile storage behavior

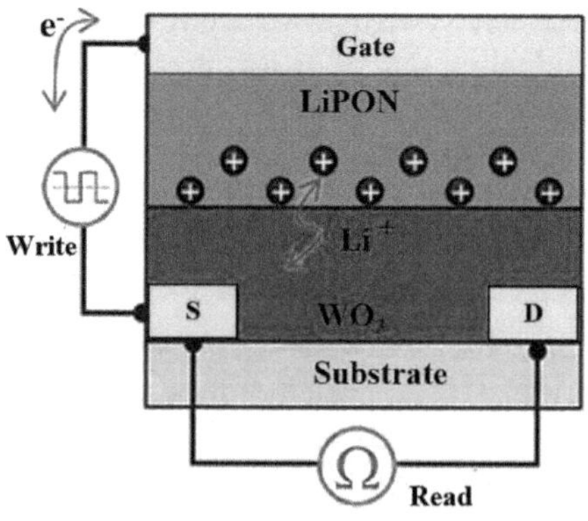

Figure 6.17 Basic structure of electrochemical random access memory.

Table 6.13 ECRAM technical specifications

Component	*Material examples*	*Function*
Gate electrode	Ag, Pt, Au	Controls ion injection
Electrolyte	LiPON, Ta_2O_5, PEO-based polymers	Transports ions (e.g., H^+, Li^+, O^{2-})
Channel material	WO_x, MoS_2, IGZO, PEDOT:PSS	Modulates conductivity via ionic doping
Source/drain	Ti/Au, Pt, ITO	Carries the current for the reading device's conductance

6.9.4 Fabrication possibility

ECRAM devices are typically fabricated using thin-film deposition techniques compatible with CMOS processes. Key steps involve the following:

- Deposition of channel material (e.g., by ALD, sputtering, or solution-based methods).
- Patterning and etching of electrodes (source, drain, and gate).
- Deposition of solid or gel electrolyte layers.
- Encapsulation for stability and reliability.

6.9.5 Challenges in fabrication

- Maintaining interface integrity between the electrolyte and the channel.
- Controlling ion mobility for predictable switching.
- Ensuring long-term stability of materials under operating conditions.
- Scaling down the devices without compromising performance.
- Achieving uniformity across large wafer areas for mass production.

6.9.6 Applications

- Neuromorphic computing (synapse emulation in artificial neural networks)
- Edge AI processors
- Reconfigurable analog circuits
- Flexible and Wearable Electronics
- In-memory computing architectures

ECRAM's superior analog tuning and efficient power consumption render it suitable for neuromorphic computing, albeit with its moderate switching speed and inadequate data retention. RRAM possesses an extended lifespan, excellent memory capabilities, and rapid switching, but is constrained by

Table 6.14 Comparison of ECRAM with other NVMs

Property	*ECRAM*	*RRAM*	*PCM*	*STT-MRAM*
Analog tuning	Excellent	Limited	Moderate	No
Power efficiency	High	Moderate	Low	High
Switching speed	Medium (μs–ns)	Fast (ns)	Medium (10 s–ns)	Fast (ns)
Retention	Medium	High	High	High
Endurance	High	High	Medium	Very high

mechanical tuning. Despite its high power consumption and limited lifespan, PCM provides substantial memory and moderate analog performance. Despite the absence of analog tuning, STT-MRAM has superior switching speed, retention, and endurance. Ultimately, each category of memory demonstrates superiority in a specific function. For example, STT-MRAM is superior for high-performance embedded systems, while ECRAM excels in artificial intelligence and in-memory processing (Table 6.14).

6.10 MEMRISTOR

A Memristor, or memory resistor, is a two-terminal nonvolatile component whose resistance depends on its voltage and current history. Due to its ability to maintain its resistance state in the absence of power, it is the optimal choice for applications related to analog signal processing, neuromorphic computation, and advanced memory technologies [24].

6.10.1 Memristor crossbars

Memristors are a form of NVM that relies on resistance changes induced by the flow of charge through them. Memristor crossbar arrays are frequently employed for high-density storage solutions. Memristor crossbars may enable exceptionally high memory density, featuring memory cells organized in a grid configuration. The primary challenge in the manufacturing process is the production of reliable and reproducible Memristor devices that maintain optimal performance over extended periods [25]. Additional issues encompass managing the switching rate and preventing crosstalk between adjacent cells.

6.10.2 Evolution of the Memristor concept

In 1971, Leon Chua introduced the Memristor, a fourth fundamental circuit component that integrates charge and flux, thereby formally defining the concept of memresistance. The phenomena were initially implemented in

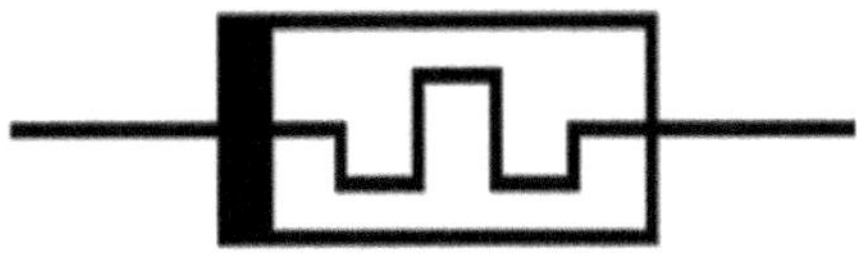

Figure 6.18 Electronic symbol of the Memristor.

2008 when HP Labs demonstrated resistive switching in TiO_2-based nano devices. Since then, devices demonstrating memory-resistance characteristics, including filamentary, phase-change, and interfacial switching mechanisms, have been collectively termed "Memristor". The electronic symbol of the Memristor is shown in Figure 6.18.

6.10.3 Construction of a Memristor

A conventional memristor is typically fabricated using a **Metal–Insulator–Metal (MIM)** structure consisting of an **upper electrode** (UE), a **switching layer**, and a **bottom electrode** (BE). The upper and bottom electrodes are commonly made of metals such as **silver (Ag), platinum (Pt), titanium (Ti), or other inert metals,** while the switching layer is composed of oxide or organic materials such as **TiO_2, HfO_2, or polymers**, as illustrated in Figure 6.19. Certain memristors are constructed from 2D materials, nanowires, or quantum dots to optimize performance and scalability These structures facilitate charge trapping or ionic migration, hence altering the device's resistance in response to input stimuli.

6.10.4 Principle of operation

Memristors work by changing their internal resistive state in response to applied voltage or current. The resistance change can be due to:

- Ionic movement (e.g., oxygen vacancies)
- Charge trapping/detrapping
- Filament formation/rupture
- Phase transitions (amorphous ↔ crystalline)

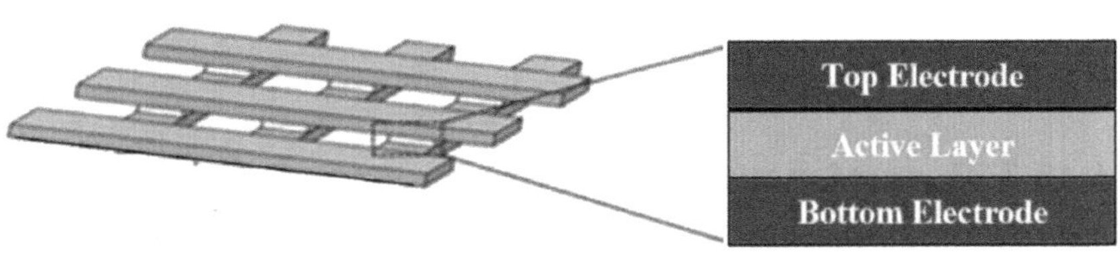

Figure 6.19 (a) Memristor crossbar array. (b) Memristor cross-sectional view.

6.10.5 Operation modes

- SET (write): Switches to LRS
- RESET (erase): Switches to HRS
- **Read:** A small voltage is applied to measure resistance

The Memristor operates based on the relationship between charge (q) and magnetic flux (φ). Unlike resistors, where resistance is constant, a Memristor's resistance—called memristance (M)—changes depending on the amount of charge that has flowed through the device.

$$V(t) = M(q) \cdot I(t)$$

- $M(q)$: Memristance depends on total charge 'q'
- $V(t)$: Voltage across the device
- $I(t)$: Current through the device

The key idea is that memristors can "remember" their resistance state even when power is removed, enabling NVM behavior.

6.10.6 Technical specifications

See Table 6.15.

Table 6.15 Memristor technical specifications

Parameter	*Specification*
Switching time	1–100 ns
Endurance	10^6–10^{12} cycles (material dependent)
Retention	>10 years
On/off resistance ratio	10^2–10^6
Operating voltage	0.2–2 V
Cell size	<$4F^2$ (ideal for crossbar architectures)
Materials	TiO_2, HfO_2, Ta_2O_5, Ag, Cu, organic materials, and graphene oxide

6.10.7 Types of memristive switching

See Table 6.16.

Table 6.16 Memristor switching modes

Type	*Description*
Bipolar	SET and RESET use opposite voltage polarities (e.g., +V for SET and −V for RESET)
Unipolar	SET and RESET occur with the same polarity, but different magnitudes
Volatile	Resistance reverts when power is removed (useful for selectors, not memory)
Nonvolatile	Resistance is retained after power-down (used for memory and neuromorphic use)

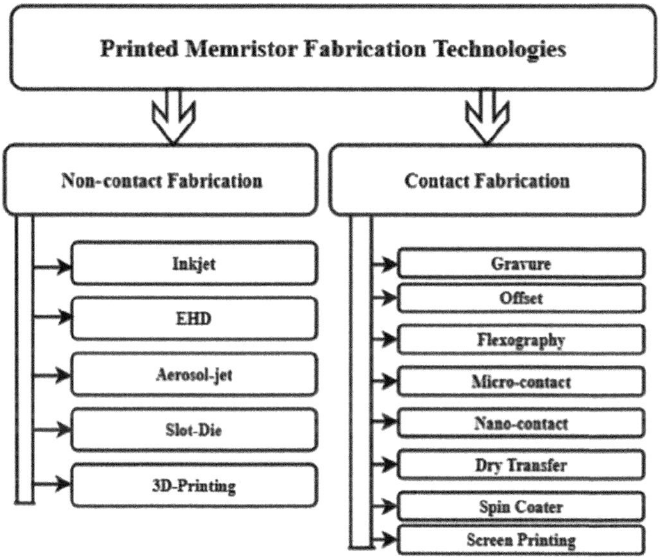

Figure 6.20 Printed Memristor fabrication technologies.

6.10.8 Fabrication challenges

Scalability and material compatibility are crucial for printed memristors. Figure 6.20 illustrates the printed Memristor fabrication technology modes. There exist trade-offs between ionic mobility (rapid switching) and data retention. The stability of the device is considerably affected by exposure to oxygen and humidity. For practical applications, issues such as low endurance, repeatability, and crosstalk in extensive arrays must be resolved.

6.10.9 Applications

- NVM
- Neuromorphic computing (synaptic emulation)

- Reconfigurable logic
- Analog signal processing
- Secure hardware and cryptographic primitives

REFERENCES

1. Jacob, B., Ng, S., & Wang, D. T. (2008). *Memory Systems: Cache, DRAM, Disk*. Morgan Kaufmann.
2. Keeth, B., Prince, R., & Brehmer, G. (2013). *DRAM Circuit Design: Fundamental and High-Speed Topics*. Wiley-IEEE Press.
3. Wong, H. S. P., & Salahuddin, S. (2015). Memory leads the way to better computing. *Nature Nanotechnology*, 10(3), 191–194.
4. Saravanan, M. et al. (2022). *Impact of Semiconductor Materials and Architecture Design on TFET Device Performance, Book: Emerging Low-Power Semiconductor Devices: Applications for Future Technology Nodes*. CRC Press. pp. 79–106.
5. Saravanan, M., et al. (2023). Biomedical devices adopting energy-harvesting schemes. *Energy Harvesting Trends for Low Power Compact Electronic Devices*, Nella, A., Bhowmick, A., Kumar, C., Rajagopal, M. (eds.). Springer International Publishing. pp. 63–80.
6. Ikeda, S., et al. (2010). A perpendicular-anisotropy CoFeB–MgO magnetic tunnel junction. *Nature Materials*, 9, 721–724.
7. Kent, A. D., & Worledge, D. C. (2015). A new spin on magnetic memories. *Nature Nanotechnology*, 10(3), 187–191.
8. Dieny, B., et al. (2020). Opportunities and challenges for spintronics in the microelectronics industry. *Nature Electronics*, 3, 446–459.
9. Bhatti, S., Sbiaa, R., Hirohata, A., Ohno, H., Fukami, S., & Piramanayagam, S. N. (2017). Spintronics based random access memory: A review. *Materials Today*, 20(9), 530–548.
10. Tehrani, S., et al. (2003). Magnetoresistive random access memory using magnetic tunnel junctions. *Proceedings of the IEEE*, 91, 703–714.
11. Ielmini, D. (2016). Resistive switching memories based on metal oxides. *Semiconductor Science and Technology*, 31, 063002.
12. Mehonic, A., & Kenyon, A. J. (2016). Emulating the electrical activity of the neuron using RRAM. *Advanced Materials*, 10, 57.
13. Pan, F., et al. (2014). Nonvolatile resistive switching memories–Characteristics, mechanisms and challenges. *Progress in Natural Science: Materials International*, 20, 1–15.
14. Raoux, S., et al. (2008). Phase-change random access memory: A scalable technology. *IBM Journal of Research and Development*, 52, 465–479.
15. Wong, H. S. P., et al. (2010). Phase change memory. *Proceedings of the IEEE*, 8, 2201–2227.
16. Loke, D., et al. (2012). Breaking the speed limits of phase-change memory. *Science*, 336, 1566–1569.
17. Schroeder, U., et al. (2022). The fundamentals and applications of ferroelectric HfO_2. *Nature Reviews Materials*, 7, 653–671.

18. Mikolajick, T., et al. (2020). The past, the present, and the future of ferroelectric memories. *IEEE Transactions on Electron Devices*, 67(4), 1434–1443.
19. Yurchuk, E., et al. (2015). Origin of the endurance degradation in ferroelectric HfO_2. *2014 IEEE International Reliability Physics Symposium.*
20. Tian, A. A., et al., (2024). Electrochemical random-access memory: Progress, perspectives. *Chemical Reviews*, 125(4), 1962–2008.
21. Kang, H., et al. (2022). Analysis of electrochemical RAM synaptic array for energy-efficient weight update. *Frontiers in Nanotechnology*, 4, 1034357.
22. Chen, P., et al. (2023). Open-loop analog programmable electrochemical memory array. *Nature Communications*, 14, 6184.
23. Strukov, D. B., et al. (2008). The missing memristor found. *Nature*, 453, 80–83.
24. Prezioso, M., et al. (2015). Training and operation of an integrated neuromorphic network based on metal-oxide memristors. *Nature*, 521, 61–64.
25. Wang, Z., et al. (2020). Resistive switching materials for information processing. *Nature Reviews Materials*, 5, 173–195.

Chapter 7

Recent trends in light-emitting diodes with applications

R. Manaswini, G. Dhivyasri, M. Manikandan, J. Ajayan, and S. Sreejith

7.1 INTRODUCTION

Figure 7.1 depicts the structure of the p–i–n semiconductor junction concept. While the semiconductor on the right contains holes in the valence band, the semiconductor on the left is n-doped, which means it has additional electrons in the conduction band. The junction sustains the electric field in the non-doped area between the two doped areas. Applying a bias, as shown, allows electrons to flow from the left to the right, indicating that holes are traveling from the right to the left, allowing electrons and holes to recombine, thereby releasing photons in the semiconductor's intrinsic area. This is the principle of a semiconductor light-emitting diode.

The reduction in the feature size of semiconductor devices is a major concern, as it must be achieved without affecting the performance of the device. The manipulation of various trade-off parameters is the primary area of concern. The popular materials used for manufacturing light-emitting diodes (LEDs) are nitride, arsenide, phosphide, and alloys of these elements. Phosphides and arsenides were widely used to design and manufacture green

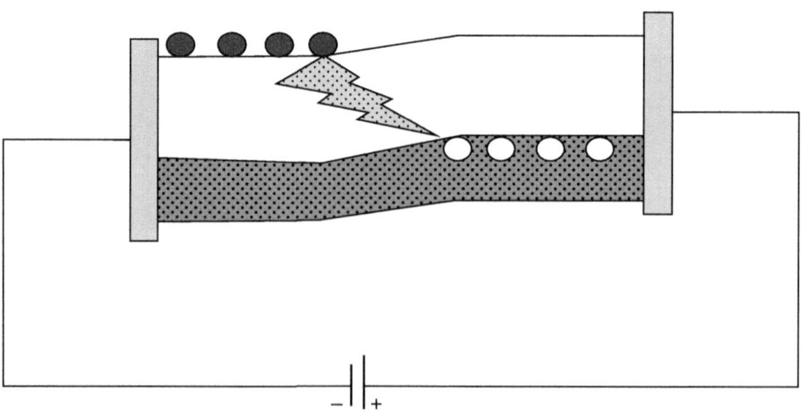

Figure 7.1 p–i–n semiconductor diode.

DOI: 10.1201/9781003635123-7

and red LEDs, but the materials used to manufacture bright blue LEDs were not available in abundance. In the present scenario, group III direct wide-energy band gap materials have excellent electrical and optical properties, high thermal and electrical conductivity, considerable ratios of carrier mobility, stable chemical and physical properties, and low dielectric constants. These attributes make them highly efficient in fabricating and manufacturing blue-light LEDs with higher performance efficiency and these perceptions till the early 1990s [1].

In 1990, Professor Shuji Nakamura developed the world's first GaN-based LED, with good performance and higher brightness, with a chip dimension of 0.6×0.5 mm^2. It was after this invention that the commercial usage and manufacturing of LEDs were initiated [2, 3]. In addition, the period also saw the parallel growth and development of the commercial manufacturing processing techniques of photoelectronic devices. The very first array-based micro-LED with a feature size of 0.5×0.5 mm^2 was implemented in the year 2000. In the literature, it is recorded that 10×10 pixels with an integrated chip dimension of 12 μm were implemented by Jiang et al. in 2001 [4]. After this milestone, the research on micro-LED gained the potential as a result new innovative design of LED with three terminals presented by Memon et al., the concept of transferable wafers for LED was experimented with by Chang et al., and the nanowire-based luminescence was included in the research finding Pandey et al. [5–7]. Figure 7.2 indicates that the size of the LED was reduced by suitably decreasing the feature's size, thereby increasing the number of LEDs on the same chip area. This rapid decrease in the size of LEDs was advantageous for various applications. Simultaneously, the manufacturing industry faced new challenges in fabrication and mass production. Improvements in micro-LED designs have led to lower power consumption, increased brightness, extended lifespan, and higher resolution. Time of response has paved the way for new applications in the fields of display systems, wearable computing, smart devices, visible light communications, and biomedical and medical device manufacturing [8–11]. Apart from these, the promising demand in the commercial market has attracted investors to set up new plants and manufacturing sites. In various statistical data analyses, it is also estimated that the market value for micro-LEDs is believed to go up to 17 billion USD by the year 2025. Also, research demand and publications increased threefold between 2016 and 2019 [12]. Table 7.1 shows the properties of different semiconductor compound materials.

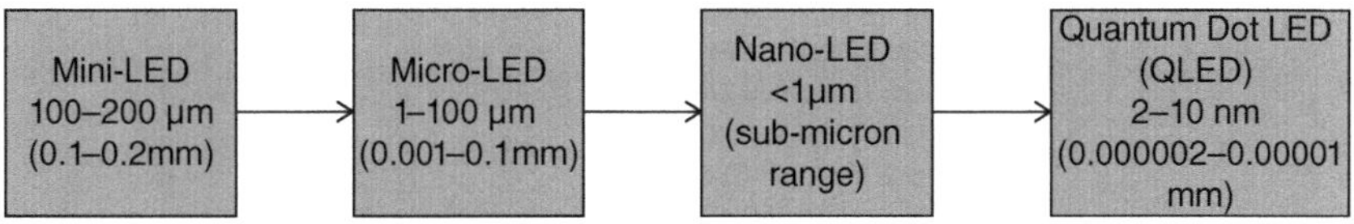

Figure 7.2 III group nitride LED types.

(Refs. [3, 4, 13].)

Table 7.1 Different semiconductor compound materials' properties

Material	*Bandgap (eV)*	*Electron mobility (cm^2/Vs)*	*Hole mobility (cm^2/Vs)*	*Critical field E_c (V/cm)*	*Thermal conductivity (W/m-K)*	*Coefficient of thermal expansion (ppm/K)*
InSb	0.17 D	77,000	850	1000	18	5.37
InAs	0.35 D	44,000	500	40,000	27	4.50
GaSb	0.73, D	3000	1000	50,000	32	7.75
InP	1344 D	5400	200	500,000	68	4.6
GaAs	1424 D	8500	400	400,000	55	5.73
GaN	344 D	900	10	3,000,000	110–200	5.4–7.2
Ge	0.661 I	3900	1900	100,000	58	5.9
Si	1.12 I	1400	450	300,000	130	2.6
GaP	2.26 I	250	150	1,000,000	110	
SiC (3C,b)	2.36 I	300–900	10–30	1,300,000	700	2.77
SiC (6H,a)	2.86 I	330–400	75	2,400,000	700	5.12
SiC (4H,a)	3.25 I	700	—	3,180,000	700	5.12
C (diamond)	5.46–5.6 I	2200	1800	600,000	1300	0.8

7.2 EVOLUTIONARY HISTORY OF LEDs

The researcher's extensive review indicates that stage-wise improvements are noticed in the development of ultraviolet, blue, green, and red LEDs for various applications. The information on infrared LEDs is purposefully limited to keep the presentations crisp and reader-friendly.

7.3 EVOLUTION OF UV LEDs

The reason for a rapid increase in the implementation and fabrication of UV-based LEDs is applications like fluorescence-based chemical sensing, flame detectors, high-efficiency lighting, optical sensors and detectors, and storage. The majority of these applications need UV LEDs designed for a 300 nm wavelength. The easy fabrication process can depend on the wavelength, as it is a function of the wavelength of light to which it is designed. When designing devices with shorter wavelengths, the bandgap of materials used for fabrication poses different challenges. The contents of the Al requirement are more in shorter wavelength-based devices, and in the case of wide bandgap materials, high Al concentration increases the dislocation density, which reduces the efficiency of the material. Hence, the amount of Al is to be optimized to obtain an optimized wavelength. A section of the research article examines efficiency of different UV LEDs considered above and below 300 nm.

7.3.1 Structure of UV LED

In the year 2020 (Li et al. [14]), the researchers designed a UV LED with an efficiency of 60% and a wavelength of 395 nm. The design includes an epitaxial layer on a substrate of Si, which is approximately 4 inches. AlGaN and AlN were grown on it, along with a layer of undoped GaN, which was used as a buffer layer. An n-type GaN layer was included with multiple quantum wells (MQWs) of 9 periods per unit. The electron-blocking layer was fabricated on p-type GaN of 200 nm thickness. The spatial overlap and carrier concentration were considered. The usage of GaN layers in the intermediate structure has shown an increase in the concentration of carriers and spatial overlap in the multiple quantum well (MQW). This resulted in a change in the polarization and energy band gap properties of the semiconducting materials and the design structures, as indicated in Figure 7.3. The UV area, ranging from 100 to 400 nm, is separated into three bands: UV-A (315–400 nm), UV-B (280–315 nm), and UV-C (100–280 nm), concerning their applications.

Manikandan et al. [15] have discussed implanting multiple quantum wells in UV-based LEDs with AlGaN/BGaN/AlGaN in the active layers, as shown in Figure 7.4. In the BGaN quantum well, the concentration and

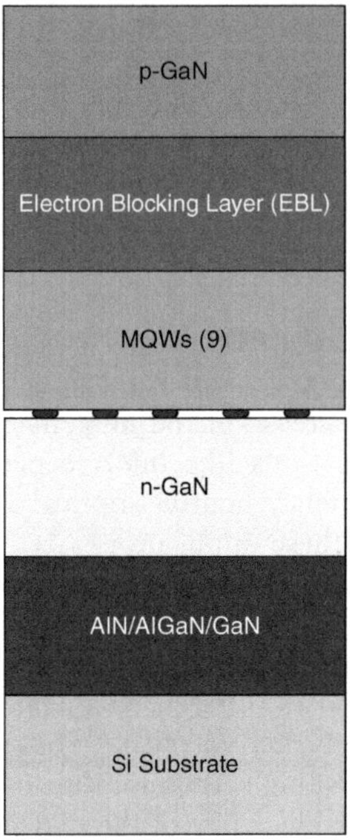

Figure 7.3 Design of UV LED with EBL layer.

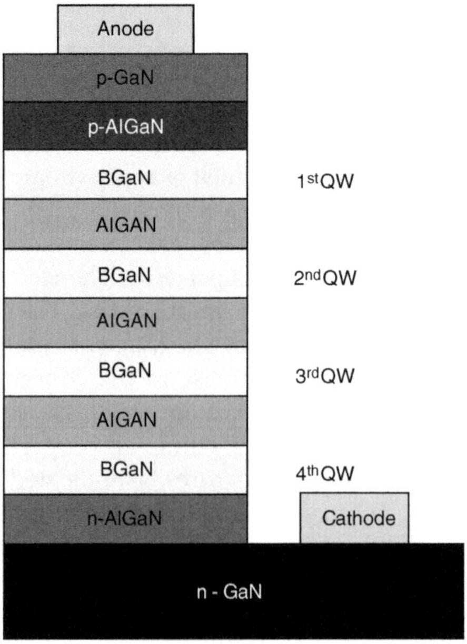

Figure 7.4 Design of UV LED with BGaN quantum well [15].

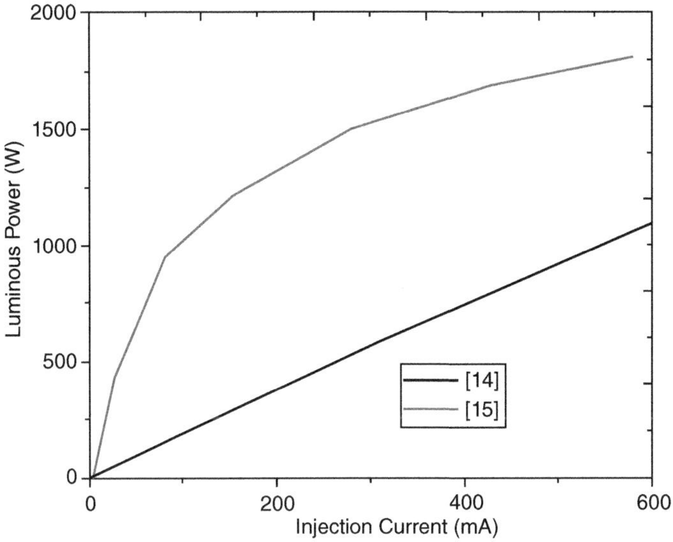

Figure 7.5 Injection current vs luminous power.

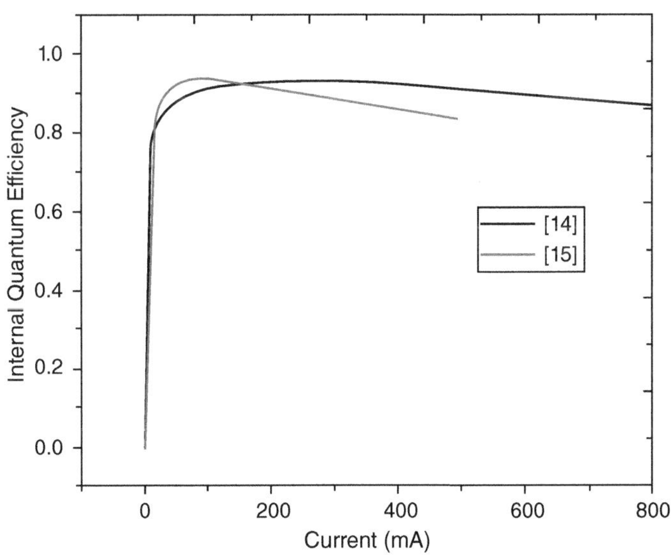

Figure 7.6 Current versus internal quantum efficiency.

thickness are fixed at 10% and 3 nm, respectively. The usage of BGaN material in UV-based LEDs has improved the performance of LEDs in terms of luminous power and internal quantum efficiency (IQE), as shown in Figures 7.5 and 7.6.

7.3.2 Applications of UV LED

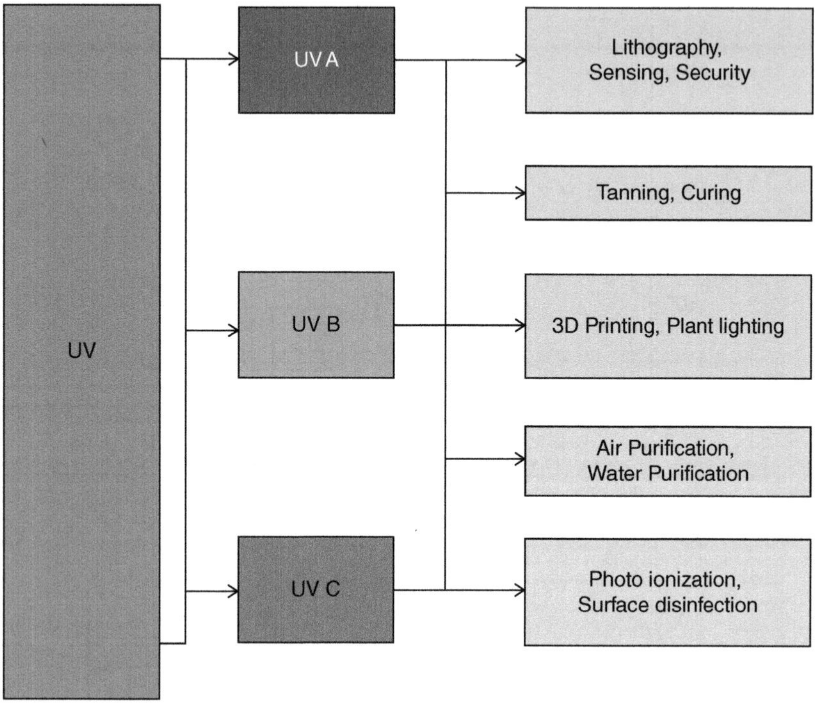

Figure 7.7 Classification and applications of UV LEDs.

7.4 EVOLUTION OF BLUE LEDs

Semiconducting materials with an E_g of 2.6 eV or more are required for the design and development of LED devices and are capable of emitting wavelengths in the range of 455–485 nm [16]. The wide bandgap materials of the direct transition type are used to design blue-light LEDs, as depicted in Figure 7.8. A highly efficient semiconducting material with a single crystal and an E_g value above 2.6 eV is used for designing brighter blue LEDs. The crystal quality of single-crystal GaN has improved the brightness of blue-colored LEDs. Semiconductors like silicon carbide (3.4 eV E_g) and zinc selenide (2.7 eV E_g) are preferred for the designing and manufacturing of blue-light LEDs.

In the current scenario, the researchers aim to improve the blue LEDs' efficiency, in particular with challenges concerned with the usage of chloride-based perovskites, which degrade the efficiency [17]. In the case of blue LEDs, the record efficiencies have increased with the usage of nanocrystal-based perovskite, with the highest quantum efficiency at 6.2% and 2.4% for the wavelengths in the range of 487 and 465 nm, respectively [18]. These

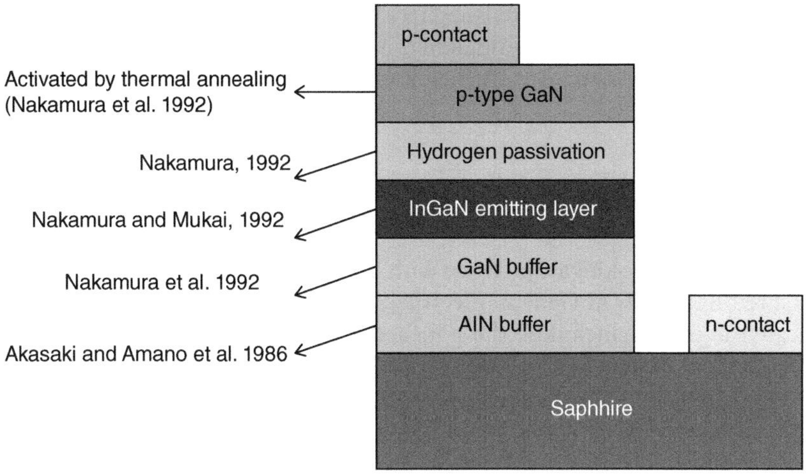

Figure 7.8 Structure of blue LEDs.

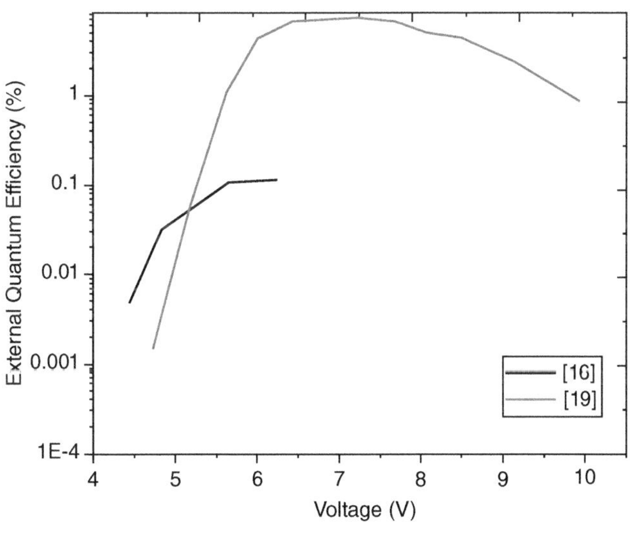

Figure 7.9 External quantum efficiency vs. voltage.

developments, that is, the usage of nanocrystal perovskite, have improved the efficiency of LED technology, as shown in Figure 7.9.

Jing Peng [19] investigated the improvements in blue LED-based perovskites and elucidated them based on the different dimensions, considering 3D and 2D. The different methods, like doping of ions, reduction in the feature size, passivation of surface defects in semiconducting materials, and anion exchange treatments, like using anti-solvents, are employed to optimize the properties of the semiconducting materials to improve the emission of different wavelengths. The perovskite used in blue LEDs is used for different applications, particularly in display devices, wearable devices, biomedical applications, bioimaging, medical instruments, and diagnostics, because of its performance efficiency, tunable wavelengths, and lower fabrication costs.

7.5 EVOLUTION OF RED LEDs

In 1962, Nick Holonyak Jr, along with his team, designed and developed the first red-light LED at General Electric [22]. The material Ga was used to fabricate the diodes, including the polished active layer of dimension 10^{-3} cm^2. The device, designed using the process vapor density method of epitaxy, emitted light at approximately 650 nm and had a performance efficiency of less than 0.1 Lm/W. The main focus of red-light LEDs is to fabricate the micro- and mini-LEDs by decreasing the feature size of the LEDs. But the main challenges are during the manufacturing process, which reduce the frequency of LED production. An ultra-small LED less than 10 μm was demonstrated in the year 2020. A 632 nm red-colored LED with a heterojunction on InGaN was fabricated on a wafer with a quantum efficiency exceeding 0.2%, which was identified as a milestone for efficient mini-display applications [23]. The performance efficiency of a red LED is related to the usage of materials used to build the LED and the techniques used in fabrication. In the research papers, it is recorded that the quantum efficiency of the device has been enhanced in the InGaN-based well structure by 10% in 640 nm [24]. Later in 2022, a 621 nm wavelength with InGaN material with a single quantum well structure and an external quantum efficiency value around 4.3% at 10.1 A/cm^2, was designed. This LED had higher improvements in its red-light gain [25]. Many other techniques were used to enhance the performance efficiency and LED output, including enhancing MQW, EBL, and lattice matching of the materials used in fabrication. Implementing organometal halide perovskite between the larger bandwidth layers effectively improves electron accumulation and hole recombination [26]. It is also indicated in the literature that the modulation concerning doping in the quantum well will act as a barrier to suppress the recombination of particles. It also reduces the rate of electrons, which improves the LED efficiency [27]. In the literature, the concept of heterostructure and quantum dots is clearly explained in the paper by Bae et al. [28, 29]. It is closely observed that both the mitigation of Auger recombination and the achievement of a balanced ratio of hole and electron injection are factors that enhance the quality of the LED. It also explains that the simultaneous surface passivation of perovskite nanocrystals will enhance the photoluminescence and increase stability and quantum yield, thereby increasing the performance of the LED, as shown in Figure 7.10. All these techniques, including interfacial control, increased charge transport, and geometrical optimization, will collectively enhance the LED performance efficiencies and capabilities of the total output. The different applications of red LEDs are depicted in Figure 7.11.

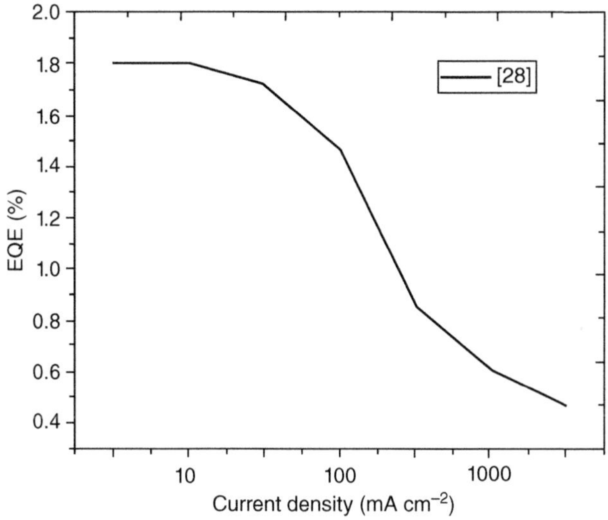

Figure 7.10 Current density versus EQE.

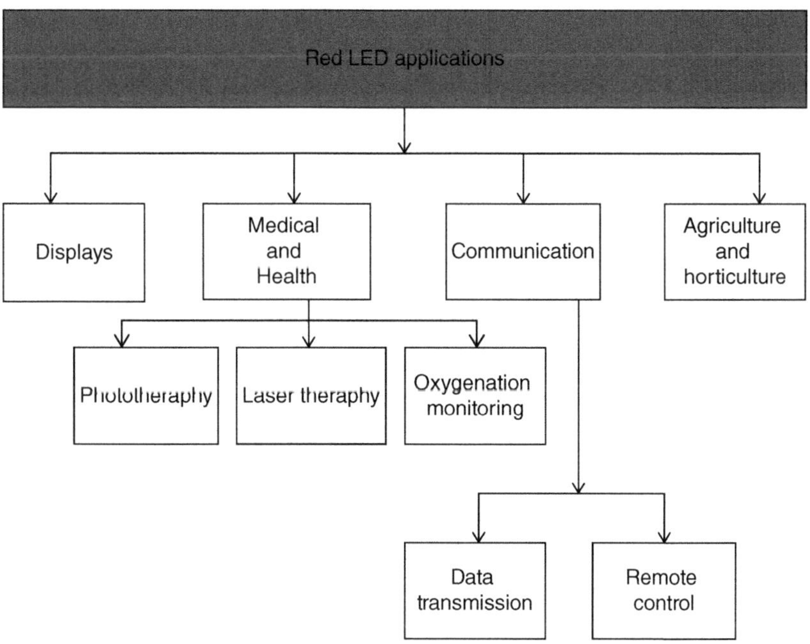

Figure 7.11 Applications of red LEDs.

7.6 EVOLUTION OF GREEN LED

The first green light LED was designed and built by Egon Loebner in the year 1958. For the fabrication of green LEDs, an alloy of lead and antimonide was used for p-type Ge. Till the 1970s, the researchers were not aware of these green LEDs. In the 1960s, Nicholas Holonyak Jr was researching the GaAsP material combinations, particularly with the material GaAs, and considering different ratios of As/P for the GaAsP material for the design and development of tunnel diodes. Research on these types of materials was the initial point for the development of research into heterostructures and junctions of semiconducting materials. In 1967, Bush S et al. [20] successfully invented a green light LED made of semiconducting material GaAsP with GaAs as a substrate, although its luminous efficiency was very low. In 2024, Guolin Zhang designed a green-colored LED, achieving a higher quantum efficiency of approximately 17%, a maximum luminance of 29,480 cd m^{-2}, and a photoluminescence efficiency achieved of 75% [21]. Figures 7.12 and 7.13 depict the various applications of green LEDs.

7.7 OLED LIGHTING

Organic light-emitting diodes (OLEDs) are efficiently used for applications requiring ultra-thin and lightweight devices, as well as a homogeneous type of light emission. This OLED-based technology is very promising for creating solid state lighting (SSL)based devices, where a carbon-based layer is

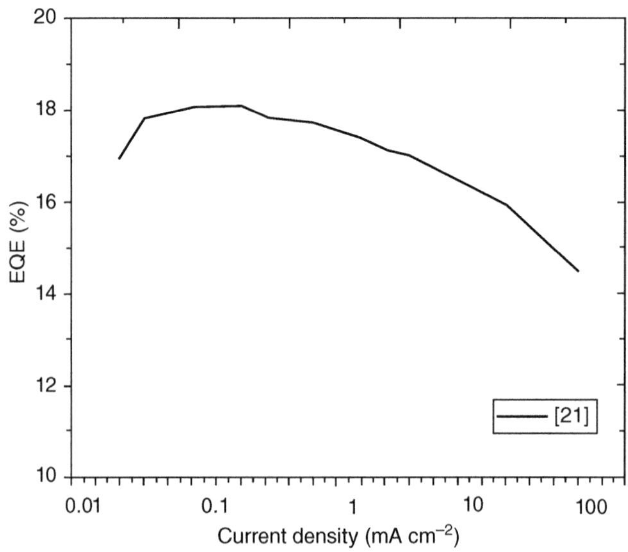

Figure 7.12 EQE vs. current density.

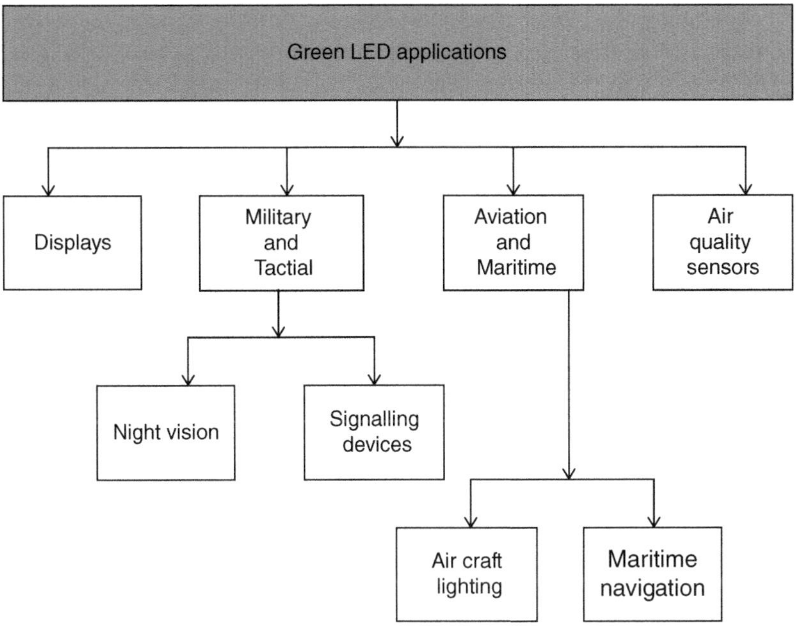

Figure 7.13 Applications of green LEDs.

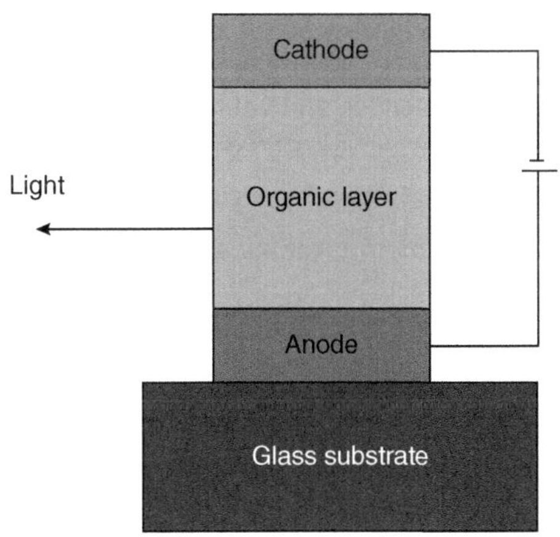

Figure 7.14 Basic structure of OLED.

sandwiched between two electrodes [30]. A DC supply is applied to excite the organic molecules, and holes are injected from the anode while electrons are injected from the cathode. The type of organic molecules decides the wavelength of the emitted light. Figure 7.14 indicates the flow of basic features of an OLED-based device used for lighting applications.

OLED light bulbs are flexible and easily customized, making these devices suitable for large-scale use in businesses and home appliances as well as for commercial usage. OLEDs use a plastic or glass substrate with organic materials, but in the case of LEDs, inorganic materials are used along with various substrates. Compared to LEDs, OLEDs offer diffuse illumination, which often demands additional optical apparatus. OLEDs can be used for manufacturing energy-efficient decorative lighting (both interior and exterior), as well as for creating light that enhances effects and changes the ambiance of the situations, whereas LEDs are primarily used for street, commercial, and residential lighting. OLED-based technology is promising for light emission-based applications, is well known for device reliability, and lastly, charge transport. The main foundation of OLEDs is the size of the organic molecules used in device manufacturing. The lower the molecular weight, the higher the precision in terms of chemical structures, which is also convenient for the implementation of the vacuum-based thermal evaporation technique. The materials used to build the OLED devices primarily include the organic compounds, small molecules, polymers, phosphorescent, and fluorescent materials [31]. The cost of these materials is less flexible, and their controlled preparation abilities are an ideal feature for OLEDs. The third generation of emissive OLED materials includes emitters based on thermally activated delayed fluorescent (TADF), which are highly preferred active materials for the efficient performance of OLEDs[32]. The stability of the deep-blue fluorescent is a key factor that has made it indispensable for the design of OLEDs [33]. The usage of materials based on thiophene has shown improvement in the performance of OLEDs due to their excellent physical and thermal properties [34]. Thus, the usage of these different materials has enhanced the performance of OLED devices, with different properties.

7.8 WHITE LEDs

The three mechanisms used for the generation of white light are: optimizing blue LEDs; mixing the three primary colors; and the phosphor method. These methods produce white light with the help of single-colored LEDs; by using short-wavelength LEDs like deep blue with a yellow phosphor coating; and through UV LEDs [35, 36]. In the research timeline, with the discovery of white light with the usage of blue LEDs, the development of SSK has progressed. In 1972, the technology to generate white light was developed in Japan. The basic structure consists of an LED chip and a lighting system part, which is a fluorescent element [37]. Following HDI lamps, fluorescent light and LED light technologies are promising, technologies that are advancing with improvement in efficiency and reduction in cost, and with enhancement in performance.

LED technology has greater benefits compared with conventional sources of light in terms of zero toxicity, performance, efficiency, portability, lifespan, durability, compactness, and negligible content of UV and IR. The LED-based devices have specific spectral contents, a simple design, ease of implementation, and the required biophysical properties. The generation of white light requires the optimum color compositions, which can be compared with natural light. If the color generated is not comparable to natural sunlight or the light emitted by a filament bulb, this results in eye difficulties as the light deviates from the original natural light [38]. The advancement of III nitride-based materials in manufacturing materials in the semiconductor industry has brought in revolutionary aspect in the manufacturing of LED and SSL technology, also emphasizing the need for highly durable and efficient materials. The recombination of the electron–hole pair was initiated by the injection of an electron into the conduction band. This recombined pair of electrons released the photon energy, matching different bandgap materials.

Currently, the development of SSL is focusing on energy management, and the devices harming the environment are of greater concern. The main features of the SSL device include shock resistance and vibrancy in nature, which enhance various design possibilities [39]. Due to the wide range of intensity and colors available, the range of applications is wider. They find applications in indoor and outdoor decorative items, streetlights, home lighting, and commercial usage. The LED market is estimated to experience an annual growth rate of about 10.5% from 2022 to 2030, which was approximately USD ~60 billion in 2022, and USD ~133 billion in 2020 [40]. The increase in the commercial market is primarily due to low energy consumption and the concept of energy conservation—even government standards demand the same.

7.9 FLEXIBLE MICRO-LEDs

The increased improvements in wearable devices, virtual reality, augmentation-based systems, and cutting-edge technology-based electronic devices have created a huge demand for versatile, flexible, and adaptive display devices of any dimension and shape [41–43]. In this scenario, another type of LED, called flexible LEDs of micro-scale, had evolved. These LEDs are promising with improved performance in technologies like displays based on absorptive and reflective smart pixels, offering flexibility in display device characteristics [44–48]. The introduction of LEDs with dimensions in the range of micrometers for display devices includes different techniques that relate to a wide range of pixel densities and panel sizes [41]. The basic integration methods are usually employed for devices with large displays ranging from 103 to 101 inches, for applications like laptops, TVs, and wall displays, as well as displays that operate with lower PPI ranges [49]. The next type is bonding-based integration, which is essentially used for displays of medium size, utilized for the applications that include projectors, mobile displays, and VR-based devices with 102–103 inches PPI [50]. The next type is growth-based integration, which is used for applications of micro- and ultra-resolution displays like nano LED-based displays, and AM and MR devices, which have the ability above the range of 104 PPI. Different integration technologies have different structural characteristics. The transfer integration includes the adhesion layer to place the LED of PGB colors on the backplane of TTF [51]. Bonding integration implants the LEDs using a transparent substrate method with a layered substrate using CMOS materials. In the growth integrations, the LEDs are grown with drive circuits, which are positioned on the substrate. The integration methods used for different applications are optimized for the applications they serve, particularly in MR/AR/VR domains, where the LED displays based on ultra-high resolution are gaining interest. The transition from larger to micro-scale display sizes has represented the correlation between required pixel densities and device dimensions. The advanced application requires higher values of PPI. The evolution of design technology benefits MR/AR/VR, in which higher pixel densities result in improved experiences and remove the screen door effects. The potential to attain higher values of PPI with sustainable values of brightness, power efficiencies, and contrast identifies the micrometer-based technology as suitable for upcoming advanced display technology-based applications. The LEDs based on flexible technology have certain key characteristics that make them better compared to displays based on OLED and LCD technology [52–55]. The characteristics include excellent performance with higher brightness, quick response time, and greater luminance [56–58].

These devices offer a wide range of characteristics that are preferred by different applications, including low power consumption, more durability, and reliability across different operating conditions [59, 60].

In micro-LED, the epitaxial growth phase is the last phase on the rigid surface of the substrate, like silicon (Si), sapphire, and silicon carbide (SiC). This substrate provides higher crystallinity and is highly thermally stable for applications of LEDs, with more efficient performance [60–63]. In micro-LED, the separation phase of the micro-LEDs from the substrate is called the lift-off stage; these procedures can be implemented using chemical lift-off, laser lift-off, or mechanical lift-off [64–65]. In the LLO method, a high-energy laser is used to differentiate the LED layers. While this provides control, a need for optimization is required to reduce damage. In CLO, selective etching of the sacrificial layer is a time-consuming procedure more suitable for LED-based devices. The physical detaching is performed in MLO for different layers of LED, which enables speedy processing, and a careful process can be enabled to avoid damage to the delicate structure. The transfer phase integrates the micro-LED into a flexible substrate using methods like roll-to-roll, pick-and-place, transfer printing, or self-assembly [66–69]. The method of pick-and-place provides more precision, but the time duration required for large-scale fabrications is significant. The transfer printing method includes steps for the transfer of multiple LEDs one after the other, along with an increase in the throughput. The roll-to-roll method of transfer includes continuous processing, which is essentially used for large-scale fabrication and production of flexible displays. In the self-assembly method, like fluid self-assembly for micro-LEDs, fluid dynamics plays an important role for predefined locations, which offers the ability to design for larger displays. The current research includes novel methods to improve the fabrication processing techniques for micro-LEDs. The inclusion of two-dimensional materials like graphene or hexagonal boron nitride as the layer between layers will facilitate the process of lift-off [70–76].

The two-dimensional materials will enable the removal of micro-LEDs, including the optoelectronic or structural materials. Advanced transfer techniques aim to achieve large-scale integration and precise positioning of micro-LEDs onto a flexible substrate, though they also face key challenges involved in scalability and manufacturability. These interns enable improved cost-effectiveness and efficiency in the flexible micrometer LED fabrications and production, which is part of next-generation display devices. The micrometer LED is designed with advanced fabrication techniques that include the selection of material types and the creation of devices for different applications. These flexible displays include foldable screens, smartphones, wearable devices, and tablets, which enable higher brightness and efficient energy conservation [77–78]. The research on micro-LED design will enhance the research possibilities in optogenetics for medical and biomedical sensor applications, thereby increasing the research opportunities for the treatment of phototherapy as depicted in Figure 7.15. In IoT-based applications and smart devices, these LEDs can easily be included to enhance system performance, sensor technology-based devices, and wearable

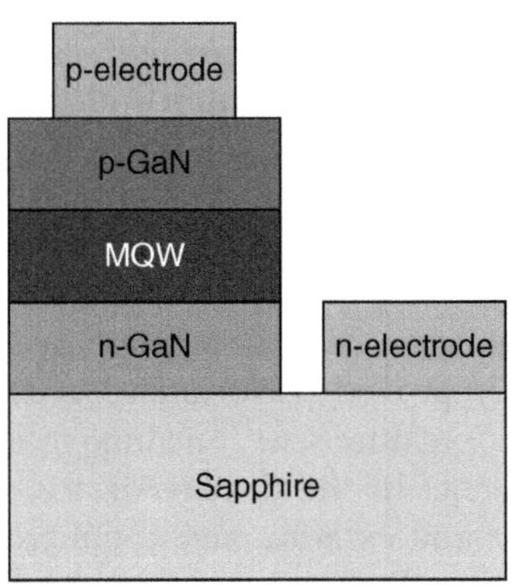

Figure 7.15 Lateral micro-LED.

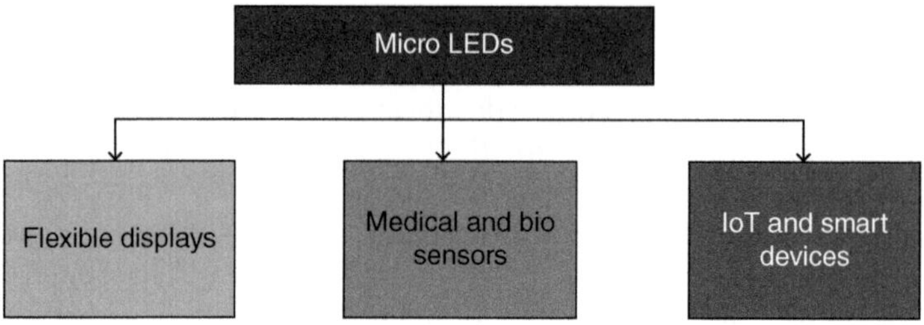

Figure 7.16 Applications of micro-LEDs.

technologies. These LEDs are also used to improve performance. These LEDs are suitable for systems based on visible light. New fabrication techniques enable a distant blend of performance, flexibility, and efficiency, making the flexible micrometer-based LED an important transformative technology for different innovative applications for users in the domains of healthcare, electronics, and communication. For a flexible, micrometer-based LED, the epitaxial growth of a green LED is obtained using the metal-organic chemical vapor deposition technique: i) epitaxial growth on the graphene substrate; ii) lift-off laser for separation; iii) mechanical lift-off; iv) roll-to-roll transfer technique; v) self-assembly of fluid transfer; vi) span multiple-layered LED patches; vii) flexible displays; viii) Micro-LED arrays; ix) e-skin; and x) adaptable technology.

Microlayered LED offers exceptional durability, greater brightness, and less power consumption, and is versatile and adapted to any application [72, 75–76]. These LEDs are integrated into flexible displays, making them more suitable for applications of wearable devices, as shown in Figure 7.16. Also, these displays can be rolled into different dimensions and different shapes, and hence are adaptable to a wide range of applications [77–78]. Micro-LEDs also find applications in medical and biomedical industries; these devices offer greater solutions to healthcare technology. The micro-LEDs are also compatible with IoT-interfaced smart devices; these are innovative and efficient performance devices that are interconnected in the ecosystem.

7.10 CONCLUSION

The innovative developments in LED technology have opened up different applications, with improved performance accuracy, fast computation speed, and reduced device feature sizes. The innovation has brought in many changes and advancements in the display device and lighting industries, including home interior and exterior, and commercial large-scale usage of lighting. The biomedical and automotive domains have undergone a revolutionary change. The evolution of LEDs, including the designs of UV-C

LEDs, Micro-LEDs, and QLEDs, as well as continuous growth in the design and manufacturing of LEDs, brings assurance in functionality of displays and lighting systems, efficiency, and performance. The future of LED display devices and applications will feature innovative combinations of more versatile and adaptive features. These applications will make human life better and sustainable, with the concept of energy management for real-time applications.

REFERENCES

1. Cai, Q., You, H., Guo, H., Wang, J., Liu, B., Chen, D., Lu, H., Zheng, Y., Zhang, R. Progress on AlGaN-based solar-blind ultraviolet photodetectors and focal plane arrays. *Light Sci. Appl.* 2021, 10, 94.
2. Nakamura, S., Mukai, T., Senoh, M. High-power GaN P-N junction blue-light-emitting diodes. *Jpn. J. Appl. Phys.* 1991, 30, L1998–L2001.
3. Nakamura, S., Senoh, M., Mukai, T. P-GaN/N-InGaN/N-GaN double-heterostructure blue-light-emitting diodes. *Jpn. J. Appl. Phys.* 1993, 32, L8–L11.
4. Jiang, H.X., Lin, J.Y. Advances in III-nitride micro-size light emitters. *III–Vs Rev.* 2001, 14, 35–37. [CrossRef] Memon, M.H., Yu, H., Luo, Y., Kang, Y., Chen, W., Li, D., Luo, D., Xiao, S., Zuo, C., Gong, C., et al. A three-terminal light emitting and detecting diode. *Nat. Electron.* 2024, 7, 279–287.
5. Chang, W., Kim, J., Kim, M., Lee, M.W., Lim, C.H., Kim, G., Hwang, S., Chang, J., Min, Y.H., Jeon, K., et al. Concurrent self-assembly of RGB microLEDs for next-generation displays. *Nature* 2023, 617, 287–291.
6. Pandey, A., Min, J., Reddeppa, M., Malhotra, Y., Xiao, Y., Wu, Y., Sun, K., Mi, Z. An ultrahigh efficiency excitonic micro-LED. *Nano Lett.* 2023, 23, 1680–1687. [CrossRef]
7. Memon, M.H., Yu, H., Luo, Y., Kang, Y., Chen, W., Li, D., Luo, D., Xiao, S., Zuo, C., Gong, C., et al. A three-terminal light emitting and detecting diode. *Nat. Electron.* 2024, 7, 279–287.
8. Day, J., Li, J., Lie, D.Y.C., Bradford, C., Lin, J.Y., Jiang, H.X. III-Nitride full-scale high-resolution microdisplays. *Appl. Phys. Lett.* 2011, 99, 031116.
9. Bonar, J.R., Valentine, G.J., Gong, Z., Small, J., Gorton, S. High-brightness low-power consumption microLED arrays. *Int. Soc. Opt. Photonics* 2016, 9768, 97680Y.
10. Xiong, J.H., Hsiang, E.L., He, Z.Q., Zhan, T., Wu, S.T. Augmented reality and virtual reality displays: Emerging technologies and future perspectives. *Light Sci. Appl.* 2021, 10, 216.
11. Poher, V., Grossman, N., Kennedy, G.T., Nikolic, K., Zhang, H.X., Gong, Z., Drakakis, E.M., Gu, E., Dawson, M.D., French, P.M.W., et al. Micro-LED arrays: A tool for two-dimensional neuron stimulation. *J. Phys. D Appl. Phys.* 2008, 41, 094014. [CrossRef]
12. Lin, J.Y., Jiang, H.X. Development of microLED. *Appl. Phys. Lett.* 2020, 116, 100502.
13. Shin, J., Sundaram, H.K.S., Jeong, J., Park, B., Chang, C.S., Choi, J., Kim, T., Saravanapavanantham, M., Lu, K., Kim, S., et al. Vertical full-colour micro-LEDs via 2D materials-based layer transfer. *Nature* 2023, 614, 81–87.

14. Li, Y., Xing, Z., Zheng, Y., Tang, X., Xie, W., Chen, X., Wang, W., Li, G. High-efficiency near-UV light-emitting diodes on Si substrates with InGaN/GaN/AlGaN/GaN multiple quantum wells. *J. Mater. Chem. C* 2020, 8, 883–888.
15. Manikandan, M., Nirmal, D., Ajayan, J., Arivazhagan, L., Prajoon, P., Dhivyasri, G., Jagadeeswari, M. Physics based modeling of AlGaN/BGaN quantum well based ultra violet light emitting diodes. *Optical and Quantum Electronics* 2022, 54(3), 168.
16. Akasaki, I. GaN-based P–N junction blue-light-emitting devices. *Proc. IEEE* 2013, 101(10), 2200–2210.
17. Hou, S., Gangishetty, M.K., Quan, Q., Congreve, D.N. Efficient blue and white perovskite light-emitting diodes via manganese doping. *Joule* 2018, 2, 2421–2433.
18. Vashishtha, P., Ng, M., Shivarudraiah, S.B., Halpert, J.E. High efficiency blue and green light-emitting diodes using Ruddlesden Popper inorganic mixed halide perovskites with butylammonium interlayers. *Chem. Mater.* 2018, 31, 83–89.
19. Peng, J., Wang, F.F., Luo, P., Dong, X.Y., Zhang, C., Zang, S.Q. Blue perovskite LEDs: A comprehensive review of emission control strategies for enhanced performance. *Coord. Chem. Rev.* 2025, 528, 216425.
20. Bush, S. 50 year history of the LED. *Electronics Weekly*. 22 September 2010. Available online: https://www.electronicsweekly.com/news/products/led/50-year-history-of-the-led-2-2010-09/#:~:text=With%20his%20PhD%20from%20Illinois
21. Zhang, G., Lu, R., Liu, Z., Ni, K., Jiang, T., Tao, X., Zheng, G., Yi, C., Wang, J. Pure-phase perovskite quantum well for green light-emitting diodes. *ACS Appl. Mater. Interfaces* 2024, 16(38), 51195–51200.
22. Holonyak, N., Bevacqua, S.F. Coherent (visible) light emission from Ga $(As_{1-x}P_x)$ junctions. *Appl. Phys. Lett.* 1962, 1, 82–83.
23. Pasayat, S.S., Gupta, C., Wong, M.S., Ley, R., Gordon, M.J., DenBaars, S.P., Nakamura, S., Keller, S., Mishra, U.K. Demonstration of ultra-small (0.2%) for mini-displays. *Appl. Phys. Express*. 2020, 14, 011004.
24. Dussaigne, A., Le Maitre, P., Haas, H., Pillet, J.C., Barbier, F., Grenier, A., Michit, N., Jannaud, A., Templier, R., Vaufrey, D., et al. Full InGaN red (625 nm) micro-LED (10 μm) demonstration on a relaxed pseudo-substrate. *Appl. Phys. Express* 2021, 14, 092011.
25. Iida, D., Kirilenko, P., Velazquez-Rizo, M., Zhuang, Z., Najmi, M.A., Ohkawa, K. Demonstration of 621-nm-wavelength InGaN based single-quantum-well LEDs with an external quantum efficiency of 4.3% at 10.1 A/cm^2. *AIP Adv.* 2022, 12, 065125.
26. Tan, Z., Moghaddam, R., Lai, M., Docampo, P., Higler, R., Deschler, F., Price, M., Sadhanala, A., Pazos, L.M., Credgington, D., et al. Bright light-emitting diodes based on organometal halide perovskite. *Nat. Nanotech.* 2014, 9, 687–692.
27. Wang, N., Mei, T., Yang, X., Li, H., Zhu, N., Zhang, M. Local doping modulation for improving quantum efficiency in gan-based light-emitting diodes. *Jpn. J. Appl. Phys.* 2013, 52, 112102.
28. Bae, W., Park, Y., Lim, J., Lee, D., Padilha, L., McDaniel, H., Robel, I., Lee, C., Pietryga, J., Klimov, V. Controlling the influence of auger recombination on the performance of quantum-dot light-emitting diodes. *Nat. Commun.* 2013, 4, 2661.

29. Lu, M., Zhang, X., Zhang, Y., Guo, J., Shen, X., Yu, W., Rogach, A. Simultaneous strontium doping and chlorine surface passivation improve luminescence intensity and stability of cspbi3 nanocrystals enabling efficient light-emitting devices. *Adv. Mater.* 2018, 30, 1804691.
30. Tang, C.W., VanSlyke, S.A. Organic electroluminescent diodes. *Appl. Phys. Lett.* 1987, 51, 913–915.
31. Kulkarni, A.P., Tonzola, C.J., Babel, A., Jenekhe, S.A. Electron transport materials for organic light-emitting diodes. *Chem. Mater.* 2004, 16, 4556–4573.
32. Cole, C.M., Kunz, S.V., Baumann, T., Blinco, J.P., Sonar, P., Barner-Kowollik, C., Yambem, S.D. Flexible inkjet printed polymer light-emitting diodes using a self-hosted non-conjugated TADF polymer. *Macromol. Rapid Commun.* 2023, 44, 2300015.
33. Huang, C.C., Xue, M.M., Wu, F.P., Yuan, Y., Liao, L.S., Fung, M.K. Deep-blue and hybrid-white organic light emitting diodes based on a twisting carbazole-benzofuro [2, 3-b] pyrazine fluorescent emitter. *Molecules* 2019, 24, 353.
34. Xu, Z., Yu, D., Yu, M. Synthesis and photophysical properties of novel 4-aryl substituted thiophene derivatives with a vinyl- quinoline unit. *J. Chem. Res.* 2013, 37, 441–443.
35. Cho, I.H., Anoop, G., Suh, D.W., Lee, S.J., Yoo, J.S. On the stability and reliability of Sr1-xBaxSi2O2N2:Eu2+ phosphors for white LED applications. *Opt. Mater. Express* 2012, 2, 1292–1305.
36. Anoop, G., Kim, K.P., Suh, D.W., Cho, I.H., Yoo, J.S. Optical characteristics of Sr2Si3O2N4:Eu2+ Phosphors for White Light Emitting Diodes. *Electrochem. Solid-State Lett.* 2011, 14, J58.
37. Taguchi, T. Technical trends and future view of white LED lighting, headline review. *OHM* 2004, 5, 8–9.
38. Taguchi, T. Present status of energy saving technologies and future prospect in white LED lighting. *IEEJ Trans. Electr. Electron. Eng.* 2007, 3, 21–26.
39. Bergh, A., Craford, G., Duggal, A., Haitz, R. The promise and challenge of solid-state lighting. *Phys. Today* 2001, 54, 42–47.
40. LED lighting market size, share & growth report, 2020–2030. Industry Report.
41. Chen, D., Chen, Y.C., Zeng, G., Zhang, D.W., Lu, H.L. Integration technology of Micro-LED for Next-Generation display. *Research* 2023, 6, 0047.
42. Zhang, S., Zheng, H., Zhou, L., Li, H., Chen, Y., Wei, C., Wu, T., Lv, W., Zhang, G., Zhang, S. Research progress of micro-LED display technology. *Crystals* 2023, 13, 1001.
43. Lin, J., Jiang, H. Development of microled. *Appl. Phys. Lett.* 2020, 116, 100502.
44. Chen, H.-W., Lee, J.-H., Lin, B.-Y., Chen, S., Wu, S.-T. Liquid crystal display and organic light-emitting diode display: Present status and future perspectives. *Light Sci. Appl.* 2018, 7, 17168.
45. Tian, P., McKendry, J.J., Gong, Z., Zhang, S., Watson, S., Zhu, D., Watson, I.M., Gu, E., Kelly, A.E., Humphreys, C.J. Characteristics and applications of micro-pixelated GaN-based light emitting diodes on Si substrates. *J. Appl. Phys.* 2014, 115, 033112.
46. McKendry, J.J., Massoubre, D., Zhang, S., Rae, B.R., Green, R.P., Gu, E., Henderson, R.K., Kelly, A., Dawson, M.D. Visible-light communications using a CMOS-controlled micro-light-emitting diode array. *J. Lightwave Technol.* 2011, 30, 61–67.

47. Carreira, J., Xie, E., Bian, R., Chen, C., McKendry, J., Guilhabert, B., Haas, H., Gu, E., Dawson, M. On-chip GaN-based dual-color micro-LED arrays and their application in visible light communication. *Opt. Express.* 2019, 27, A1517–A1528.
48. Moon, C.W., Kim, Y., Hyun, J.K. Active electrochemical high contrast gratings as on/off switchable and color tunable pixels. *Nat. Commun.* 2022, 13, 3391.
49. Liu, Y., Wang, G., Feng, F., Zhanghu, M., Yuan, Z., Li, Z., Xu, K., Kwok, H.S., Liu, Z. Ultra-low-defect homoepitaxial micro LEDs with enhanced efficiency and monochromaticity for high PPI AR/MR displays. *PhotoniX* 2024, 5, 23.
50. Horng, R.-H., Chen, Y.-F., Wang, C.-H., Chen, H.-Y. Development of metal bonding for passive matrix micro-LED display applications. *IEEE Electron. Device Lett.* 2021, 42, 1017–1020.
51. Ma, T., Chen, J., Chen, Z., Liang, L., Hu, J., Shen, W., Li, Z., Zeng, H. Progress in color conversion technology for Micro LED. *Adv. Mater. Technol.* 2023, 8, 2200632.
52. Geffroy, B., Le Roy, P., Prat, C. Organic light-emitting diode (OLED) technology: Materials, devices and display technologies. *Polym. Int.* 2006, 55, 572–582.
53. Kobayashi, S., Mikoshiba, S., Lim, S. *LCD Backlights*. Wiley. 2009.
54. Qin, F., Liu, C., Wu, W., Peng, W., Huo, S., Ye, J., Gu, S. Ink jet printed quantum Dots color conversion layers for full-color micro-LED displays. *Electron. Mater. Lett.* 2023, 19, 19–28.
55. Jana, R., Rajaitha, P.M., Hajra, S., Kim, H.J. Advancements in visible-light-driven double perovskite nanoparticles for photo degradation. *Micro Nano Syst. Lett.* 2023, 11, 3.
56. Liu, Y., Zhang, K., Hyun, B.-R., Kwok, H.S., Liu, Z. High brightness InGaN/GaN micro-LEDs with secondary peak effect for displays. *IEEE Electron. Device Lett.* 2020, 41, 1380–1383.
57. Lee, V.W., Twu, N., Kymissis, I. Micro-LED technologies and applications. *Inf. Disp.* 2016, 32, 16–23.
58. James Singh, K., Huang, Y.-M., Ahmed, T., Liu, A.-C., Chen, H., Liou, S.-W., Wu, F.-J., Lin, T., Chow, C.-C., Lin, G.-R. Micro-LED as a promising candidate for high-speed visible light communication. *Appl. Sci.* 2020, 10, 7384.
59. Hsiang, E.-L., He, Z., Huang, Y., Gou, F., Lan, Y.-F., Wu, S.-T. Improving the power efficiency of micro-LED displays with optimized LED chip sizes. *Crystals* 2020, 10, 494.
60. Lee, S., Kim, J., Oh, J., Ryu, J., Hwang, K., Hwang, J., Kang, S., Choi, J.H., Sim, Y.C., Cho, Y.-H. A discrete core-shell-like micro-light-emitting diode array grown on Sapphire nano-membranes. *Sci. Rep.* 2020, 10, 7506.
61. Xue, J., Kim, J., Mestre, A., Tan, K., Chong, D., Roy, S., Nong, H., Lim, K., Gray, D., Kramnik, D. Low-voltage, high-brightness silicon micro-LEDs for CMOS photonics. *IEEE Trans. Electron. Devices.* 68, 3870–3875 (2021)
62. Vlaskina, S. Silicon carbide LED. *Semicond. Phys. Quantum Electron. Optoelectron.* 2002, 5, 71–75
63. Jung, S.Y., Choi, H.-J., Lee, J.Y., Kim, M.-S., Ning, R., Han, D.-H., Kim, S.K., Won, S.O., Lee, J.H., Jang, J.-S. Evolution of microcracks in epitaxial CeO_2 thin films on YSZ-Buffered Si. *Electron. Mater. Lett.* 2024, 20, 484–490.
64. Lin, C.-F., Dai, J.-J., Wang, G.-M., Lin, M.-S. Chemical lift-off process for blue light-emitting diodes. *Appl. Phys. Express.* 2010, 3, 092101.

65. Wu, F.-L., Ou, S.-L., Kao, Y.-C., Chen, C.-L., Tseng, M.-C., Lu, F.-C., Lin, M.-T., Horng, R.-H. Thin-film vertical-type AlGaInP leds fabricated by epitaxial lift-off process via the patterned design of Cu substrate. *Opt. Express.* 2015, 23, 18156–18165.
66. Manjula, V., Karamagi, R. Automatic pick and place robot manipulation using a microcontroller. *J. Appl. Comput. Math.* 2018, 7, 1000408.
67. Zhou, X., Tian, P., Sher, C.-W., Wu, J., Liu, H., Liu, R., Kuo, H.-C. Growth, transfer printing and colour conversion techniques towards full-colour micro-LED display. *Prog. Quantum Electron.* 2020, 71, 100263.
68. Gao, W., Ota, H., Kiriya, D., Takei, K., Javey, A. Flexible electronics toward wearable sensing. *Acc. Chem. Res.* 2019, 52, 523–533.
69. Liu, Y., Yao, D., Zhang, H. Self-assembly driven aggregation induced emission of copper nanoclusters: A novel technology for lighting. *ACS Appl. Mater. Interfaces* 2017, 10, 12071–12080.
70. Shin, J., Kim, H., Sundaram, S., Jeong, J., Park, B.-I., Chang, C.S., Choi, J., Kim, T., Saravanapavanantham, M., Lu, K. Vertical full-colour micro-LEDs via 2D materials-based layer transfer. *Nature* 2023, 614, 81–87.
71. Tran, T., Kassem, A., Ottapilakkal, V., Vuong, P., Gujrati, R., Bourras, M., Srivastava, A., Perepeliuc, A., Moudakir, T., Gautier, S. Transferrable GaN-based Micro-LED heterostructures grown on h-BN for optogenetic applications. *J. Cryst. Growth.* 2025, 650, 127979.
72. Anwar, A.R., Sajjad, M.T., Johar, M.A., Hernández-Gutiérrez, C.A., Usman, M., Łepkowski, S. Recent progress in micro-LED based display technologies. *Laser Photonics Rev.* 2022, 16, 2100427.
73. Niu, J., Bai, X., Wang, J., Chen, Y., Zhao, B., Sha, W., Long, Y., Wang, Z., Hu, W. Fully flexible All-in-One electronic display skin with seamless integration of microled and hydrogel battery. *Adv. Funct. Mater.* 2024, 35, 2411916.
74. Xu, H., Zhang, J., Davitt, K., Song, Y., Nurmikko, A. Application of blue–green and ultraviolet micro-LEDs to biological imaging and detection. *J. Phys. D: Appl. Phys.* 2008, 41, 094013.
75. Wu, T., Sher, C.-W., Lin, Y., Lee, C.-F., Liang, S., Lu, Y., Chen, S.-W.H., Guo, W., Kuo, H.-C., Chen, Z. Mini-LED and micro-LED: Promising candidates for the next generation display technology. *Appl. Sci.* 2018, 8, 1557.
76. Bandari, V.K., Schmidt, O.G. A bright future for micro-LED displays. *Light Sci. Appl.* 2024, 13, 1–3.
77. Lee, H.E., Shin, J.H., Park, J.H., Hong, S.K., Park, S.H., Lee, S.H., Lee, J.H., Kang, I.S., Lee, K.J. Micro light-emitting diodes for display and flexible biomedical applications. *Adv. Funct. Mater.* 2019, 29, 1808075.
78. Wu, Y., Ma, J., Su, P., Zhang, L., Xia, B. Full-color realization of micro-LED displays. *Nanomaterials* 2020, 10, 2482.

Chapter 8

Piezoelectric nanogenerators

Vidhya Bhojan, K.J. Niveditha, Raju Nandhakumar, Rajesh Swaminathan, and Velusamy Maheskumar

8.1 PIEZOELECTRIC EFFECT

Some materials can generate an electric charge when subjected to mechanical pressure—a phenomenon known as the piezoelectric effect. The term "piezoelectric" comes from Greek roots: *piezein*, meaning "to press," and *piezo*, meaning "push." As illustrated in Figure 8.1, when a piezoelectric crystal is placed between two conductive plates, the application of force causes the material to become electrically polarized. This polarization leads to a buildup of electric charge on the metal surfaces, producing a voltage (often represented by a lightning symbol). This process is known as the direct piezoelectric effect and essentially enables the material to act like a miniature power source. This effect is widely used in sensors like microphones, pressure detectors, and underwater sound devices (hydrophones).

Through electromechanical harvesting, mechanical energy from everyday sources, such as bodily motion or vibrations, can be transformed into

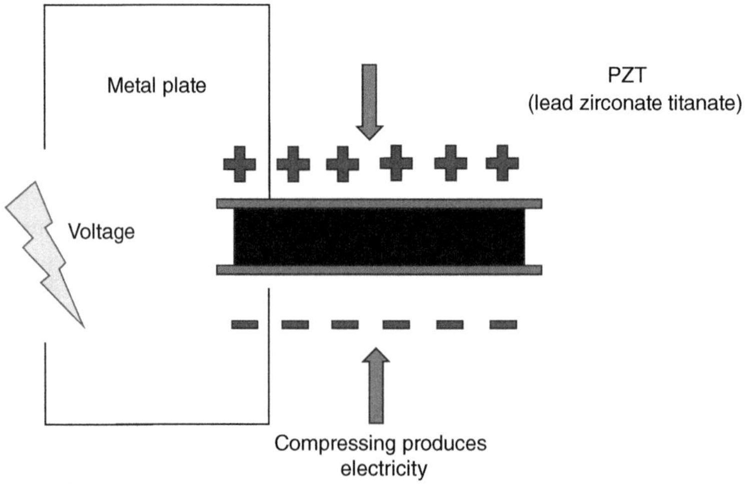

Figure 8.1 Piezoelectric effect.

DOI: 10.1201/9781003635123-8

electrical energy. This generated electricity is suitable for powering small-scale electronic devices. Recent advancements have notably enhanced the capabilities of nanogenerators (NGs). These developments have increased their energy output, enabling them to replace conventional energy-harvesting processes. Traditional methods often rely on materials like piezoelectric ceramics (e.g., lead zirconate titanate, or PZT), polymers such as polyvinylidene fluoride (PVDF), or systems utilizing electrostatic or electromagnetic mechanisms. The efficiency of an NG is generally evaluated by analyzing parameters like open-circuit voltage and short-circuit current to derive its overall power output.

8.2 PIEZOELECTRIC MATERIALS

Piezoelectric materials possess an asymmetric crystal structure and lack a center of symmetry, enabling them to uniquely transform mechanical force into electrical signals. Upon applying mechanical pressure, these materials generate an internal electric charge as a result of induced polarization. The efficiency of this energy conversion depends largely on the material's intrinsic piezoelectric properties—stronger piezoelectric responses generally lead to better device performance. These materials are commonly grouped into three main categories: single crystals, perovskite-based compounds, and polymers (as illustrated in Figure 8.2). This section explores the various piezoelectric materials used in fabricating piezoelectric nanogenerators (PENGs) and highlights their key characteristics.

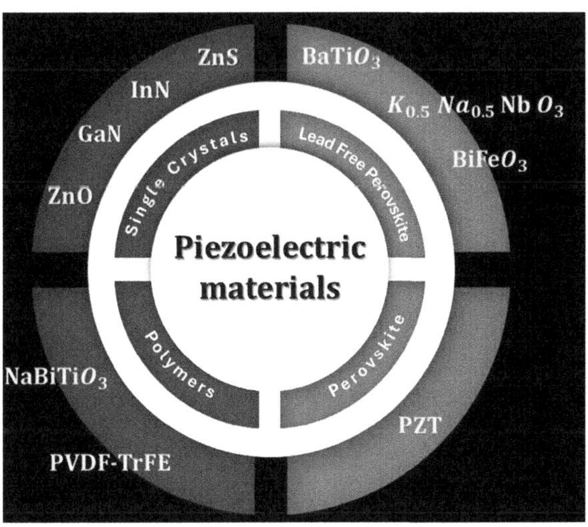

Figure 8.2 Piezoelectric materials classification and examples.

8.2.1 Single-crystal materials

Single-crystal compounds with a wurtzite structure, such as zinc oxide (ZnO), gallium nitride (GaN), indium nitride (InN), and zinc sulfide (ZnS), form the first category of piezoelectric materials. Among these, ZnO has been widely researched for its potential in PENGs due to its natural abundance and biocompatibility. However, the initial ZnO nanowire-based PENGs produced relatively low electrical output. In early designs, a single ZnO nanowire was mounted laterally on a flexible base, and the ends were connected to electrodes. When the substrate was bent, the resulting strain in the nanowire generated a piezoelectric potential. Repeated bending and releasing actions produced an electrical signal, representing the piezoelectric response [1–3]. This type of PENG could produce output voltages ranging from 20 to 50 millivolts and short-circuit currents ranging from 400 to 750 picoamperes. These lateral nanowire-based generators were even used to harvest energy from small-scale human activities, such as heartbeats and finger taps.

To enhance the energy output, researchers introduced two new configurations: vertically integrated nanogenerators (VING) and laterally integrated nanogenerators (LING), as shown in Figure 8.3. These architectures featured arrays of ZnO nanowires, assembled either perpendicular (VING) or parallel (LING) to the substrate surface. Both are collectively referred to as integrated nanowire NGs. These improved designs achieved significantly better performance, with maximum recorded open-circuit voltages reaching 1.26 volts and short-circuit currents up to 28.8 nanoamperes. Despite this progress, the inherent limitations of ZnO—particularly its relatively weak piezoelectric properties—led researchers to seek alternative materials with higher dielectric constants and stronger piezoelectric coefficients to further improve device efficiency.

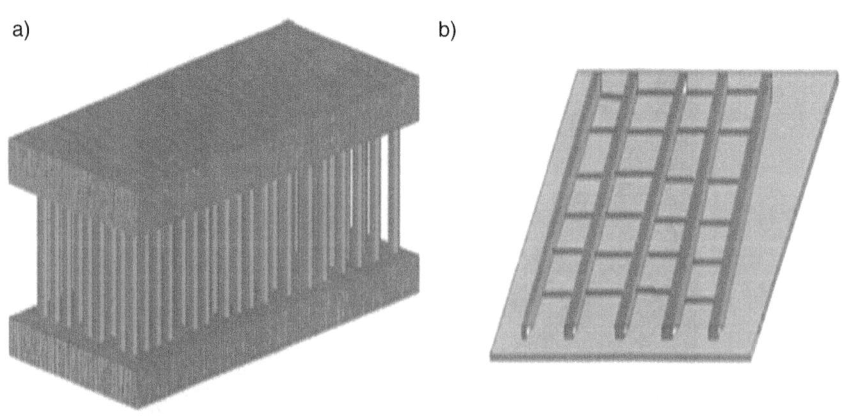

Figure 8.3 Schematic of (a) VING and (b) LING.

8.2.2 Perovskites

Lead-perovskite materials and lead-free perovskite materials are the two major classifications of perovskites.

8.2.2.1 *Lead-perovskite materials*

Upon recognizing the limited performance of zinc oxide (ZnO)-based PENGs, researchers shifted their focus to lead zirconate titanate ($Pb[Zr_xTi_{1-x}O_3]$ or PZT) ceramic materials. PZT ceramics demonstrate excellent crystallinity, a high dielectric constant, and a significant piezoelectric constant. These favorable properties have captured the interest of many researchers, leading to significant advancements in PZT-based PENGs. Despite its high piezoelectricity, the fabrication of thin films for PZT poses challenges due to its brittle nature. The intricate synthesis process and brittleness are major hurdles in creating PZT-based PENGs. Researchers have attempted to overcome these challenges by experimenting with various morphologies and employing unique processes. However, progress in PZT-based PENGs has slowed considerably, primarily due to the hazardous nature of lead. Consequently, attention is now shifting toward exploring lead-free perovskite materials for use in PENGs [4, 5].

8.2.3 Lead-free perovskite materials

Due to the harmful and environmentally hazardous properties of lead, there is a growing focus among researchers on developing lead-free perovskite alternatives for use in high-efficiency PENGs. These environmentally friendly perovskite ceramics are generally polycrystalline, consisting of numerous small, irregular grains. They are most often fabricated through solid-state synthesis followed by a sintering process. The piezoelectric behavior of these materials is significantly affected by their phase transition temperatures, as shifts in crystal structure—such as transitions from tetragonal to orthorhombic or monoclinic and further to rhombohedral phases—allow polarization vectors to reorient, thereby improving piezoelectric properties [6]. Structurally, these compounds typically follow the general formula ABO_3, where the A and B cation sites produce oxide classes like $AIIBIVO_3$, $AIIIBIIIO_3$, or $AIBVO_3$, depending on the elements involved.

8.2.3.1 *ABO₃*

Figure 8.4 depicts an ideal cubic perovskite ABO_3 structure, showing corner-shared BO_6 octahedra and cation A at the interstitial position, and cation B and BO_6 octahedra at the center of the unit cell.

Barium titanate ($BaTiO_3$ or BTO) is among the most extensively investigated lead-free perovskite materials, owing to its notable ferroelectric and

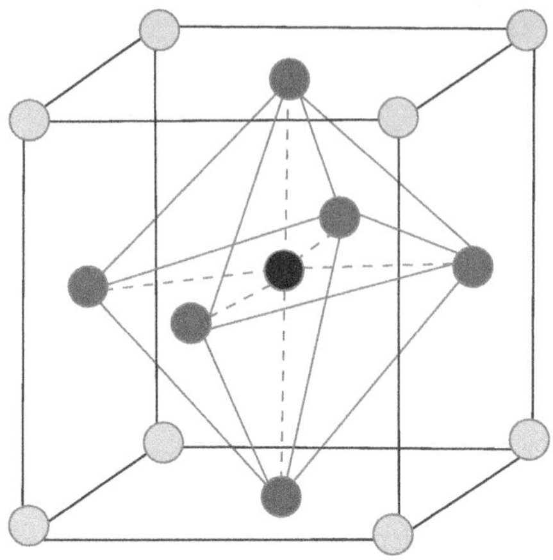

Figure 8.4 Structure of perovskite, ABO_3, where A (*gray*) is a big ion, B (*black*) is a small ion; O is *blue*.

piezoelectric properties at the nanoscale [7]. It has served as a benchmark system in numerous studies aimed at understanding these functional behaviors. As BTO is cooled, it undergoes a sequence of structural phase changes—from a tetragonal to an orthorhombic phase, and subsequently to a rhombohedral configuration—each influencing its electrical characteristics [8, 9]. In 2009, researchers reported a major advancement involving a composite based on barium titanate (BTO) exhibiting an exceptionally high piezoelectric coefficient (around $d_{33} \approx 620$ pC/N) [10]. A notable subsequent development was the formulation of a composite combining barium zirconate titanate and barium calcium titanate, identified as 0.5 $Ba(Ti_{0.8}Zr_{0.2})$–$(Ba_{0.7}Ca_{0.3})TiO_3$ (BZT-BCT). This material showed substantial enhancement in its functional properties. Through a two-step hydrothermal process, BZT-BCT nanowires were synthesized, resulting in an open-circuit voltage output of 6.25 V and a power density reaching 2.25 $\mu W/cm^3$ [11]. To further improve these characteristics, scientists have investigated various approaches, including modifying the morphology of the materials. Techniques such as solvothermal synthesis, molten salt-based processing, and the use of supercritical fluids have been applied to fabricate BZT-BCT structures [12–14]. Nanostructured materials may exhibit piezoelectric behaviors that differ significantly from those of bulk materials, largely due to their differing surface-to-volume ratios [15].

Sodium bismuth titanate, often referred to as NBT or BNT and represented by the formula $Na_{1/2}Bi_{1/2}TiO_3$, adopts a perovskite-type crystal structure under standard conditions. When the temperature is reduced from around 540°C down to 300°C, the material undergoes a series of structural transformations—transitioning successively from a cubic phase to a tetragonal phase, and finally settling into a rhombohedral configuration. This material is known for its robust ferroelectric behavior, characterized by a notable remnant polarization of approximately 38 μC/cm^2. Several ferroelectric compositions derived from NBT have also demonstrated promising electrostrictive effects and elevated piezoelectric coefficients, making them suitable candidates for high-efficiency, lead-free PENGs. Despite these advantages, NBT-based materials face limitations, including relatively low depolarization temperatures and high leakage currents, which constrain their broader practical applications [16].

8.2.3.2 *AIII–BIII–O$_3$*

Bismuth ferrite ($BiFeO_3$, BFO) is a prominent example of a lead-free perovskite oxide belonging to the AIII–BIII–O_3 structural family. This material demonstrates piezoelectric behavior and possesses a relatively high Curie temperature, generally falling in the range of 809°C–830°C, making it well-suited for high-temperature applications. When combined with barium titanate ($BaTiO_3$, BTO) or lead titanate ($PbTiO_3$), it can generate rhombohedral-tetragonal phase boundaries, which are known to boost its piezoelectric response. A distinctive feature of BFO is its simultaneous exhibition of ferroelectric and magnetic properties at room temperature, making it a strong candidate for modifying the ferroelectric performance of solid solutions derived from BTO and strontium titanate ($SrTiO_3$) [17].

8.2.3.3 *AIBVO$_3$*

Materials belonging to the AIBVO_3 family are recognized for their advantageous properties, including a Curie temperature of approximately 420°C [18] and notable ferroelectric behavior, evidenced by a remnant polarization of approximately 3.3 μC/cm^2 [19]. Additionally, they exhibit strong electrochemical coupling performance. These attributes make $A^{I}B^{V}O_3$ compounds highly suitable as environmentally benign and biocompatible alternatives to traditional lead-based piezoelectric systems [20]. Among these, potassium sodium niobate (KNN) ($K_xNa_{1-x}NbO_3$) is frequently applied in PENG devices. Pristine KNN undergoes phase transitions from a cubic (paraelectric) phase through tetragonal and orthorhombic phases to a final rhombohedral (ferroelectric) phase; the orthorhombic phase is typically observed at room temperature [21]. Different KNN morphologies, such as nanoparticles, nanofibers, nanotubes, and nanopellets, have been synthesized through various physical and chemical methods [22–24].

8.2.4 Polymer materials

Due to their inherent rigidity, single-crystal and perovskite materials are unsuitable for flexible applications, making polymeric materials the preferred candidates for developing PENGs [25, 26]. The introduction of PVDF has considerably advanced piezopolymer applications in sensor and actuator systems, owing to their mechanical flexibility, cost-effectiveness, and ease of fabrication [27]. Polymers such as PVDF and its copolymer P(VDF-TrFE) also offer excellent biocompatibility, making them ideal for biomedical piezoelectric applications. Piezopolymers are generally classified into bulk types, void-charged layers, and composite systems [28]. Void-charged types generate internal dipoles via trapped gases within their polymer matrix, while bulk materials like PVDF exhibit semicrystalline structures with multiple phases—α, β, γ, δ, and ε—of which the α and β phases are particularly crucial for piezoelectricity [29].

Although commercial PVDF films are readily available, their output power remains suboptimal. Techniques such as electrospinning produce PVDF nanofibers that are more easily aligned into the piezoelectric β-phase, improving performance by up to fourfold [30–33]. However, the power output of these devices still falls short of the requirements for critical uses like powering pacemakers. To enhance their performance, researchers have incorporated nanofillers into PVDF matrices. These include ceramic-based nanostructures such as $BaTiO_3$ [34], $NaNbO_3$ [35, 36], and ZnO [37, 38], and carbon-based nanomaterials like graphene and carbon nanotubes [39, 40]. These additions increase the β-phase content, thereby improving the electromechanical response. Recent innovations also involve hybrid filler systems, which synergize the benefits of different nanomaterials. Notably, PVDF films embedded with ZnO and multi-walled carbon nanotubes (MWCNTs) have demonstrated output voltages as high as 3.22 ± 0.24 V [41]. Additionally, PVDF/AlO–rGO composites have achieved power densities reaching 27.97 $\mu W/cm^3$ [42], illustrating the significant potential of these materials in next-generation flexible energy-harvesting devices. (Table 8.1) shows a comparison of different piezoelectric materials and its applications.

8.3 APPLICATIONS

NGs are emerging as innovative solutions for harvesting energy from untapped ambient sources, including solar radiation, wind, thermal gradients, mechanical pressure, human motion, and environmental vibrations.

8.3.1 Self-powered nano/micro devices

NGs hold significant potential as standalone or auxiliary power sources for nano- and microscale devices that operate with minimal energy requirements, especially when there is a continuous supply of mechanical motion. A well-known demonstration of this concept was presented in 2010 by the research team led by Professor Zhong Lin Wang. In their work, a VING was integrated with a self-powered sensor for detecting pH or ultraviolet light,

Table 8.1 Comparison of modified piezoelectric materials: voltage enhancement and application spectrum

Material	*Additive*	*Output voltage*	*Applications*	*Reference*
PVDF	i. $BaTiO_3$	6 V	• Energy harvesting • Biomedical sensors • Actuators • Environmental monitoring • Smart textiles	[43]
	ii. Gr/HNT/GO	Output voltage ~1.3 (1.6 wt% Gr), ~1.5 (1.6 wt% HNT) ~ 2 (1.6 wt% GO)	• Hydrostatic sensor • Energy harvesting	[44]
	iii. CNT	8.5 mV	• Portable electronics • Implantable biomedical devices • Wireless sensors	[45]
	iv. MWCNT	6 V		[46]
	v. ZnO	The electrical power output of hybrid was found to be enhanced ~17 times from a pristine PVDF.	• Energy harvesting	[47]
P(VDF-TrFE)	$BaTiO_3$	13.2 V.	• Self-powered sensors • Wearable electronics • Consumer electronics • Medical sciences • Smart clothes	[48]
$BaTiO_3$@P(VDF-TrFE)	Polydopamine (Pdop)	6V	• Self-powered wearable sensors • Energy harvesting • Real-time monitoring • Flexibility and versatility • Potential for smart textiles	[49]

(*Continued*)

Table 8.1 (Continued)

Material	*Additive*	*Output voltage*	*Applications*	*Reference*
GaN	—	1.2 V	• NGs with high output to drive small electric devices. • Energy harvesting	[50]
[P(VDF-HFP)]	PANI-ZnS	Maximum **3 V** output voltage	• Energy harvesting • Energy storage devices	[51]
PVA	ZnS	Wearable nanogenerator **(~2 V/Pa)**	• Portable electronics and wearable technology.	[52]
PVDF	ZnO@ZnS	PVDF—10 wt% ZnO @ ZnS composite resulted in a sensitivity of 0.153 V/N·mm^3	• Wearable self-powered electronic devices.	[53]
ZnO nanowires (NWs) on carbon fibers Au-coated ZnO NWs on paper	—	18 mV 35 nA (current)	• Power sources that are flexible, foldable, and customizable in shape, making them suitable for integrating into fabrics and wearable textiles.	[54]
PVDF	MWCNTs	6 V	• Implantable biomedical devices • Wireless sensors • Portable electronics.	[46]
PVDF/rGO-TNL pressure sensor was prepared by the solution casting method.	—	—	• Sensing applications.	[55]

PVDF	CNT and Fe-rGO	Open-circuit output voltage of 2.5 V and 1.2 V for CNT and Fe-rGO-based devices, respectively.	• Energy harvesting	[56]
PDMS/$BaTiO_3$ hydrothermal method	$BaTiO_3$	V_{oc}—5.5 V and I_{SC}—350 nA, a current density of 350 nA/cm^2	• Highly transparent and useful for a large-scale (11 × 11 cm) fabrication of NG.	[57]
PVDF	PZT	Output voltage of 62.0 V and power of 136.9 μW for composite film thickness of 220 μm.	• Low-power electric devices	[58]
PDMS	PZT	An output voltage of 1.1 V and an output current of 1.4 μA for a thickness of 80 μm and an area of 8 cm^2.	• Energy harvesting	[59]
PVDF	$BaTiO_3$	The highest piezoelectric power output was obtained as 0.243 μW.	• Energy harvesting under mechanical vibration	[60]
Chitosan	$BaTiO_3$-nanorods	The combination of 7 wt% lead-free piezoelectric $BaTiO_3$ nanorods (BT-NRs) in the chitosan matrix, (V_{oc}) of ~111.4 V, (I_{sc}) of ~21.6 μA/cm^2, and also output power density of 756 μW/cm^2.	• Provided a platform for an eco-friendly TENG.	[61]

generating an output voltage in the range of 20–40 millivolts, sufficient for sensor functionality.

Although NGs feature a novel and compact design, their energy output remains relatively low, which limits their capacity to independently power nano- and microscale electronic devices. Consequently, they are often employed as supplementary power sources in conjunction with conventional batteries. To overcome this challenge, current research focuses on hybrid energy-harvesting systems that integrate NGs with technologies such as solar cells and bioenergy harvesters. These combined approaches hold promise for enabling self-powered systems in autonomous applications, including emerging technologies like smart dust.

Incorporating piezoelectric fibers into textiles presents a promising avenue for advancing smart wearable technology. These specialized fibers are capable of transforming mechanical energy generated through body motion into usable electrical energy, which can then be harnessed to power small-scale devices, such as wearable health sensors. Furthermore, VINGs developed for footwear have opened new possibilities for capturing energy from walking, enhancing the practicality of energy-harvesting wearables. Another innovative application involves artificial skin capable of generating electricity. This concept was demonstrated by Professor Zhong Lin Wang's team, who developed a flexible single-wire generator that produced alternating current voltages of up to 100 millivolts when attached to a running hamster. This experiment illustrates the potential for developing self-powered wearable and biomedical electronic systems [62].

8.3.2 Transparent and flexible piezoelectric devices

These devices are being developed using nanostructures fabricated on diverse substrates, including flexible, optically clear organic materials. Research groups led by Professor Sang-Woo Kim at SKKU and Dr. Jae-Young Choi at SAIT have successfully developed an NG that is both transparent and flexible, offering promising applications in self-powered tactile sensing technologies. This innovation has the potential to support the advancement of energy-efficient touch-screen interfaces. Ongoing studies aim to enhance the optical transparency and reduce the production costs of these devices by substituting traditional indium–tin-oxide electrodes with graphene-based alternatives [63].

8.3.3 Implantable telemetric energy receiver

An NG built using ZnO nanowires is particularly suitable for use in implantable medical devices, owing to ZnO biocompatibility and its capability to grow on organic materials. This compatibility ensures that the entire system remains safe for use within the human body. When integrated into an implantable setup, the NG can utilize ultrasonic vibrations transmitted from outside the body. These mechanical signals are then transformed into electrical energy through the piezoelectric effect of the ZnO nanowires, providing a power source for the implanted device.

8.3.4 Biomedical applications

In the expanding landscape of the Internet of Things (IoT), the sensitivity of PENG devices to mechanical stimuli and their capacity for autonomous energy harvesting make them a suitable option for biomedical and health-monitoring systems. One of the major challenges in developing miniaturized devices depends on the conventional power sources, such as batteries. On the other side, self-sustaining power solutions offer clear advantages, including reduced maintenance and simplified system architecture.

8.3.5 Health monitoring

PENGs, when used in association with biomedical sensors, address the challenge of powering portable and compact medical devices. They play a significant role in the advancement of self-sustaining technologies for health monitoring.

Ravikumar et al. [64] reported a new PENG fabricated using a polymer matrix composite based on a metal-phenolic coordination framework (MPCF). The composite material was prepared by combining copper oxide nanospheres with tannic acid (CuO-TA) and copper-tannic acid (Cu-TA) nanocubes within a PVDF polymer matrix. The PENG device was constructed using the MPCF composite placed between copper sheet electrodes (2 × 2 cm) and encapsulated by Polydimethylsiloxane (PDMS). The CuO-TA/PVDF composite exhibited a measured output voltage of 40.20 V and a current of 4.40 μA, whereas Cu-TA/PVDF composite showed 8.90 V and 3.80 μA, respectively. Thus, a notable difference in electrical output was observed. This PENG system was successfully tested in monitoring physiological signals as well as harvesting energy.

These PENG systems are highlighted for their capability in continuous, real-time tracking of heart rate and arterial pulse, aiming to serve as an early warning system [65]. To assess its performance, the device was tested on a healthy 22-year-old subject. Heart rate measurements were approximately 76 BPM at rest and increased to 145 BPM following exercise, closely matching those obtained from a conventional commercial monitor. These results demonstrated the sensor's accuracy and sensitivity. The integration of this PENG with IoT platforms further enabled remote pulse monitoring, extending its usability in practical healthcare settings. All prototypes developed were characterized by their flexibility, energy autonomy, and high sensitivity, making them ideal for wearable health-monitoring technologies.

In a related innovation, Rana et al. [66] developed a wireless, self-powered cardiac sensor using porous PVDF, as shown in Figure 8.5(a–c). This

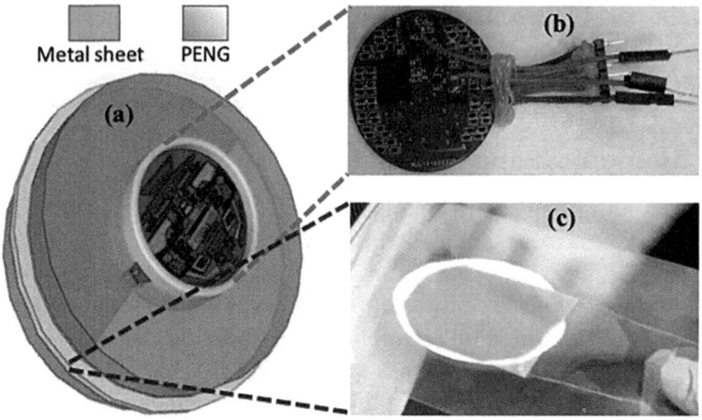

Figure 8.5 Schematic of a **wireless, self-powered cardiac sensor system**. (a) Schematic illustration of the functional components of a piezoelectric nanogenerator (PENG); (b) the integrated circuit; (c) an as-fabricated porous PVDF-based PENG device.

(Reproduced with permission from [66].)

auto-functioning, wireless PENG demonstrated excellent durability and was tailored for operation in variable environments. The device achieved a peak output of 22 μA and 84.5 V, while also emphasizing environmental sustainability. A complete self-powered sensing network was also constructed to support broader health-monitoring applications.

8.3.6 Bio-implant

Implantable medical devices are increasingly exploring the use of PENGs, particularly due to the biodegradable and biocompatible nature of some piezoelectric materials. In this context, Azimi et al. [67] proposed a novel self-powered cardiac pacemaker based on PVDF, a material known for both its piezoelectric and biocompatible properties. Conventional pacemakers depend on batteries that eventually require replacement, thereby limiting their long-term use. To overcome this, the research team designed a PENG incorporating PVDF, ZnO, and reduced graphene oxide (rGO) fabricated via electrospinning. The device was evaluated in a canine model as shown in Figure 8.6, where it was implanted on the lateral wall of the heart's left ventricle. It was able to harvest 0.487 μJ of energy from each heartbeat. This device marks a major milestone toward developing intelligent, energy-autonomous implantable medical devices.

Wu et al. [68] designed a PENG aimed at supporting nerve tissue regeneration. The device, constructed from KNN, poly(L-lactic acid), and poly(3-hydroxybutyrate-co-3-hydroxyvalerate), serves as a self-powered electrical stimulator. They utilized either poly(lactic acid) or poly(ε-caprolactone) for

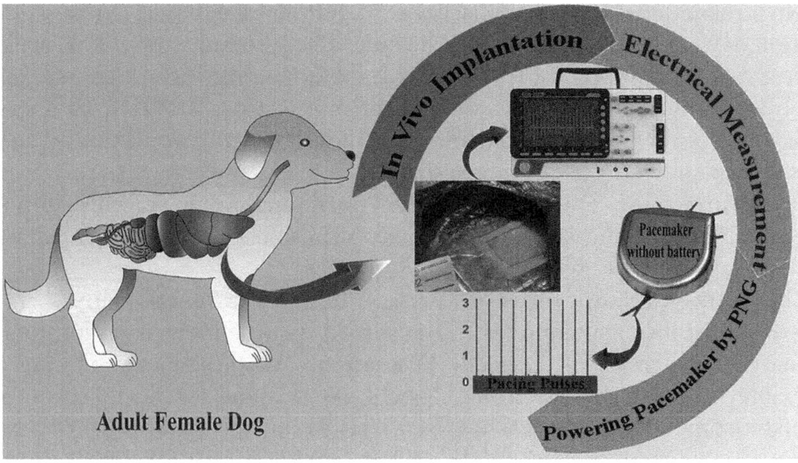

Figure 8.6 Electrospun PVDF/ZnO/rGO-based PENG implanted in an animal model.

(Reproduced with permission from [67].)

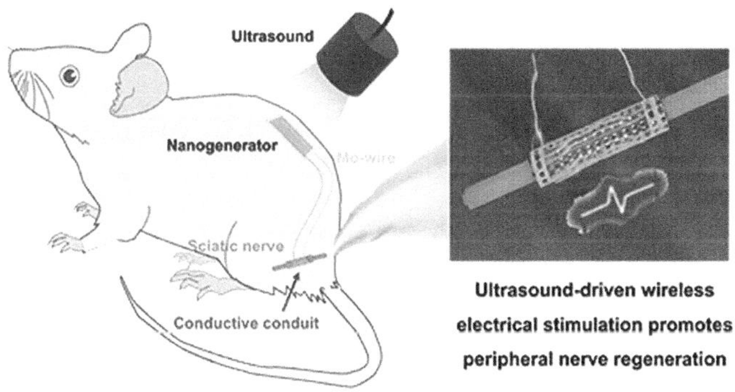

Figure 8.7 Peripheral nerve regeneration simulated by ultrasound. Biocompatibility study of the system and electrical simulation without any external rectifying circuitry.

(Reproduced with permission from [68].)

encapsulation. The performance efficiency and biodegradability of the system were verified under various environmental conditions, including water and cell culture environments. The device was also implanted into a lab rat for three weeks (Figure 8.7(a)), during which no organ damage was detected, confirming its biocompatibility for both in vitro and in vivo applications. Importantly, the device was able to deliver electrical stimulation to promote nerve regeneration without requiring any external rectifying circuitry (see Figure 8.7(b).

In a related investigation, Sharma et al. [69] designed an electrical stimulation device to accelerate wound healing. The system featured a piezoelectric hydrogel composed of carbonized polydopamine, polydopamine, and polyacrylamide, which was combined with an electrospun PVDF membrane. This hydrogel exhibited excellent mechanical strength and flexibility, making it well-suited for use during activities involving movement, such as walking or stretching. The composite film facilitated tissue regeneration by converting mechanical motions into electrical signals and also offered antimicrobial properties to protect the wound area.

In recent years, driven by advances in design and functionality, PENGs have gained increasing interest. The current research focus is on improving their energy-harvesting efficiency. Furthermore, with the increased usage of flexible and wearable electronic devices, the demand for flexible, durable, and biocompatible materials has risen. This has encouraged the use of piezoelectric polymers such as PVDF and its copolymer PVDF-TrFE. Various optimization strategies, namely doping, composite formation, morphological tuning, and sophisticated synthesis methods to precisely tailor the properties of these materials, are being explored in order to achieve high-output efficiency. Further, advanced manufacturing techniques like 3D printing are contributing to improved performance of PENGs. Enhancements in designs and interface engineering also play a crucial role.

8.4 FUTURE PROSPECTS OF PIEZOELECTRIC NANOGENERATORS (PENGs)

The progress in piezoelectric nanogenerators (PENGs) is poised to impact several critical fields in the near future:

1. **Biomedical Applications**: PENGs exhibit great promise as power supplies for miniaturized biomedical devices and self-powered bio-implants, whereas conventional batteries face challenges such as high expenses, recyclability issues, and limited life span. In contrast, PENGs offer a cost-effective, safer, and maintenance-free alternative. The compact size of these devices enables seamless integration within biomedical systems.
2. **Enhancing Performance**: One of the prime research areas aims at improving the efficiency of PENGs. Advances in fabrication methods—including doping and surface engineering—have significantly elevated device output. For instance, Liu et al. [70] highlighted how interfacial engineering can substantially enhance performance.
3. **Novel Materials**: With the advancement in the easy and scalable preparation methods, such as solid-state reactions, several new nanoscale materials, such as graphitic carbon nitride (g-C_3N_4) and boron nitride

nanosheets (BNNs), have gained attention for superior piezoelectric characteristics due to their crystalline structures.

4. **Hybrid Nanogenerators**: Hybrid NGs aid in simultaneous energy harnessing from different sources, enabling a significant output capability. These devices generate much higher voltages and currents due to the integration of various energy conversion mechanisms, making them potential candidates for energy harvesting.

8.5 CONCLUSION

This chapter elaborates the classification of piezoelectric materials, emphasizing their unique attributes, including piezoelectric efficiency, synthesis complexity, mechanical flexibility, and biocompatibility. Each material type brings specific advantages, underscoring the importance of selecting the right material to meet application-specific requirements. With notable progress in the development and manufacturing of PENGs, particularly for biomedical and wearable devices, current research efforts are concentrated on enhancing energy output, durability, and multifunctional capabilities. The incorporation of advanced materials, flexible designs, and hybrid energy-harvesting strategies signifies a move toward next-generation, self-sustaining compact devices.

REFERENCES

1. Wang, Z.L., Song, J. Piezoelectric Nanogenerators Based on Zinc Oxide Nanowire Arrays. *Science* 2006, *312*, 242–246.
2. Fan, F.R., Tang, W., Wang, Z.L. Flexible Nanogenerators for Energy Harvesting and Self-Powered Electronics. *Adv. Mater.* 2016, *28*, 4283–4305.
3. Xu, S., Qin, Y., Xu, C., Wei, Y., Yang, R., Wang, Z.L. Self Powered Nanowire Devices. *Nat. Nanotechnol.* 2010, *5*, 366–373.
4. Niu, X., Jia, W., Qian, S., Zhu, J., Zhang, J., Hou, X., Mu, J., Geng, W., Cho, J., He, J., et al. High-Performance PZT-Based Stretchable Piezoelectric Nanogenerator. *ACS Sustain. Chem. Eng.* 2019, *7*, 979–985.
5. Wu, W., Bai, S., Yuan, M., Qin, Y., Wang, Z.L., Jing, T. Lead Zirconate Titanate Nanowire Textile Nanogenerator for Wearable Energy-Harvesting and Self-Powered Devices. *ACS Nano* 2012, *6*, 6231–6235.
6. Budimir, M., Damjanovic, D., Setter, N. Piezoelectric Anisotropy-Phase Transition Relations in Perovskite Single Crystals. *J. Appl. Phys.* 2003, *94*, 6753–6761.
7. Lee, W.H., Su, C.Y. Improvement in the Temperature Stability of a $BaTiO_3$-Based Multilayer Ceramic Capacitor by Constrained Sintering. *J. Am. Ceram. Soc.* 2007, *90*, 3345–3348.
8. Zhang, Y., Sun, H., Chen, W. Influence of Cobalt and Sintering Temperature on Structure and Electrical Properties of $BaZr_{0.05}Ti_{0.95}O_3$ Ceramics. *Ceram. Int.* 2015, *41*, 8520–8532.

9. Acosta, M., Novak, N., Rojas, V., Patel, S., Vaish, R., Koruza, J., Rossetti, G.A., Rödel, J. $BaTiO_3$-Based Piezoelectrics: Fundamentals, Current Status, and Perspectives. *Appl. Phys. Rev.* 2017, *4*(4), 041305.
10. Liu, W., Ren, X. Large Piezoelectric Effect in Pb-Free Ceramics. *Phys. Rev. Lett.* 2009, *103*, 257602.
11. Zhou, Z., Bowland, C.C., Malakooti, M.H., Tang, H., Sodano, H.A. Lead-Free $0.5Ba(Zr_{0.2}Ti_0.8)O_3$–$0.5(Ba_{0.7}Ca_{0.3})TiO_3$ Nanowires for Energy Harvesting. *Nanoscale* 2016, *8*, 5098–5105.
12. Caruntu, D., Rostamzadeh, T., Costanzo, T., Salemizadeh Parizi, S., Caruntu, G. Solvothermal Synthesis and Controlled Self-Assembly of Monodisperse Titanium-Based Perovskite Colloidal Nanocrystals. *Nanoscale* 2015, *7*, 12955–12969.
13. Xue, P., Wu, H., Xia, W., Pei, Z., Lu, Y., Zhu, X. Molten Salt Synthesis of $BaTiO_3$ Nanorods: Dielectric, Optical Properties, and Structural Characterizations. *J. Am. Ceram. Soc.* 2019, *102*, 2325–2336.
14. Philippot, G., Elissalde, C., Maglione, M., Aymonier, C. Supercritical Fluid Technology: A Reliable Process for High Quality $BaTiO_3$ Based Nanomaterials. *Adv. Powder Technol.*, 2014, *25*(5), 1415–1429.
15. Zhao, Z., Buscaglia, V., Viviani, M., Buscaglia, M.T., Mitoseriu, L., Testino, A., Nygren, M., Johnsson, M., Nanni, P. Grain-Size Effects on the Ferroelectric Behavior of Dense Nanocrystalline $BaTiO_3$ Ceramics. *Phys. Rev. B Condens. Matter. Mater. Phys.* 2004, *70*, 024107.
16. Hong, C.H., Fan, Z., Tan, X., Kang, W.S., Ahn, C.W., Shin, Y., Jo, W. Role of Sodium Deficiency on the Relaxor Properties of $Bi_{1/2}Na_{1/2}TiO_3$-$BaTiO_3$. *J. Eur. Ceram. Soc.* 2018, *38*, 5375–5381.
17. Pan, H., Li, F., Liu, Y., Zhang, Q., Wang, M., Lan, S., Zheng, Y., Ma, J., Gu, L., Shen, Y., et al. Ultrahigh-Energy Density Lead-Free Dielectric Films via Polymorphic Nanodomain Design. *Science* 2019, *365*, 578–582.
18. Deng, Y., Wang, J., Zhang, C., Ma, H., Bai, C., Liu, D., Wu, F., Yang, B. Structural and Electric Properties of MnO_2-Doped KNN-LT Lead-Free Piezoelectric Ceramics. *Crystals* 2020, *10*, 705.
19. Swain, S., Kumar, P., Agrawal, D.K., Sonia. Dielectric and Ferroelectric Study of KNN Modified NBT Ceramics Synthesized by Microwave Processing Technique. *Ceram. Int.* 2013, *39*, 3205–3210.
20. Kwak, J., Kingon, A.I., Kim, S.H. Lead-Free $(Na_{0.5},K_{0.5})NbO_3$ Thin Films for the Implantable Piezoelectric Medical Sensor Applications. *Mater. Lett.* 2012, *82*, 130–132.
21. Wu, J., Xiao, D., Zhu, J. Potassium-Sodium Niobate Lead-Free Piezoelectric Materials: Past, Present, and Future of Phase Boundaries. *Chem. Rev.* 2015, *115*(7), 2559–2595.
22. Khorrami, G.H., Kompany, A., Khorsand Zak, A. Structural and Optical Properties of $(K,Na)NbO_3$ Nanoparticles Synthesized by a Modified Sol-Gel Method Using Starch Media. *Adv. Powder Technol.* 2015, *26*, 113–118.
23. Ghasemian, M.B., Rawal, A., Shahrbabaki, Z., Zhang, Q., Lu, T., Liu, Y., Wang, D. Evidence of Phase Coexistence in Hydrothermally Synthesized $K_{0.5}Na_0\cdot_5NbO_3$ nanofibers. *J. Mater. Chem. A Mater.* 2020, *8*, 8731–8739.
24. Zhou, D., Gu, H., Hu, Y., Tian, H., Wang, Z., Qian, Z., Wang, Y. Synthesis, Characterization and Ferroelectric Properties of Lead-Free $K_{0.5}Na_0\cdot_5NbO_3$ Nanotube Arrays. *J. Appl. Phys.* 2011, *109*, 11410.

25. Brown, L.F., Carlson, R.L., Sempsrntt, J.M. *Spin-Cast P(VDF-TrFE) Films for High Performance Medical Ultrasound Transducers*, IEEE: Belfast, UK, 2006.
26. Jung, I., Shin, Y.H., Kim, S., Choi, J.Y., Kang, C.Y. Flexible Piezoelectric Polymer-Based Energy Harvesting System for Roadway Applications. *Appl. Energy* 2017, *197*, 222–229.
27. Ye, L., Chen, L., Yu, J., Tu, S., Yan, B., Zhao, Y., Bai, X., Gu, Y., Chen, S. High-Performance Piezoelectric Nanogenerator Based on Electrospun ZnO Nanorods/P(VDF-TrFE) Composite Membranes for Energy Harvesting Application. *J. Mater. Sci. Mater. Electron.* 2021, *32*, 3966–3978.
28. Sappati, K.K., Bhadra, S. Piezoelectric Polymer and Paper Substrates: A Review. *Sensors* 2018, *18*, 3605.
29. Sharma, M., Srinivas, V., Madras, G., Bose, S. Outstanding Dielectric Constant and Piezoelectric Coefficient in Electrospun Nanofiber Mats of PVDF Containing Silver Decorated Multiwall Carbon Nanotubes: Assessing through Piezoresponse Force Microscopy. *RSC Adv.* 2016, *6*, 6251–6258.
30. Abolhasani, M.M., Naebe, M., Hassanpour Amiri, M., Shirvanimoghaddam, K., Anwar, S., Michels, J.J., Asadi, K. Hierarchically Structured Porous Piezoelectric Polymer Nanofibers for Energy Harvesting. *Adv. Sci.* 2020, *7*, 2000517.
31. Abolhasani, M.M., Shirvanimoghaddam, K., Naebe, M. PVDF/Graphene Composite Nanofibers with Enhanced Piezoelectric Performance for Development of Robust Nanogenerators. *Compos. Sci. Technol.* 2017, *138*, 49–56.
32. Abolhasani, M.M., Shirvanimoghaddam, K., Khayyam, H., Moosavi, S.M., Zohdi, N., Naebe, M. Towards Predicting the Piezoelectricity and Physiochemical Properties of the Electrospun P(VDF-TrFE) Nanogenerators Using an Artificial Neural Network. *Polym. Test.* 2018, *66*, 178–188.
33. Fang, J., Wang, X., Lin, T. Electrical Power Generator from Randomly Oriented Electrospun Poly(Vinylidene Fluoride) Nanofibre Membranes. *J. Mater. Chem.* 2011, *21*, 11088–11091.
34. Lee, C., Wood, D., Edmondson, D., Yao, D., Erickson, A.E., Tsao, C.T., Revia, R.A., Kim, H., Zhang, M. Electrospun Uniaxially-Aligned Composite Nanofibers as Highly Efficient Piezoelectric Material. *Ceram. Int.* 2016, *42*, 2734–2740.
35. Zeng, W., Tao, X.-M., Chen, S., Shang, S., Chan, H.L.W., Choy, S.H. Highly Durable All fiber Nanogenerator for Mechanical Energy Harvesting. *Energy Environ. Sci.* 2013, *6*, 2631–2638.
36. Briscoe, J., Jalali, N., Woolliams, P., Stewart, M., Weaver, P.M., Cain, M., Dunn, S. Measurement Techniques for Piezoelectric Nanogenerators. *Energy Environ. Sci.* 2013, *6*, 3035–3045.
37. Parangusan, H., Ponnamma, D., Al-Maadeed, M.A.A. Stretchable Electrospun PVDFHFP/Co-ZnO Nanofibers as Piezoelectric Nanogenerators. *Sci. Rep.* 2018, *8*, 754.
38. Bafqi, M.S.S., Bagherzadeh, R., Latifi, M. Fabrication of Composite PVDF-ZnO Nanofiber Mats by Electrospinning for Energy Scavenging Application with Enhanced Efficiency. *J. Polym. Res.* 2015, *22*, 130.
39. Garain, S., Jana, S., Sinha, T.K., Mandal, D. Design of In Situ Poled Ce^{3+}-Doped Electrospun PVDF/Graphene Composite Nanofibers for Fabrication of Nanopressure Sensor And Ultrasensitive Acoustic Nanogenerator. *ACS Appl. Mater. Interfaces* 2016, *8*, 4532–4540.

40. Lu, X., Qu, H., Skorobogatiy, M. Piezoelectric Micro- and Nanostructured Fibers Fabricated from Thermoplastic Nanocomposites Using a Fiber Drawing Technique: Comparative Study and Potential Applications. *ACS Nano* 2017, *11*, 2103–2114.
41. Xu, Z., Jin, C., Cabe, A., Escobedo, D., Hao, N., Trase, I., Closson, A.B., Dong, L., Nie, Y., Elliott, J., Feldman, M., Chen, Z., Zhang, J.X.J. Flexible Energy Harvester on a Pacemaker Lead Using Multibeam Piezoelectric Composite Thin Films. *ACS Appl. Mater. Interfaces* 2020, *12*, 34170–34179.
42. Karan, S.K., Bera, R., Paria, S., Das, A.K., Maiti, S., Maitra, A., Khatua, B.B. An Approach to Design Highly Durable Piezoelectric Nanogenerator Based on Self-Poled PVDF/AlO-rGO Flexible Nanocomposite with High Power Density and Energy Conversion Efficiency. *Adv. Energy Mater.* 2016, *6*, 1601016.
43. Hussein, A.D., Sabry, R.S., Abdul Azeez Dakhil, O., Bagherzadeh, R. Effect of Adding $BaTiO_3$ to PVDF as Nano Generator. *J. Phys.* 2019. doi:10.1088/1742-6596/1294/2/022012
44. Abbasipour, M., Khajavi, R., Yousefi, A.A., Yazdanshenas, M.E., Razaghian, F., Akbarzadeh, A. Improving Piezoelectric and Pyroelectric Properties of Electrospun PVDF Nanofibers Using Nanofillers for Energy Harvesting Application. *Polym. Adv. Technol.* 2019, *30*(2), 279–291. doi:10.1002/pat.4463
45. Chang C., Fuh Y.K., Lin L. A Direct-Write Piezoelectric PVDF Nanogenerator. *Proceedings of the TRANSDUCERS 2009-15th International Conference Solid-State Sensors, Actuators Microsystems*, Denver, CO, June 21–25 2009, 2009, 485–488.
46. Yu, H., Huang, T., Lu, M., Mao, M., Zhang, Q., Wang, H. Enhanced Power Output of an Electrospun PVDF/MWCNTs-Based Nanogenerator by Tuning Its Conductivity. *Nanotechnology* 2013, *24*(40). doi:10.1088/0957-4484/24/40/405401
47. Choi, M., Murillo, G., Hwang, S., Kim, J.W., Jung, J.H., Chen, C.Y., Lee, M. Mechanical and Electrical Characterization of PVDF-ZnO Hybrid Structure for Application to Nanogenerator. *Nano Energy* 2017, *33*, 462–468.
48. Chen, X., Li, X., Shao, J., An, N., Tian, H., Wang, C., Han, T., Wang, L., Lu, B. High-Performance Piezoelectric Nanogenerators with Imprinted P(VDF-TrFE)/$BaTiO_3$ Nanocomposite Micropillars for Self-Powered Flexible Sensors. *Small* 2017, *13*(23). doi:10.1002/smll.201604245
49. Guan, X., Xu, B., Gong, J. Hierarchically Architected Polydopamine Modified $BaTiO_3$@P(VDF-TrFE) Nanocomposite Fiber Mats for Flexible Piezoelectric Nanogenerators and Self-Powered Sensors. *Nano Energy* 2020, *70*. doi:10.1016/j.nanoen.2020.104516
50. Lin, L., Lai, C.-H., Hu, Y., Zhang, Y., Wang, X., Xu, C., Snyder, R.L., Chen, L.J., Wang, Z.L. High Output Nanogenerator Based on Assembly of GaN Nanowires. *Nanotechnology* 2011, *22*(47). doi:10.1088/0957-4484/22/47/475401
51. Parangusan, H., Bhadra, J., Al-Thani, N. Flexible Piezoelectric Nanogenerator Based on [P(VDF-HFP)]/PANI-ZnS Electrospun Nanofibers for Electrical Energy Harvesting. *J. Mater. Sci. Mater. Electron.* 2021, 32(5), 6358–6368. doi:10.1007/s10854-021-05352-4
52. Sultana, A., Alam, M.M., Biswas, A., Middya, T.R., Mandal, D. Fabrication of Wearable Semiconducting Piezoelectric Nanogenerator Made with Electrospun-Derived Zinc Sulfide Nanorods and Poly(Vinyl Alcohol) Nanofibers. *Transl. Mater. Res.* 2016, *3*(4), 045001. doi:10.1088/2053-1613/3/4/045001

53. Ali, N., Kenawy, E.R., Wadoud, A.A., Elhadary, M.I. Wearable Electrospun Piezoelectric Mats Based on a PVDF Nanofiber–ZnO@ZnS Core–Shell Nanoparticles Composite for Power Generation. *Nanomaterials* 2023, *13*(21). doi:10.3390/nano13212833
54. Liao, Q., Zhang, Z., Zhang, X., Mohr, M., Zhang, Y., Fecht, H.J. Flexible Piezoelectric Nanogenerators Based on a Fiber/ZnO Nanowires/Paper Hybrid Structure for Energy Harvesting. *Nano Res.* 2014, *7*(6), 917–928. doi:10.1007/s12274-014-0453-8
55. Al-Saygh, A., Ponnamma, D., AlMaadeed, M.A.A., Poornima Vijayan, P., Karim, A., Hassan, M.K. Flexible Pressure Sensor Based on PVDF Nanocomposites Containing Reduced Graphene Oxide-Titania Hybrid Nanolayers. *Polymers (Basel)* 2017, *9*(2), 33. doi:10.3390/polym9020033
56. Pusty, M., Sharma, A., Sinha, L., Chaudhary, A., Shirage, P. Comparative Study with a Unique Arrangement to Tap Piezoelectric Output to Realize a Self Poled PVDF Based Nanocomposite for Energy Harvesting Applications. *ChemistrySelect* 2017, *2*(9), 2774–2782. doi:10.1002/slct.201602046
57. Lin, Z.H., Yang, Y., Wu, J.M., Liu, Y., Zhang, F., Wang, Z.L. $BaTiO_3$ Nanotubes-Based Flexible and Transparent Nanogenerators. *J. Phys. Chem. Lett.* 2012, *3*(23), 3599–3604. doi:10.1021/jz301805f
58. Du, X., Zhou, Z., Zhang, Z., Yao, L., Zhang, Q., Yang, H. Porous, Multi-Layered Piezoelectric Composites Based on Highly Oriented PZT/PVDF Electrospinning Fibers for High-Performance Piezoelectric Nanogenerators. *J. Adv. Ceram.* 2022, *11*(2), 331–344. doi:10.1007/s40145-021-0537-3
59. Lee, H., Kim, H., Kim, D.Y., Seo, Y. Pure Piezoelectricity Generation by a Flexible Nanogenerator Based on Lead Zirconate Titanate Nanofibers. *ACS Omega* 2019, *4*(2), 2610–2617. doi:10.1021/acsomega.8b03325
60. Güçlü, H., Kasım, H., Yazici, M. Investigation of the Optimum Vibration Energy Harvesting Performance of Electrospun PVDF/$BaTiO_3$ Nanogenerator. *J. Compos. Mater.* 2023, *57*(3), 409–424. doi:10.1177/00219983221144696
61. Pongampai, S., Charoonsuk, T., Pinpru, N., Pulphol, P., Vittayakorn, W., Pakawanit, P., Vittayakorn, N. Triboelectric-Piezoelectric Hybrid Nanogenerator Based on $BaTiO_3$-Nanorods/Chitosan Enhanced Output Performance with Self-Charge-Pumping System. *Compos. B Eng.* 2021, *208*. doi:10.1016/j.compositesb.2020.108602
62. Momeni, K., Odegard, G. M., Yassar, R. S. Nanocomposite Electrical Generator Based on Piezoelectric Zinc Oxide Nanowires. *J. Appl. Phys.* 2010, *108*(11): 114303–114307. doi:10.1063/1.3517095
63. Qin, Y., Wang, X., Wang, Z.L. Microfibre–Nanowire Hybrid Structure for Energy Scavenging. *Nature* 2009, *457*(7227): 340. doi:10.1038/nature07628
64. Ravikumar, R., Natraj, V., Sivalingam, Y., Velappa Jayaraman, S., Zheng, F., Wang, H., Du, Q., Liu, N. Piezoelectric Nanogenerator Induced Work Function on Metal Phenolic Coordination Framework from Copper Oxide Nanosphere for Efficient Biomechanical Energy Harvesting and Physiological Monitoring. *J. Mater. Chem. C Mater.* 2022, *10*, 16492–16505.
65. Kappiarukudil, K.J., Ramesh, M.V. Real-time monitoring and detection of "heart attack" using wireless sensor networks. In Proceedings of the 4th International Conference on Sensor Technologies and Applications, Venice/Mestre, Italy, 18–25 July 2010, 2010, pp. 632–636.

66. Rana, M.M., Khan, A.A., Huang, G., Mei, N., Saritas, R., Wen, B., Zhang, S., Voss, P., Abdel-Rahman, E., Leonenko, Z., et al. Porosity Modulated High-Performance Piezoelectric Nanogenerator Based on Organic/Inorganic Nanomaterials for Self-Powered Structural Health Monitoring. *ACS Appl. Mater. Interfaces* 2020, *12*, 47503–47512.
67. Azimi, S., Golabchi, A., Nekookar, A., Rabbani, S., Amiri, M.H., Asadi, K., Abolhasani, M.M. Self-Powered Cardiac Pacemaker by Piezoelectric Polymer Nanogenerator Implant. *Nano Energy* 2021, *83*, 105781.
68. Wu, P., Chen, P., Xu, C., Wang, Q., Zhang, F., Yang, K., Jiang, W., Feng, J., Luo, Z. Ultrasound-Driven in Vivo Electrical Stimulation Based on Biodegradable Piezoelectric Nanogenerators for Enhancing and Monitoring the Nerve Tissue Repair. *Nano Energy* 2022, *102*, 107707.
69. Sharma, A., Panwar, V., Mondal, B., Prasher, D., Bera, M.K., Thomas, J., Kumar, A., Kamboj, N., Mandal, D., Ghosh, D. Electrical Stimulation Induced by a Piezo-Driven Triboelectric Nanogenerator and Electroactive Hydrogel Composite, Accelerate Wound Repair. *Nano Energy* 2022, *99*, 107419.
70. Liu, C., Yu, A., Peng, M., Song, M., Liu, W., Zhang, Y., Zhai, J. Improvement in the Piezoelectric Performance of a ZnO Nanogenerator by a Combination of Chemical Doping and Interfacial Modification. *J. Phys. Chem. C* 2016, *120*, 6971–6977.

Chapter 9

Nanofabrication techniques and lithography process

Bipul Kumar Mahato and Rajdeep Singh Rawat

9.1 INTRODUCTION

Driven by Moore's law—which predicts the doubling of transistors on integrated circuits approximately every two years—the relentless miniaturization of complementary metal oxide semiconductor (CMOS) transistors continues to push the boundaries of modern technology. Nanodevice fabrication lies at the heart of this progress, serving as a cornerstone for the rapidly advancing fields of nanotechnology. Enabling the creation of devices at the nanometer scale (10^{-9} m) opens pathways to breakthroughs in electronics, medicine, energy storage, sensing, and beyond. The ability to engineer and manipulate matter with atomic-scale precision not only fuels innovation but also redefines the performance limits of next-generation systems. This chapter explores the fundamental methods and techniques of nanodevice fabrication, discussing key materials, fabrication strategies, and the challenges involved in building devices capable of operating at the scale of individual molecules or atoms.

9.2 NANOFABRICATION: OVERVIEW AND TECHNIQUES

Nanofabrication[1–5] is the process of designing, creating, and integrating structures and devices at the nanometer scale, where at least one dimension is typically smaller than 100 nm. Due to quantum confinement effects, materials at this scale often exhibit unique physical, chemical, electrical, and optical properties that can be harnessed for advanced functionality. The goal is not only to achieve extreme miniaturization but also to control structure and composition with atomic or molecular precision, enabling the development of highly specialized devices.

Producing such features requires advanced methods capable of precisely modifying, patterning, and assembling materials. These methods fall broadly into two categories: **top-down** approaches, which sculpt or etch bulk materials into nanoscale structures, and **bottom-up** approaches, which assemble

DOI: 10.1201/9781003635123-9

structures atom-by-atom or molecule-by-molecule. Each approach offers distinct advantages and challenges, but in many modern applications, they are combined to exploit the strengths of both.

9.3 TOP-DOWN VERSUS BOTTOM-UP APPROACHES

Top-down fabrication begins with a bulk material and shapes it into nanoscale structures through processes such as grinding, etching, or precision patterning. Common techniques include photolithography, electron beam lithography (EBL or E-beam), and focused ion beam (FIB)[6] milling, which allow the direct sculpting of intricate nanoscale features. We will focus more on top-down approach, as it is largely used in research and industries.

Bottom-up fabrication, in contrast, constructs structures atom-by-atom or molecule-by-molecule, typically using controlled chemical or physical processes. Representative techniques include chemical vapor deposition (CVD), molecular beam epitaxy (MBE), self-assembly, and atomic layer deposition (ALD), each capable of producing highly ordered nanostructures with exceptional precision.

Both strategies have unique strengths and limitations: top-down methods are mature, reproducible, and well-suited for large-scale manufacturing, while bottom-up approaches offer superior control at the molecular level and the potential to create novel architectures that are not easily achievable through subtractive methods. In many cutting-edge applications, hybrid approaches integrate both strategies to harness their respective advantages.

9.4 MAJOR FABRICATION TECHNIQUES

9.4.1 Hard mask

Pattern transfer techniques utilizing hard-mask were established much earlier. A hard mask is a prefabricated, opaque pattern—typically fabricated on a glass plate—used in photolithography for rapid exposure and high-throughput or mass-production processes. The opaque regions block light, while the transparent areas allow it to pass through, thereby defining the pattern on the photoresist. This process is analogous to creating alternating shadow and illuminated regions on the resist, as shown in Figure 9.1. Hard masks are not employed in EBL.

9.4.2 Soft mask

A soft mask is a computer-generated pattern file that defines exposure regions using contrast differences, enabling a lithography system to accurately transfer the design onto a resist-coated substrate. A digitally designed

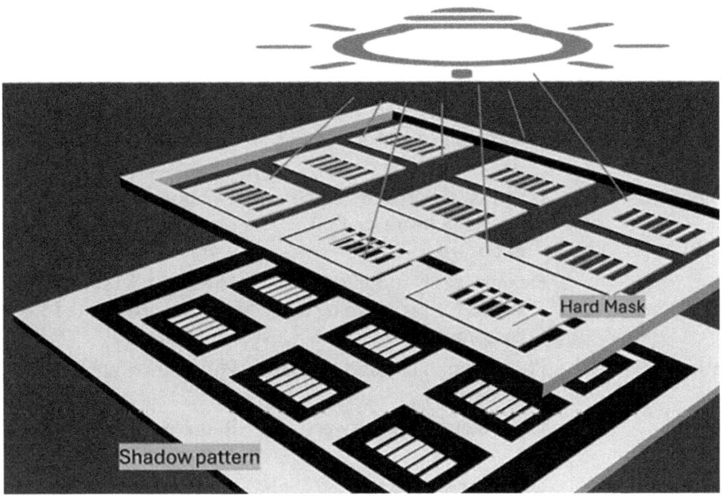

Figure 9.1 Formation of shadow pattern from a hard mask.

pattern is created using software such as AutoCAD, *K*-layout, or even Microsoft Paint, in which the regions to be exposed are assigned a contrasting shade or color from the unexposed regions. The lithography system recognizes this contrast, using it to control the passage of light or the electron beam and thereby transferring the pattern onto the resist-coated substrate. A typical example of a soft mask is shown in Figure 9.2.

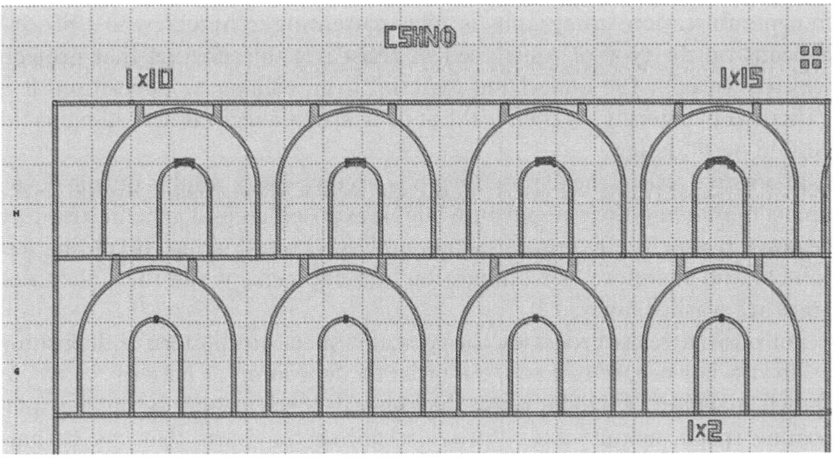

Figure 9.2 Part of a soft mask made using *K*-layout software. Two types of regions include thick lines with deeper color, and the rest of the area with lighter color. The lithography machine will allow either region to pass light while blocking the rest of the region.

Table 9.1 Differences between soft masks and hard masks

Feature	*Soft mask*	*Hard mask*
Form	Digital file created in design software (e.g., AutoCAD, K -layout, and MS Paint)	Prefabricated, opaque pattern printed on a transparent substrate (usually glass)
Pattern definition	Contrast differences in the file guide beam or light exposure	Opaque and transparent regions physically block or transmit light
Usage	Flexible, easily modified; suited for prototyping and custom designs	High-speed, repeatable exposures; suited for mass production
Application	Photolithography and e-beam lithography	Photolithography only
Modification	Easily editable in software	Requires re-fabrication for changes
Cost & turnaround	Low cost and fast turnaround	Higher cost and longer preparation time
Durability	Not physical—depends on storage of digital file	Physical plate and reusable for many exposures

To distinguish between hard and soft masks, a comparison is given in Table 9.1.

The following are some of the major techniques employed in the patterning steps of nanodevice fabrication.

9.4.3 Lithography steps in general

In nanofabrication, lithography is a key patterning technique whose process depends on the type of resist used. A **resist** is a thin coating that protects selected areas of the underlying material from etching or deposition. It is patterned by altering its chemical bonds in specific regions according to the desired mask design.

In **positive resists,** the exposed regions become more soluble during development, allowing their removal, while unexposed regions remain intact. In **negative resists,** the exposed regions undergo cross-linking, becoming less soluble and therefore remaining after development, while the unexposed areas are washed away.

For positive resist processes, the typical sequence is: pattern → deposition → lift-off. For negative resists, the order is: deposition → pattern → etch, as shown in Figure 9.3. The resist thickness depends primarily on the spin-coating speed; manufacturers provide calibration charts that can be fine-tuned for specific etch requirements. Development time also requires periodic optimization, as it can drift over time. Baking the resist before exposure removes air bubbles, improves adhesion, and stabilizes its properties. Yellow-light cleanrooms are used to prevent unintentional exposure.

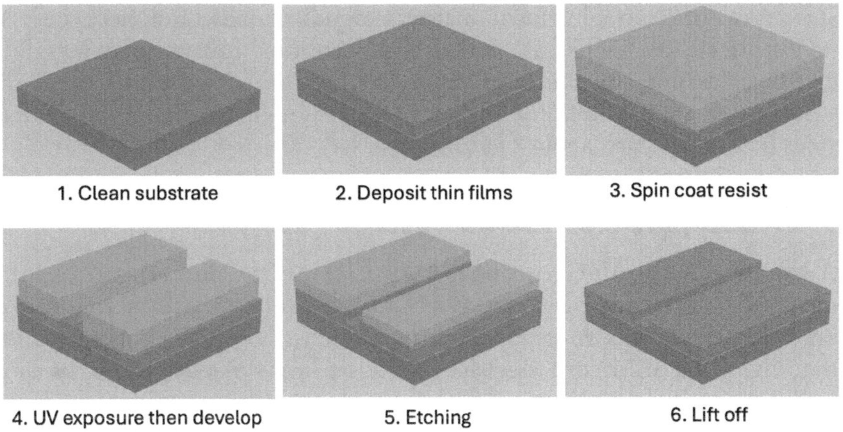

Figure 9.3 Fabrication steps using negative resist, a step-by-step photolithography process. 1. Clean the substrate > 2. Deposit a thin film of the desired material > 3. Spin coat resist (positive or negative (mask design will vary accordingly)) > 4. Develop the pattern on resist > 5. Etch the material according to the pattern in the mask > 6. Lift-off the residual resists to get the pattern fabricated on the thin film.

Patterning precision depends on resist properties, the exposure beam's wavelength, the developer chemistry, and (for negative resists) the etching process. Shorter wavelengths offer higher resolution due to reduced diffraction limits, but also correspond to higher photon energy. In photolithography, deep-ultraviolet (UV) light marks the practical resolution limit (~1 µm resolution), as shorter wavelengths like X-rays are highly penetrating and hazardous. For higher resolution, **EBL** is employed, where a focused electron beam, generated from a heated filament and shaped using magnetic lenses, directly writes nanoscale patterns with superior control.

9.4.4 Photolithography

Photolithography[7–9] is a fundamental top-down fabrication technique widely used in microelectronics and nanotechnology to define patterns on a substrate. In this process, a light-sensitive material called a **photoresist** is coated onto the surface, and UV light is projected through a mask to expose specific regions. The exposed resist undergoes chemical changes that enable selective removal during development, revealing the desired pattern for subsequent etching or deposition.

The resolution of photolithography is primarily limited by the wavelength of the exposure light; shorter wavelengths allow finer feature sizes by reducing diffraction effects. Recent advances, such as **extreme ultraviolet (EUV) lithography**, have extended this capability into the deep-nanometer regime. In practice, photolithography is often used to create relatively larger structures—such as alignment markers, bonding pads, and electrodes in nanoscale devices—whereas smaller, high-resolution features may be fabricated using complementary techniques like EBL.

It is necessary to learn the component systems used in the nanofabrication steps before we learn about the EBL process.

9.4.5 Scanning electron microscope (SEM)

A scanning electron microscope (SEM) is a powerful and versatile instrument that has transformed our ability to visualize and analyze the micro- and nanoscale world. By employing a finely focused beam of electrons rather than light, the SEM surpasses the resolution limits of optical microscopy, enabling detailed examination of surface structure, composition, and other material properties, while providing remarkable depth of field and clarity. Its ability to produce highly detailed, three-dimensional images has made it indispensable across a wide range of disciplines, including materials science, biology, semiconductor technology, and forensic science.

9.4.5.1 Principle of operation

The SEM operates by raster-scanning a focused electron probe across the surface of a specimen. As the beam interacts with the sample, it produces a variety of signals—such as secondary electrons (SEs), backscattered electrons (BSEs), and characteristic X-rays. These are collected by specialized detectors to form images or spectra. SEs provide fine surface detail and a three-dimensional appearance, while BSEs reveal compositional contrast based on atomic number differences. Meanwhile, X-ray detection enables elemental analysis through techniques like energy-dispersive X-ray spectroscopy (EDS/EDX), combining high-resolution imaging with chemical composition mapping.

9.4.5.2 Typical components of an SEM

To perform these functions, a modern SEM integrates several key subsystems:

A. **Electron Source (Gun)**: Generates the beam; can be a thermionic emitter (tungsten filament, or LaB_6 crystal) or a field emission source (field emission gun (FEG)), providing higher brightness and resolution.
B. **Electron Optics**: A series of condenser and objective lenses that focus and control the beam's spot size and current. These are equipped with apertures and stigmators to fine-tune the beam.
C. **Scanning Coils**: Deflect the electron beam across the specimen in a raster pattern for imaging or analysis.
D. **Specimen Chamber and Stage**: A vacuum chamber containing a motorized stage to position, tilt, or rotate the specimen for versatile imaging.

E. **Vacuum System:** Maintains the high-vacuum environment required for unimpeded electron travel and detector performance. Typically, a combination of roughing pumps and turbomolecular or ion pumps.

F. **Detectors:**

  i. SE detector (Everhart-Thornley type) for topography.
  ii. BSE detector for composition or crystallographic contrast.
  iii. X-ray detectors (EDS/WDS) for elemental analysis.
  iv. Additional detectors (CL, STEM-in-SEM) may also be integrated.

G. **Control and Display System:** Dedicated electronics and software for beam scanning, signal collection, image generation, and data analysis.

9.4.5.3 Applications and significance

In practice, SEM serves as a bridge between theory and reality, making the invisible visible and quantifiable. It can reveal nanoscale structures on a butterfly's wing, analyze fracture surfaces of engineering alloys, inspect defects in semiconductor wafers, or detect asbestos fibers in building materials. This combination of high-resolution imaging, compositional analysis, and versatility makes the SEM a cornerstone of modern science, technology, and quality assurance (Figure 9.4).

9.4.6 Electron beam lithography (EBL)

Electron beam lithography (EBL)[10, 11] is a high-resolution nanofabrication technique that utilizes a focused beam of accelerated electrons to directly pattern electron-sensitive resist layers on a substrate, circumventing the diffraction limits inherent in optical photolithography. Unlike conventional photolithography, which is constrained by the wavelength of UV or deep-ultraviolet (DUV) photons (typically above 193 nm), **EBL leverages electrons with de Broglie wavelengths** on the order of picometers, enabling patterning with critical dimensions down to the sub-10 nm scale and, in optimized conditions, approaching a few nanometers. A typical diagram representing the working principles of a high-resolution EBL system is shown in Figure 9.5. There are a lot of similarities between an SEM and an EBL system. **Many commercial EBL systems are built on a modified SEM platform.** They retain the basic electron column and imaging capabilities but are enhanced with the critical additional components required for lithography: a high-specification pattern generator, a high-precision stage, and a beam blanker. This allows them to both align to existing features on a substrate (using SEM mode) and write new patterns.

In EBL, an electron beam generated by a field emission gun or thermionic source is precisely controlled and raster-scanned across the resist-coated substrate using electromagnetic lenses and deflectors. The incident electrons interact with the resist material (commonly **polymethyl methacrylate**

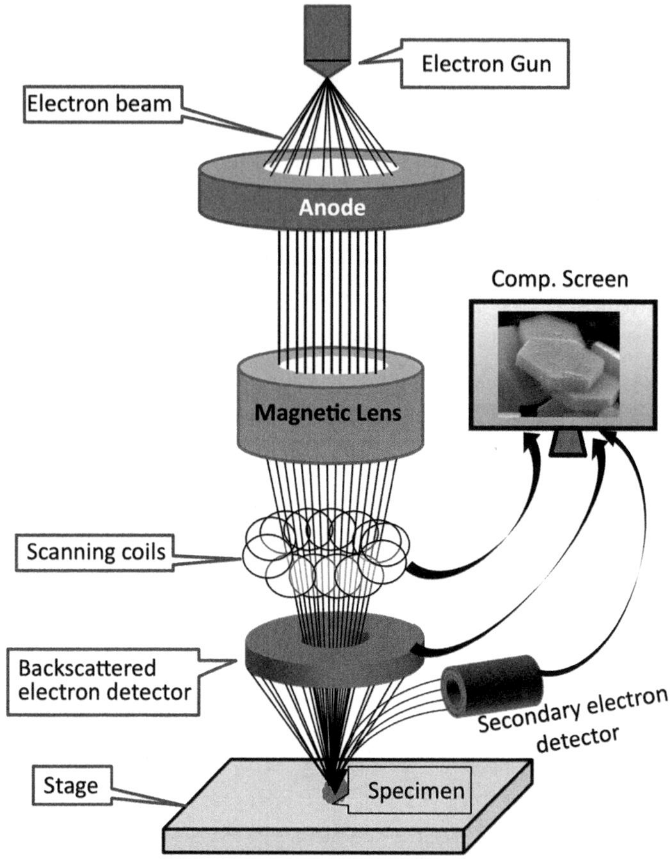

Figure 9.4 A linear diagram displaying the working principle of a scanning electron microscope.

(**PMMA**) or chemically amplified resists) via inelastic scattering processes, inducing localized chain scission or cross-linking depending on the resist chemistry. This electron–resist interaction modulates the solubility of the resist in subsequent development steps, effectively transferring the designed pattern with nanometric fidelity.

The **technique's maskless**, direct-write nature affords unparalleled flexibility for fabricating arbitrary and highly complex nanoscale geometries, eliminating the need for photomasks, which are costly and time-consuming to produce at comparable resolutions. However, the electron scattering within the resist and substrate (proximity effect) necessitates computational correction algorithms to mitigate pattern distortion and maintain dimensional accuracy. Furthermore, the **serial writing approach of EBL imposes throughput limitations**, making it inherently slower than parallel exposure techniques such as photolithography, thereby limiting its practicality for high-volume semiconductor manufacturing.

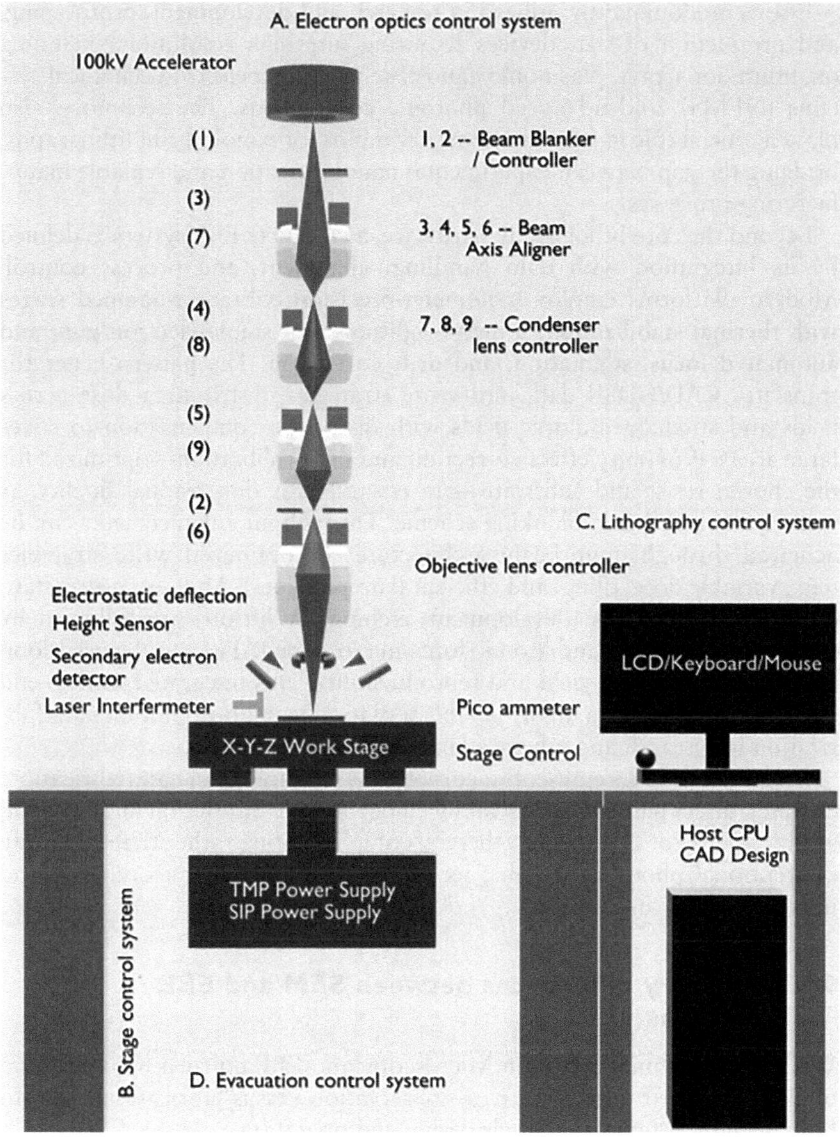

Figure 9.5 A diagram of high-resolution electron beam lithography (EBL) showing four main systems (a) Electron optics control system. (b) Stage control system. (c) Lithography control system. (d) Evacuation control system.

EBL is predominantly utilized in **research and development**, prototyping, and production of nanodevices requiring ultrahigh resolution, including quantum dot arrays, plasmonic nanostructures, nanoelectromechanical systems (NEMS), and advanced photonic components. The technique also plays a crucial role in fabricating master molds for nanoimprint lithography, bridging the gap between experimental nanofabrication and scalable manufacturing processes.

Beyond the core lithography hardware, a complete EBL system is defined by **its integration with data handling, alignment, and process control.** Modern platforms employ **nanometer-precision, vibration-damped stages with thermal stabilization; a high-brightness and stable electron gun; and automated focus, stigmation, and drift correction**. The pattern generator translates CAD/GDSII data into write strategies, distributing dose across fields and stitching multiple fields with distortion compensation to cover large areas. **Proximity effect correction and dose calibration**—optimized for the chosen resist and substrate—are essential for dimensional fidelity, as well as a robust beam blanking scheme. Throughput enhancements can be achieved through multi-beam architectures or optimized write strategies (e.g., variable dose, tiling, and efficient data pipelines). After exposure, standard processing steps (development, etching, or lift-off) are followed by metrology (CD-SEM and atomic force microscope (AFM)) and closed-loop feedback to maintain yield and reproducibility. This integrated, end-to-end approach makes EBL a highly flexible tool for ultrahigh-resolution nanofabrication in research and advanced manufacturing contexts.

In summary, EBL represents a cornerstone technology in nanofabrication, enabling direct patterning at scales well beyond the diffraction limit of light, at the expense of throughput, thereby complementing rather than replacing conventional photolithographic techniques in modern semiconductor and nanotechnology industries.

9.4.7 Primary differences between SEM and EBL systems

While both Scanning Electron Microscopy and EBL utilize a focused beam of electrons, their core objectives—observation versus fabrication—lead to fundamental differences in their design and operation.

A. **Purpose and core function**
   - **SEM (Scanning Electron Microscopy): Imaging and Characterization.** Its primary function is to obtain high-resolution images for analyzing a sample's surface morphology, composition, and topography. It achieves this by scanning the beam and detecting signals (e.g., SEs and BSEs) generated from electron-sample interactions.
   - **EBL (Electron Beam Lithography): Patterning and Fabrication.** Its primary function is to directly "write" nanoscale patterns onto a substrate. It uses the electron beam to precisely expose a

radiation-sensitive polymer (resist) coated on the substrate, defining a pattern for subsequent processing steps like etching or metal deposition.

B. **Beam parameters and optimization**
- SEM: Optimized for **imaging contrast and resolution**. It typically operates with relatively **low beam currents (picoamperes to nanoamperes)** and accelerating voltages ranging from **0.5 to 30 kV** to avoid damaging the sample while achieving clear imaging.
- EBL: Optimized for **pattern fidelity and exposure dose control**. It requires extremely fine control over beam spot size, current, and dwell time (the time the beam spends on each pixel). It often operates at **higher beam currents (nanoamperes to microamperes)** to expose resist efficiently and may trade some ultimate resolution for critical beam stability.

C. **Control and scanning system**
- SEM: The beam scanning is relatively straightforward. **It rasterized over a rectangular area** with fixed pixel dwell times to form an image. The stage position is generally fixed during imaging.
- EBL: The beam movement is dictated by a dedicated **pattern generator**, not just imaging software. This system interprets CAD/GDSII design files to guide the beam along arbitrary, complex shapes. **Precise dosage control** and synchronization with a **high-precision, continuously moving stage** are critical for stitching together pattern fields over a large area without errors.

D. **Key hardware components**
- SEM: Equipped with **signal detectors** (e.g., for SEs, BSEs, and X-rays), which are essential for creating images and performing compositional analysis.
- EBL: Its defining hardware is the **beam blanker** (to instantly turn the beam on/off) and the **pattern generator**. While EBL systems often include detectors for substrate alignment and inspection, these are secondary to the patterning function.

E. **Substrate and sample preparation**
- SEM: Samples require minimal special preparation. Nonconductive samples may need a thin conductive coating (e.g., gold or carbon) to prevent charging, but many modern SEMs can image uncoated samples at low voltages.
- EBL: Requires substrates to be meticulously coated with an **electron-sensitive resist** (e.g., PMMA, ZEP, or HSQ). The process is incomplete after exposure; the sample must subsequently undergo **chemical development** and further steps (etching and lift-off) to transfer the pattern into the desired material.

F. **Throughput**

- **SEM: Relatively fast.** Acquiring an image involves scanning a defined area, which can be done in seconds to minutes.
- **EBL: Inherently slow and serial.** Because each feature must be exposed point-by-point in a direct-write process, throughput becomes a major limitation for large-area patterning and mass production.

9.4.7.1 Summary and key relationship

- An **SEM is an observation tool** designed for imaging with optimized detectors and scanning systems.
- An **EBL system is a fabrication tool** designed for nanofabrication with specialized hardware for beam control, patterning, and resist processing.

A crucial point of connection is that many commercial EBL systems are built on a **modified SEM platform**. They retain the basic electron column and imaging capabilities but are enhanced with the critical additional components required for lithography: a high-specification pattern generator, a high-precision stage, and a beam blanker. This allows them to align to existing features on a substrate using SEM mode, and to write new patterns.

We have now returned to learn the process of nanofabrication using EBL. The first thing we need for top-down fabrication is a thin film or an engineered stack of thin films. An example of a film stack is shown in Figure 9.6. Following that, we go through various steps to fabricate the film into a desired device architecture.

Figure 9.7 illustrates an EBL-based patterning workflow on a thin-film multilayer (~58 nm total) deposited on a silicon substrate. In (a), the substrate for the stack is prepared and cleaned, establishing a smooth base for subsequent processing; in (b), a **negative resist such as ma-N-2403** is

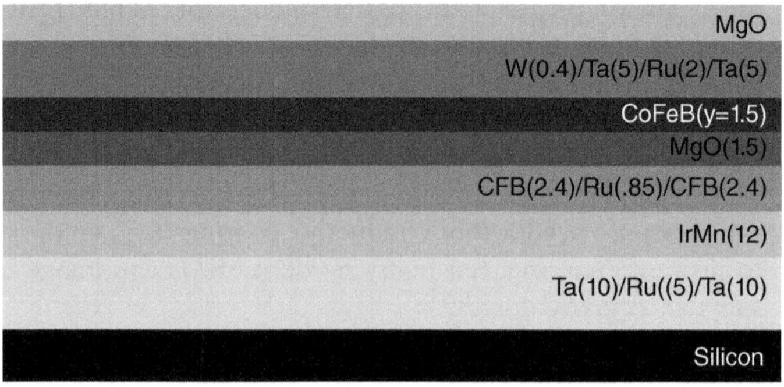

Figure 9.6 A typical stack of thin films deposited layer by layer for magnetic tunnel junction device fabrication.

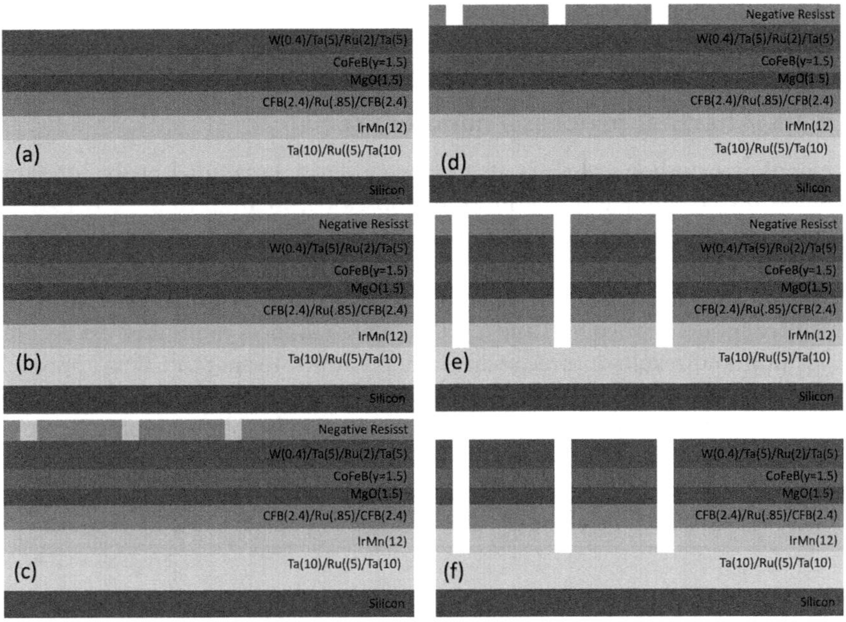

Figure 9.7 (a) An example of thin-film multilayer thin-film sample (~58 nm) in total; (b) a negative resist (ex. ma-N-2403) of a few microns is spun-coated on top of it; (c) a suitable dose of electron beam is exposed through the patterned mask on the sample, which makes the exposed area of the resist loosened from the rest of the resist, and it floats on the substrate surface; (d) a suitable developer (ex. ma-D-525) is used for a suitable time to develop the pattern on the sample surface; (e) a calibrated ion miller is used to etch the sample till the required depth; and (f) the resist is then lifted off from the sample using acetone. A fresh pattern on the sample is thus made.

spin-coated to a thickness of several hundred nanometers to a few micrometers and soft-baked to remove solvent and improve adhesion; in (c), the resist is exposed with a calibrated electron beam through the patterned mask, **causing cross-linking in the exposed regions,** so they remain insoluble during development, while the unexposed regions are dissolved away; in (d), development with a suitable **developer (e.g., ma-D-525)** reveals the patterned resist with well-defined features, where the exposure dose, development time, and agitation are tuned to control the sidewall profile and fidelity; in (e), a calibrated **ion milling (see Section 9.6.2.3)** step etches the substrate to the required depth according to the pattern, using appropriate endpoint detection and etch chemistries to minimize redeposition and preserve pattern integrity; finally, in (f), the remaining resist is stripped with acetone (or another suitable solvent), leaving a clean, fully patterned sample ready for further processing. This sequence showcases EBL's high-resolution, direct-write capability and the end-to-end workflow, from resist coating to pattern transfer, including critical considerations for dose optimization, undercut control, and dimensional metrology.

9.4.8 Thermal process—nano Frazer

NanoFrazor lithography[12] is an advanced thermal scanning probe lithography (tSPL) technique that utilizes a nanoscale heated tip to directly pattern heat-sensitive resist materials, such as polyphthalaldehyde, a commonly used positive-tone resist. A photo of the tSPL system is shown in Figure 9.8. The method operates by locally heating the resist with a tip temperature typically ranging between 800 and 1200 K, which **induces rapid thermal decomposition** and results in precise nanoscale pattern formation. This approach achieves critical dimensions on the order of tens of nanometers, significantly surpassing the resolution limits of conventional photolithography.

While the overall lithographic **workflow—substrate preparation, resist coating, exposure, development, and post-processing**—remains similar to other resist-based techniques (as shown in Figure 9.9), NanoFrazor

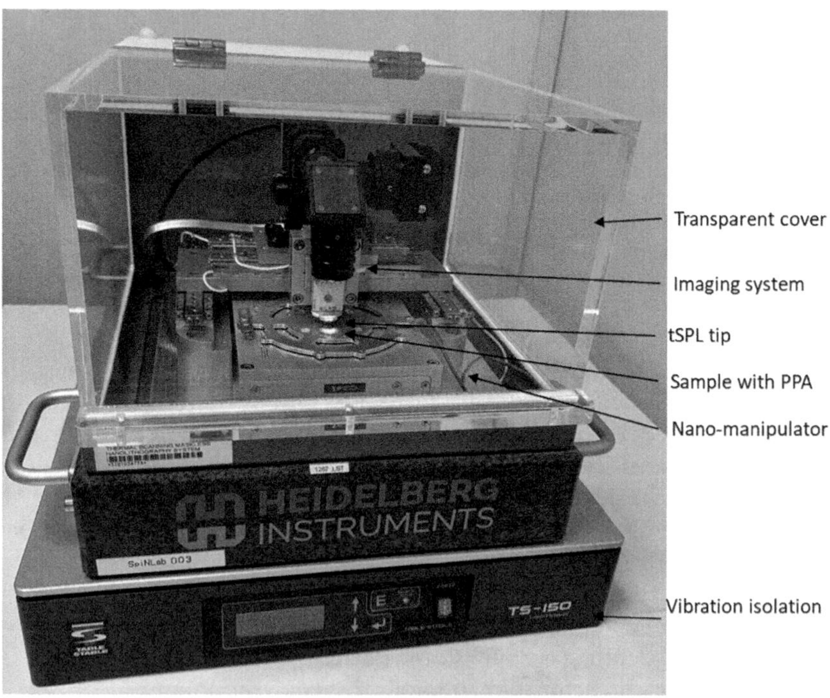

Figure 9.8 Photo of a thermal scanning probe lithography (tSPL) system made by Heidelberg Instruments. Arrows indicate different parts of the machine.

Figure 9.9 (a) A spun-coated PPA (positive resist) layer is patterned using a hot tip. After developing the PPA using a solution (DI water: acetone = 8: 92), SiO_2 is deposited on top of it. (b) An ion beam is used to etch the sample till a required depth (in this case, depth equals to the SiO_2 thickness) on the sample. (c) The resist is then lifted off using acetone, and we get the intended pattern on silicon.

distinguishes itself by employing localized thermal energy delivered via a scanning probe rather than photons or electrons. The tip raster scans the resist surface, heating it pixel by pixel, which causes selective vaporization of the resist in exposed regions. Due to this serial writing mechanism, NanoFrazor's throughput is slower than both photolithography and EBL, but its **direct-write capability** offers unmatched flexibility for rapid prototyping and the fabrication of complex nanostructures.

When compared to other scanning probe lithography (SPL) methods, NanoFrazor offers unique advantages. For instance, mechanical SPL techniques rely on physically scratching or indenting the resist with a stylus tip to pattern features, which can cause resist damage and tip wear, potentially limiting resolution and repeatability. In contrast, NanoFrazor's thermal decomposition process is nonmechanical, **preserving tip integrity and enabling finer, more controlled patterning.** The **tSPL tips** are integrated with thermal actuators, whereas dip-pen nanolithography (DPN) uses an AFM tip to deliver molecular "inks" onto a substrate, enabling chemical patterning but generally limited to softer, organic materials and often lacking the resolution and speed required for semiconductor applications.

Thermal SPL variants that heat the tip without vaporizing the resist, such as thermomechanical lithography, typically rely on softening or melting the resist, resulting in lower resolution and less precise feature edges. NanoFrazor's direct resist cleavage method, utilizing localized **thermal decomposition**, achieves sharper, cleaner features with reduced proximity effects often encountered in EBL.

Furthermore, NanoFrazor operates in ambient or controlled atmospheric conditions, unlike EBL, which requires high-vacuum environments, simplifying equipment design and reducing operational costs. Its gentle thermal process also minimizes substrate damage and allows patterning on heat-sensitive and multilayer substrates, thereby broadening the range of compatible materials.

In summary, NanoFrazor lithography combines the high resolution and direct-write flexibility of SPL with the benefits of nonmechanical, thermally induced resist patterning. It stands out among SPL techniques by offering superior feature definition, minimal tip wear, and compatibility with heat-sensitive materials, making it an ideal choice for nanoscale prototyping and specialized device fabrication, albeit with lower throughput compared to mask-based lithography methods. Recently, Marcus et al. reported single-digit nanometer fabrication using the tSPL technique.[13]

9.5 DEPOSITION TECHNIQUES

Thin films are ultrathin layers of material, typically ranging in thickness from a few nanometers to several micrometers, that are deposited onto a substrate surface to modify its physical, chemical, or electronic properties. Common substrates, including silicon (Si), silicon dioxide (SiO_2), quartz, glass, and various metals, are selected based on the intended application and compatibility with the deposition process. The fabrication of thin films is a critical step in numerous fields such as microelectronics, optics, sensor technology, and surface engineering.

The deposition techniques for thin films can be broadly categorized into three major classes: PVD, CVD, and solution-based methods. Each category encompasses a variety of processes that differ in their underlying principles, operating conditions, material compatibility, and film quality.

PVD involves the physical transfer of material from a solid source to the substrate, often in a vacuum environment. This category includes techniques such as thermal evaporation, electron beam evaporation, and sputtering. In PVD, the source material is vaporized by heating or ion bombardment and then condenses on the substrate to form a thin film. PVD methods offer precise thickness control and yield high-purity films, making them widely applicable for metals, dielectrics, and alloys.

9.5.1 Physical vapor deposition (PVD)

In PVD technique,[14,15] a solid material is vaporized in a vacuum chamber and deposited onto a nearby substrate placed in an optimized position. Some common PVD techniques are as follows.

9.5.1.1 Thermal evaporation

Thermal evaporation is a **PVD** method where a solid material is heated inside a high-vacuum chamber until it vaporizes. The vapor deposits on the nearby cooler substrate, forming a thin film on the substrate. It is one of the oldest and simplest PVD techniques, widely used in research and industry.

Process overview

1. **Material Heating:** The source material (e.g., Al, Au, and Ag) is placed in a resistive boat or crucible (tungsten, tantalum, or graphite).
2. **Vaporization:** High currents are sent through the boat, heating both the boat and the target material beyond its melting point, thus producing vapor.
3. **Deposition:** The vapor condenses on cooler substrates, forming a uniform thin film.

Advantages

a) **High Deposition Rates:** Faster than sputtering for certain metals. Deposition of electrodes of hundreds of nanometers becomes faster.
b) **Simple & Cost-Effective:** Less complex than e-beam evaporation or sputtering.
c) **High-Purity Films:** Minimal contamination due to a vacuum environment.
d) **Good for Low-Melting-Point Materials:** Ideal for Al, Au, Ag, and organic films.

Limitations

a) **Poor Step Coverage:** Line-of-sight deposition leads to uneven coating on complex shapes.
b) **Limited to Low-Melting Materials:** High-melting-point materials (e.g., W and Ta) require e-beam evaporation.
c) **Possible Alloy Segregation:** Different evaporation rates can alter film composition.

Applications

- **Microelectronics** (Al interconnects and organic light – emitting diode (OLED) electrodes)
- **Optical Coatings** (mirrors and antireflective layers)
- **Decorative Films** (gold-like coatings on jewelry)
- **Research and Development** (due to simplicity and low cost)

Thermal evaporation remains a fundamental technique in thin-film deposition technology, especially where simplicity and high deposition rates are prioritized over extreme uniformity or high-temperature compatibility, such as electrode deposition.

9.5.1.2 Electron beam (E-beam) evaporation

Electron beam (e-beam) evaporation is an advanced **PVD** method that uses a focused electron beam to vaporize high-melting-point materials in a high-vacuum environment. This technique is widely used in the semiconductor, optical, and aerospace industries due to its ability to deposit ultra-pure, high-performance thin films.

Process overview

1. **Electron Beam Generation**: A tungsten filament emits electrons, which are accelerated and focused onto the source material using magnetic fields.
2. **Localized Heating**: The high-energy beam melts and vaporizes the target material (e.g., SiO_2, Ti, and W) without contaminating the crucible.
3. **Film Deposition**: The vaporized material condenses onto the substrate, forming a uniform, high-purity coating.

Advantages

a) **High-Melting-Point Compatibility**: It can evaporate refractory metals (e.g., Ta and Mo) and ceramics.
b) **High Purity and Density**: This method minimizes contamination due to crucible-free heating.
c) **Precise Rate Control**: Electron beam power can be finely adjusted for optimal deposition.
d) **Excellent Film Uniformity**: Suitable for demanding applications like optical coatings.

Limitations

a) **High Equipment Cost**: More complex and expensive than thermal evaporation.
b) **Line-of-Sight Deposition**: Poor step coverage on complex 3D structures.
c) **Potential X-Ray Emission**: Requires proper shielding for operator safety.

Applications

- **Semiconductors** (gate oxides and metallization layers)
- **Optical Coatings** (laser mirrors and antireflective films)
- **Superconductors and Aerospace** (high-performance material deposition)
- **Solar Cells and MEMS** (precise thin-film integration)

E-beam evaporation is ideal for applications requiring **high-purity, high-temperature materials**, which makes it indispensable in advanced manufacturing and research.

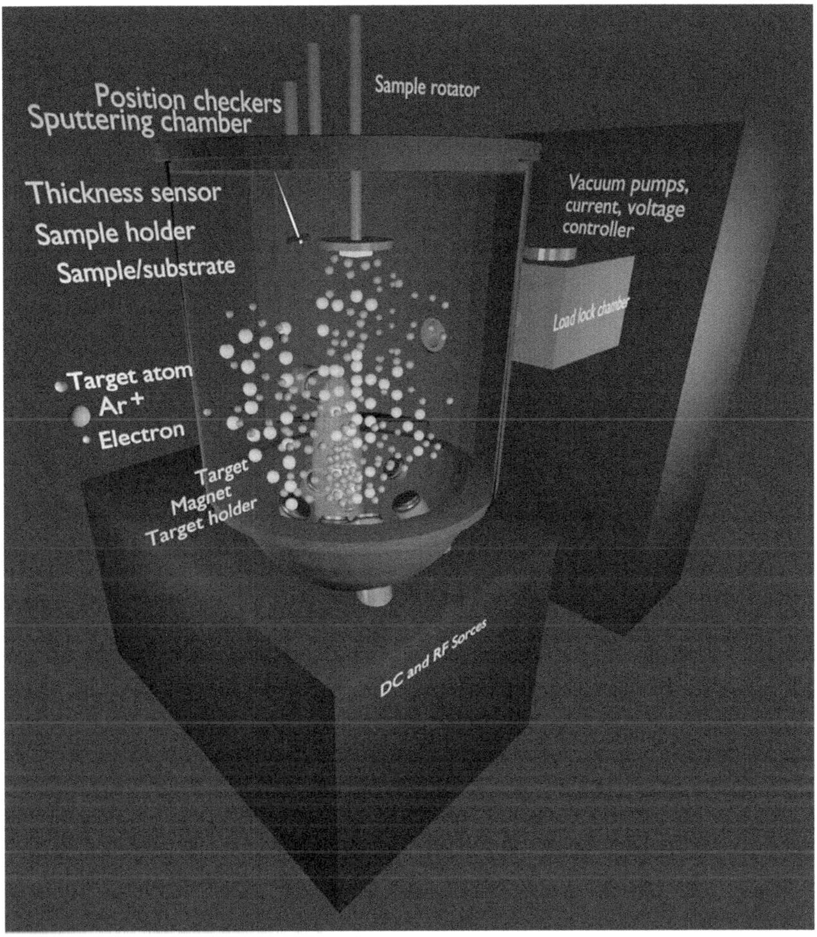

Figure 9.10 The electric field helps in the formation of the plasma state of Ar^+ ions and electrons. The electrons are captured by the magnetic field; heavier Ar^+ ions hit the targets and eject target atoms. Some of the ejected atoms get deposited on the substrate inside the sputtering chamber. The substrate is rotated at optimum speed to make the deposition more uniform. Generally, one target at a time is biased for pure metal deposition.

9.5.1.3 Sputtering

Sputtering[16] is a PVD[14] technique used to grow thin films by ejecting material from a solid target onto a substrate. In a typical sputtering system, an inert gas—most often argon—is ionized in a low-pressure plasma and accelerated toward the target by an applied electric field (Figure 9.10). A schematic of the sputtering system is shown in Figure 9.10. The momentum transfer from these energetic ions ejects atoms (sputtered species) from the

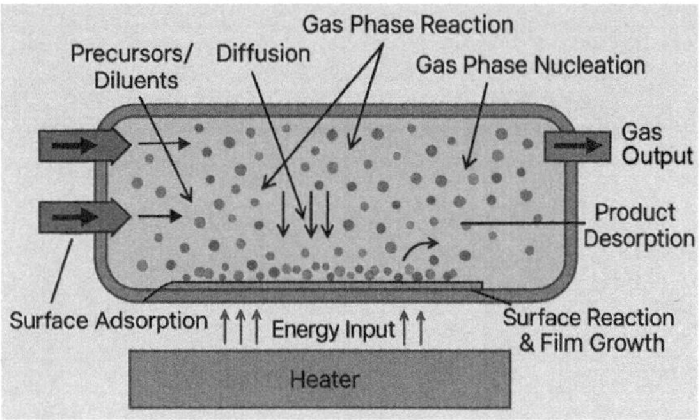

Figure 9.11 Different chemical and physical processes inside the CVD chamber are depicted.

target, which then travel through the vacuum and condense on the substrate to form a uniform film. This process offers precise control over film thickness, composition, and microstructure and is compatible with a wide range of materials, including metals, ceramics, and semiconductors.

Variants of sputtering include direct-current (DC) sputtering for conductive targets, radio-frequency (RF) sputtering for insulating materials, and magnetron sputtering, where magnetic fields confine the plasma close to the target to boost deposition rates and efficiency. Reactive sputtering introduces a reactive gas (e.g., O_2 or N_2) to form compound films such as oxides or nitrides in situ. The technique provides good step coverage, strong film adhesion, and scalability, making it a workhorse for optical, protective, and functional coatings across microelectronics, optics, and surface engineering. Key process parameters—base and working pressure, target-to-substrate distance, power, gas composition and flow, and substrate temperature—control film density, morphology, stress, and stoichiometry, and will be explored in detail in this chapter alongside common configurations, material systems, and representative applications.

9.5.1.3.1 The sputtering process

9.5.1.3.1.1 BASIC PRINCIPLE

Sputtering involves bombarding a target material with high-energy ions (typically argon) in a vacuum chamber. The collision ejects atoms from the target, which then deposit onto a substrate to form a thin film.

Step-by-step mechanism

a. **Vacuum Chamber Setup**: The system is evacuated to **10^{-6}–10^{-9} bar** to minimize contamination. Argon (Ar) gas is then introduced as the sputtering medium.

b. **Plasma Generation**: An applied high-voltage electric field ionizes argon gas, creating a **plasma**. Positively charged argon ions (Ar^+) accelerate toward the negatively charged target (cathode).
c. **Sputtering Ejection**: When Ar^+ ions strike the target, they dislodge (sputter) atoms due to momentum transfer. The ejected atoms travel in a straight line and deposit onto the substrate.
d. **Film Formation**: The sputtered atoms condense on the substrate, forming a thin, uniform coating.

9.5.1.3.2 Types of sputtering techniques

9.5.1.3.2.1 DIRECT-CURRENT MAGNETRON SPUTTERING

When targets are **conductive materials** (metals like Al, Cu, and Ti), DC magnetron sputtering is used. It is called magnetron sputtering because a magnetic field, generated by a permanent magnetic ring placed above the target, traps electrons near the target, increasing ionization efficiency.

Applications: Microelectronics, solar cells, and decorative coatings.

9.5.1.3.2.2 RF (RADIO-FREQUENCY) MAGNETRON SPUTTERING

When targets are **insulating materials** (SiO_2 and Al_2O_3), RF magnetron sputtering is used. Alternating current (13.56 MHz) prevents charge buildup on the target.

Applications: Optical coatings and semiconductor passivation.

9.5.1.3.2.3 REACTIVE SPUTTERING

Reactive sputtering introduces a reactive gas (e.g., O_2 and N_2) to form compound films (e.g., TiO_2 and TiN).

Applications: Hard coatings for cutting tools and wear-resistant films.

9.5.1.3.2.4 HIGH-POWER IMPULSE MAGNETRON SPUTTERING (HIPIMS)

High-power impulse magnetron sputtering (HiPIMS) uses short, high-power pulses to achieve **denser** and **smoother films**.

Applications: Aerospace coatings, medical implants.

9.5.1.3.2.5 ION BEAM SPUTTERING (IBS)

Ion beam sputtering (IBS) uses a separate ion source (rather than plasma) for ultra-precise deposition.

Applications: High-end optics (laser mirrors and telescope coatings).

9.5.1.3.3 Advantages of sputtering

i. **Wide Material Compatibility**: It works with metals, alloys, oxides, nitrides, and more.
ii. **Excellent Film Uniformity**: It is better than thermal evaporation for complex substrates.
iii. **High Adhesion and Density**: It produces durable, pinhole-free coatings.
iv. **Low Substrate Temperature**: It is suitable for heat-sensitive materials (e.g., plastics).
v. **Scalability**: It can be used in both lab-scale and industrial production.

9.5.1.3.4 Limitations of sputtering

i. **Slower deposition rates** compared to thermal evaporation.
ii. **Higher equipment cost because it** requires vacuum systems and power supplies.
iii. **Target utilization issues** due to nonuniform erosion, which can reduce target life.
iv. **Plasma-induced damage** may affect sensitive substrates (e.g., organic films).

9.5.1.3.5 Applications of sputtering

Semiconductor Industry: Metallization (Al and Cu interconnects) and dielectric layers (SiO_2 and Si_3N_4 for insulation).

Optical Coatings: Antireflective films (eyeglasses and camera lenses) and mirror coatings (telescopes and laser systems).

Decorative and Functional Coatings: Wear-resistant coatings (TiN for cutting tools) and decorative finishes (gold-like coatings on watches).

Energy and Aerospace: Solar cell electrodes (indium tin oxide (ITO) for transparent conductive layers) and thermal barrier coatings (jet engine components).

Medical Devices: Biocompatible coatings (Ti and Ta for implants) and antimicrobial films (Ag coatings on surgical tools) (Table 9.2).

Table 9.2 Comparison with other deposition techniques

Feature	*Sputtering*	*Thermal evaporation*	*E-beam evaporation*
Deposition rate	Moderate	Fast	Moderate
Material range	Metals, ceramics, and polymers	Low-melting metals only	High-melting metals
Film uniformity	Excellent	Poor (line-of-sight)	Good
Cost	High	Low	Very high
Best for	Complex coatings	Simple metal films	High-purity refractory films

9.5.1.3.6 *Future trends in sputtering technology*

- **Hybrid Systems** (combining sputtering with CVD or ALD).
- **Green Sputtering** (reducing argon usage with alternative gases).
- **AI-Driven Process Optimization** (real-time plasma monitoring).

9.5.1.3.7 *Conclusion*

Sputtering is a **versatile, high-performance** thin-film deposition method, offering superior film quality and material flexibility. Despite its higher costs and slower deposition rates compared to thermal evaporation, its ability to coat complex geometries and diverse materials makes it indispensable in **advanced manufacturing, optics, and the semiconductor industry**.

9.5.1.3.8 *Key takeaways*

- Works with **metals, oxides, nitrides, and polymers.**
- Produces **high-adhesion, uniform films.**
- Preferred for **semiconductors, optics,** and **wear-resistant coatings.**
- Future advancements focus on **efficiency and hybrid deposition methods.**

9.5.2 Pulsed laser deposition (PLD)

9.5.2.1 *Working principle of pulsed laser deposition (PLD)*

Pulsed laser deposition (PLD)[17] is a sophisticated thin-film growth technique that relies on laser-matter interaction to achieve highly controlled material deposition. The process can be broken down into four fundamental stages, each with distinct physical mechanisms:

9.5.2.1.1 *Laser-target interaction (ablation phase)*

When an intense pulsed laser beam (typically ns pulses with fluences of 1–10 J/cm^2) strikes the target surface, the following processes can happen:

- **Electronic Excitation**: Photons with energy exceeding the material's bandgap (UV lasers like KrF at 248 nm are common) create electron-hole pairs.
- **Thermal and Nonthermal Effects**: For metals, absorption occurs via free electrons; for insulators, it occurs via valence band excitation.
- **Phase Explosion**: The surface rapidly heats to 10,000+ K, forming a superheated melt that explosively vaporizes.

9.5.2.1.2 Plasma plume formation

The ablated material forms a highly directional plasma plume with:

- **Three Distinct Regions:**
  1. **Atomic/ionic front** (fast component, ~10^6 cm/s)
  2. **Molecular/cluster region**
  3. **Macroparticle tail** (source of droplets)
- **Ionization States:** The plume contains ions (1^+ to 3^+), excited neutrals, and electrons with temperatures of 1–100 eV.

9.5.2.1.3 Plume expansion dynamics

The plasma expansion follows:

- **Adiabatic Expansion Model:** The plume evolves in a forward-peaked cosine distribution (n ~ $\cos^n \theta$, where $n \approx 5$–20).
- **Gas Interactions:** When using background gases (e.g., O_2 for oxides):
  - **Scattering Effects:** Reduce plume velocity via collisions.
  - **Reactive Processes:** Enable stoichiometry control (critical for oxide films).

9.5.2.1.4 Film growth on substrate

Deposition occurs through:

- **Sticking Coefficient Effects:** High-energy species (10–100 eV) enhance surface mobility.
- **Nucleation Modes:**
  - **3D Island Growth** (Volmer-Weber) for dissimilar materials.
  - **Layer-by-Layer Growth** (Frank-van der Merwe) for homoepitaxy.
- **Thermodynamic Factors:** Substrate temperature critically affects crystallinity (often 500°C–800°C for oxides).

9.5.2.1.5 Key physics considerations

- **Laser Spot Size:** Determines ablation rate (typically 1–5 mm^2).
- **Pulse Repetition Rate:** Affects deposition rate (1–50 Hz typical).
- **Plasma Screening:** Later pulses interact with preceding plasma.
- **Stoichiometry Preservation:** Resulting from congruent ablation.

This multistage process enables PLD to uniquely transfer complex target compositions directly to films, making it unparalleled for oxide heterostructures and metastable phases. The high kinetic energy (20–100 eV) of deposited species often eliminates the need for post-annealing, a process required by other techniques.

9.5.2.2 *Arc vapor deposition: working principle and process dynamics*

Arc vapor deposition (AVD) is a high-energy PVD technique that utilizes an electrical arc discharge to vaporize material from a solid cathode target. This process is particularly effective for depositing ultra-hard coatings and dense metallic films.

9.5.2.2.1 Core mechanism: vacuum arc discharge

The fundamental process begins with a high-current (50–200 A), low-voltage (20–40 V) DC power supply between electrodes, separated by 1–10 mm in vacuum (10^{-2}–10^{-6} Torr), enabling mechanisms (mechanical or laser) to trigger the arc. When the arc strikes the cathode, a **spot forms,** consisting of microscopic plasma zones (1–10 μm diameter) with current densities exceeding 10^{12} A/m^2. Local temperatures instantaneously reach 8000–15,000 K at the spot. This causes **explosive vaporization** of material at an ejection rate of 10–100 μg/Coulomb.

9.5.2.2.1.1 PLASMA GENERATION CHARACTERISTICS

The arc creates a unique plasma composition:

- **Ionization States**: Typically 1^+–4^+ ions (90% ionization).
- **Particle Energies**: 10–100 eV for most metallic ions.
- **Plume Velocity**: 10^4–10^5 m/s expansion speed.
- **Macroparticle Emission**: 0.1–10 μm diameter droplets (this big size is a key challenge).

9.5.2.2.1.2 CATHODE SPOT DYNAMICS

The arc root moves erratically across the target surface at 10–100 m/s due to the following:

- **Retrograde Motion**: Opposite to conventional current flow direction.
- **Spot Splitting**: Continuous generation of new spots.
- **Crater Formation**: Leaves 1–20 μm diameter pits on the target.

9.5.2.2.1.3 DEPOSITION PROCESS PHYSICS

As the plasma reaches the substrate, an initial 1–2 nm mixed interface layer is formed, which is called ion implantation. The film then grows in either a columnar or a dense structure, depending on the applied bias field. Generally, compressive stress develops in the 1–10 GPa range. Key process control parameters are given in Table 9.3.

Table 9.3 Key process control parameters

Parameter	*Typical range*	*Effect on deposition*
Arc current	50–200 A	Higher current increases deposition rate (0.1–10 μm/min)
Bias voltage	−50 to −1000 V	Controls ion bombardment energy
Pressure	10^{-6}–10^{-2} Torr	Affects plasma scattering
Target-substrate distance	100–500 mm	Determines deposition uniformity

9.5.2.3 *Macroparticle mitigation strategies*

Several approaches address droplet contamination. The most followed methods are as follows:

- **Magnetic Filtering:** 90° bent filters using 10–100 mT fields.
- **Pulsed Operation:** μs–ms pulses reduce spot dwelling time.
- **Shielding Geometries:** Mechanical baffles and apertures.

9.5.2.4 *Reactive arc deposition variant*

When combining with reactive gases (N_2, O_2, and CH_4):

- Gas partial pressures of 10^{-4}–10^{-2} Torr are observed.
- Compound formation occurs both in plasma and on the substrate.
- Requires precise control to avoid target poisoning.

9.5.2.5 *Advantages over other PVD methods*

- **Higher Ionization:** 90% versus 1–5% in sputtering.
- **Greater Adhesion:** Due to high-energy ion bombardment.
- **Better Target Utilization:** Up to 80% versus 30% for sputtering.

9.5.2.6 *Industrial applications*

- **Tool Coatings:** TiN and AlCrN for cutting tools.
- **Aerospace:** Cr-based coatings for turbine blades.
- **Medical:** Ta coatings for implants.
- **Electronics:** Cu interconnects in semiconductors.

Arc deposition enables unique coating properties unattainable by other methods, particularly for applications requiring extreme hardness or perfect stoichiometry in compound films. Ongoing research focuses on better droplet control and scaled-up systems for industrial throughput.

9.5.3 Chemical vapor deposition (CVD)

CVD,[18,19] on the other hand, is a chemical process where volatile precursor gases react or decompose on the heated substrate surface to form a solid thin film. CVD techniques can produce highly conformal coatings with excellent uniformity over complex topographies. Variants of CVD include plasma-enhanced CVD (PECVD), low-pressure CVD (LPCVD), and ALD, each tailored to specific material systems and film properties. CVD is extensively employed for the deposition of semiconductors, oxides, nitrides, and various compound materials. Common CVD techniques are as follows:

9.5.3.1 Atmospheric pressure CVD (APCVD): working principle

Atmospheric pressure chemical vapor deposition (APCVD) is a thin-film deposition technique in which precursor gases react at ambient pressure (approximately 760 Torr) on a heated substrate, typically maintained between 300°C and 1200°C. In this process, volatile precursor gases—such as silane (SiH_4) for silicon films or tetraethyl orthosilicate (TEOS) for silicon dioxide (SiO_2)—are introduced into the reaction chamber along with a carrier gas like nitrogen (N_2) or hydrogen (H_2). The elevated substrate temperature thermally decomposes these precursor molecules into highly reactive species; for example, silane decomposes as $SiH_4 \rightarrow Si + 2H_2$.

These reactive radicals adsorb onto the substrate surface, where they undergo surface reactions and nucleation to form a solid thin film, such as polycrystalline silicon (poly-Si) during silicon deposition. Concurrently, volatile reaction byproducts—such as water vapor (H_2O) or hydrogen chloride (HCl)—are generated and efficiently removed from the deposition zone by the gas flow to prevent contamination and undesired secondary reactions.

APCVD offers several advantages, including a relatively simple and cost-effective reactor design and high deposition rates, making it attractive **for large-area coating** and industrial-scale production. However, operating at atmospheric pressure can lead to challenges that degrade film quality, such as reduced film uniformity across the substrate due to turbulent gas flow and the potential for gas-phase particulate formation or unwanted particle nucleation.

This technique finds widespread applications in the fabrication of solar cell layers, doped oxide films, and epitaxial silicon growth, where its scalability and high throughput are particularly beneficial despite some limitations in uniformity control.

Advantages: Simple setup and high growth rates.
Challenges: Poor uniformity and gas-phase particle formation.
Uses: Solar cells, doped oxides, and epitaxial Si.

9.5.3.2 Low-pressure CVD (LPCVD): working principle

LPCVD is a thin-film deposition technique carried out at sub-atmospheric pressures, typically in the range of **0.1–10 Torr**, and at elevated temperatures between **400°C and 900°C**. The process is performed in a heated, evacuated reactor—often a horizontal or vertical furnace—designed to hold multiple substrates for batch processing.

In LPCVD, gaseous precursor molecules such as silane (SiH_4) or ammonia (NH_3) are introduced into the reactor under controlled vacuum conditions. The reduced operating pressure increases the mean free path of gas molecules, minimizing intermolecular collisions and enabling more uniform precursor distribution throughout the reaction chamber. Under these conditions, mass transport is largely diffusion-dominated, not convection-dominated, which is key to achieving exceptional thickness uniformity across large batches of wafers.

Upon reaching the heated substrate surfaces, the precursor molecules adsorb and react in thermally driven surface reactions. For example, **silane decomposes as: $SiH_4 \rightarrow Si$ (solid) + $2H_2$ (gas)**. This reaction results in the deposition of a solid thin film—such as polysilicon or silicon nitride—while volatile byproducts like hydrogen are continuously removed by the vacuum system to maintain a clean reaction environment.

LPCVD offers several significant advantages: excellent film thickness uniformity, high material purity due to the low contamination risk in a vacuum, and efficient large-scale batch processing. However, it also has some limitations, including relatively slow deposition rates compared to atmospheric pressure processes and a high thermal budget, which may limit compatibility with temperature-sensitive substrates or prefabricated device layers.

LPCVD is widely used in the fabrication of high-quality silicon nitride (Si_3N_4) layers, polysilicon films for gate electrodes and interconnects, and in microelectromechanical systems (MEMS) manufacturing, where film uniformity and conformality are critical.

9.5.3.3 Plasma-enhanced CVD (PECVD): working principle

PECVD is **a low-temperature** thin-film deposition technique that **utilizes a plasma** to dissociate and activate precursor gases, enabling film growth at substrate temperatures typically between 100°C and 400°C. This makes PECVD particularly suitable for temperature-sensitive substrates, such as **polymers or prefabricated semiconductor devices.**

In PECVD, a plasma is generated within the reaction chamber by applying an electric field—commonly RF (13.56 MHz) or DC—across the precursor gases. The applied field accelerates free electrons, giving them kinetic energies in the range of 1–10 eV. These energetic electrons collide with precursor molecules (e.g., silane (SiH_4) and ammonia (NH_3)), breaking their chemical bonds through inelastic collisions. This dissociation produces reactive radicals, ions, and metastable species; for example, **SiH_4 → SiH_3 + H**. These activated species then adsorb onto the substrate surface and undergo surface reactions, such as **SiH_3 + N → SiN*x***. The presence of plasma enhances the chemical reactivity at the substrate surface, promoting film growth even at low temperatures. Ion bombardment, particularly when the substrate is biased, can improve film density, adhesion, and step coverage by rearranging surface atoms during deposition. However, excessive ion bombardment may introduce film stress or damage sensitive substrates, requiring careful process optimization.

PECVD offers several advantages, including low-thermal budget, high deposition rates, and the ability to deposit high-quality dielectric films such as silicon dioxide (SiO_2) and silicon nitride (SiN_x) with good uniformity. However, challenges include controlling intrinsic film stress, preventing plasma-induced substrate damage, and avoiding particle contamination from chamber walls or from flaking films.

PECVD is widely used in applications such as passivation layers for microelectronics, dielectric coatings for solar cells, moisture barriers for flexible electronics, and protective coatings in optoelectronic devices, requiring the essential combination of low-temperature processing and high film quality.

9.5.3.4 Metal–organic CVD (MOCVD): working principle

Metal–organic chemical vapor deposition (MOCVD) is an advanced epitaxial thin-film growth technique that utilizes **metal–organic precursors** to deposit compound semiconductors with precise control over composition, thickness, and crystal quality. The process operates in a heated reactor, typically **at substrate temperatures** ranging from 500°C to 1200°C, and is widely employed for fabricating high-performance **optoelectronic** and high-frequency **electronic** devices.

The process proceeds through several key phases:

i. **Vapor Delivery: Volatile metal–organic precursors**, such as **trimethylgallium (TMGa)** for gallium **or trimethylaluminum (TMA)** for aluminum, are carried into the reaction chamber by inert or reducing carrier gases (H_2 or N_2). Flow rates and precursor ratios are carefully controlled to achieve the desired stoichiometry in the grown film.
ii. **Thermal Decomposition (Pyrolysis)**: As the precursor-laden gas stream approaches the heated substrate, the high temperature induces pyrolytic decomposition of the precursors. For example:
iii. TMGa → Ga (adsorbed) + $3CH_3$. Liberated metal atoms become available for film growth, while organic ligands form volatile byproducts.
iv. **Surface Migration and Epitaxial Growth**: The freshly adsorbed atoms migrate across the substrate surface and align with the underlying crystal lattice through epitaxial growth. This stage is critical for

producing high-quality single-crystal films and heterostructures with atomically abrupt interfaces.

v. **Byproduct Exhaust:** Volatile organic fragments, such as methane (CH_4) and other decomposition products, are swept away from the reaction zone by the carrier gas flow and removed via the exhaust system, ensuring a clean growth environment.

Advantages:

- Atomic-level control over layer thickness and composition.
- Ability to grow complex heterostructures, quantum wells, and superlattices.
- High crystalline quality and scalability for wafer-scale production.

Challenges:

- High capital and operational costs due to sophisticated equipment and ultra-pure precursors.
- Use of hazardous and toxic chemicals, such as arsine (AsH_3) and phosphine (PH_3), requiring stringent safety protocols.

Applications:

MOCVD is the dominant technique for **producing III–V compound** semiconductors, including gallium nitride (GaN) for light-emitting diodes (LEDs) and laser diodes, gallium arsenide (GaAs) for high-efficiency photovoltaic cells, and indium phosphide (InP) for high-speed optoelectronic devices. It is also used for growing multilayer epitaxial structures in high-electron-mobility transistors (HEMTs) and other advanced electronic components.

9.5.3.5 Atomic layer deposition (ALD): working principle

ALD[20] is a vapor-phase thin-film deposition technique that achieves **atomic-scale thickness precision** by exploiting sequential, self-limiting surface reactions. This process allows the controlled growth of films, one atomic layer at a time, exhibiting exceptional conformality over high-aspect-ratio and complex 3D structures.

The ALD cycle consists of four fundamental steps:

1. **Precursor Pulse:** A gaseous metal-containing precursor—such as **trimethylaluminum (TMA)** for aluminum oxide (Al_2O_3)—is introduced into the reaction chamber. The precursor molecules **chemisorb onto available reactive sites on the substrate surface, forming a saturated monolayer**. Due to the self-limiting nature of the reaction, once all

surface sites are occupied, no further precursor adsorption occurs, preventing uncontrolled film growth.

2. **Purge**: An inert carrier gas (e.g., nitrogen or argon) is introduced to **flush out** unreacted precursor molecules and volatile byproducts, preventing gas-phase reactions in the subsequent step.
3. **Reactant Pulse: A counter-reactant**—such as water vapor (H_2O) or ozone (O_3)—is pulsed into the chamber. **This reactant chemically reacts with the chemisorbed precursor layer, completing the formation of the desired monolayer.** For example: TMA + $H_2O \rightarrow Al_2O_3$ (solid) + CH_4 (gas). Again, the reaction is self-limiting, ceasing upon the consumption of all reactive sites.
4. **Final Purge**: A second purge step removes excess reactant gases and gaseous byproducts, restoring a clean environment before the next cycle begins.

Repeating this cycle builds the film in discrete increments, with typical growth rates of 0.1–3 Å per cycle, enabling sub-nanometer thickness control.

Advantages:

- Exceptional conformality and uniformity, even over high-aspect-ratio and nanoporous surfaces.
- Precise thickness control at the atomic scale.
- Superior film quality with low defect density and excellent step coverage.

Challenges:

- Low deposition rates due to the sequential nature of the process.
- Limited by the availability and volatility of suitable precursors.
- Some processes require elevated substrate temperatures, restricting compatibility with certain materials.

Applications:

ALD is widely used for depositing **high-κ dielectric materials** (e.g., Al_2O_3 and HfO_2) in advanced semiconductor devices; **barrier and passivation layers** in MEMS and NEMS; and **nanoscale coatings** for applications in energy storage devices, flexible electronics, and optical coatings. It is also critical for conformally coating 3D NAND flash memory structures and catalytic supports in nanotechnology applications.

9.5.3.6 *Comparison among APCVD, LPCVD, PECVD, MOCVD, and ALD in terms of process regimes*

See Table 9.4.

Table 9.4 A comparison table covering APCVD, LPCVD, PECVD, MOCVD, and ALD—summarizing working principles, conditions, advantages, challenges, and applications

Technique	*Pressure and temperature*	*Working principle*	*Advantages*	*Challenges*	*Typical applications*
APCVD (atmospheric pressure CVD)	~760 Torr 300°C–1200°C	Precursor gases (e.g., SiH_4 and TEOS) flow over a heated substrate at atmospheric pressure; thermal decomposition generates reactive species that adsorb and react to form the film; byproducts removed in the gas flow.	- Simple setup - High growth rates - Scalable for large areas	- Poor uniformity due to turbulence - Gas-phase particle formation	Solar cell coatings, doped oxides, and epitaxial Si
LPCVD (low-pressure CVD)	0.1–10 Torr 400°C–900°C	Precursors (e.g., SiH_4 and NH_3) delivered under vacuum; long mean free path → diffusion-dominated transport; uniform adsorption and surface reaction; byproducts pumped away.	- Excellent uniformity - High purity - Batch processing	- Slow deposition rates - High thermal budget	Si_3N_4 films, polysilicon, and MEMS

PECVD (plasma-enhanced CVD)	0.1–5 Torr 100°C–400 °C	RF/DC plasma (13.56 MHz) ionizes precursors; energetic electrons dissociate molecules (e.g., $SiH_4 \rightarrow SiH_3$ + H); radicals adsorb and react on substrate; ion bombardment can densify film.	- Low-temperature processing - High deposition rates - Good step coverage	- Film stress control - Plasma-induced damage - Particle contamination	SiO_2 and SiN_x dielectrics, solar cells, and flexible electronics
MOCVD (metal–organic CVD)	10–760 Torr 500°C–1200°C	Metal–organic precursors (e.g., TMGa and TMAl) carried by H_2/N_2; thermal decomposition near the substrate releases metal atoms; atoms migrate and align with the crystal lattice (epitaxy); organic byproducts exhausted.	- Atomic-level composition control - Complex heterostructures - High crystalline quality	- High cost - Toxic/hazardous precursors	III–V semiconductors (GaN LEDs, and GaAs PV), HEMTs
ALD (atomic layer deposition)	0.1–1 Torr 50°C–350°C (varies)	Sequential, self-limiting surface reactions in cycles: precursor pulse → purge → reactant pulse → purge; each cycle deposits ~1 atomic layer with perfect conformality.	- Sub-nm thickness control - 100% conformality - Low defect density	- Slow growth rates - Limited precursor options	High-κ dielectrics, conformal nanoscale coatings, 3D NAND, and MEMS

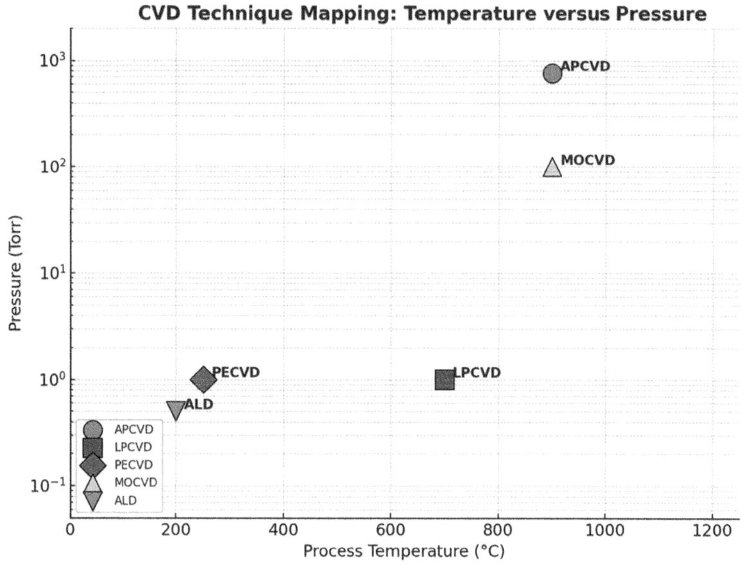

Figure 9.12 A temperature–pressure map showing where each CVD technique typically operates—making it easier to visually compare APCVD, LPCVD, PECVD, MOCVD, and ALD in terms of process regimes.

Figure 9.12 presents a temperature–pressure diagram that highlights the typical operating regimes of common CVD techniques, namely APCVD, LPCVD, PECVD, MOCVD, and ALD. The map shows how each method occupies a characteristic region in T–P space: thermal CVD methods (APCVD and LPCVD) generally require higher temperatures; LPCVD, in particular, often utilizes lower pressures to enhance film quality; PECVD enables deposition at lower temperatures with plasma assistance; MOCVD relies on metal–organic precursors at high temperatures; and ALD operates in a low-pressure, narrow temperature window to achieve highly conformal, self-limiting growth. The diagram also reveals overlaps, indicating regimes where multiple techniques can be suitable, and gaps where process constraints or substrate compatibility favor one method over others. By visualizing these regimes, the figure provides a quick reference for selecting a CVD technique based on desired film properties, deposition rate, conformality, and process limitations.

9.5.4 Solution-based deposition

Solution-based methods, such as sol–gel processing, spin coating, dip coating, and inkjet printing, involve depositing thin films from liquid-phase precursors. These techniques offer cost-effective and scalable routes for fabricating films, especially organic, polymeric, or hybrid materials. While solution-based methods typically operate under ambient or mild conditions, they may require subsequent thermal treatment to remove solvents and consolidate the film.

Common solution-based methods:

- **Spin Coating:** Liquid solution is spun at high speed (e.g., photoresists in lithography).
- **Dip Coating:** The substrate is dipped and withdrawn from a solution (e.g., sol–gel coatings).
- **Spray Pyrolysis:** Solution is sprayed onto a heated substrate (e.g., transparent conductive oxides).
- **Inkjet Printing:** Deposits droplets in precise patterns (flexible electronics).

Applications

- Organic electronics, solar cells, and antireflective coatings.

9.5.5 Electrochemical deposition (electroplating and electroless plating)

- **Electroplating**[21]: Uses electric current to deposit metal ions (e.g., Cu interconnects in chips).
- **Electroless Plating:** Chemical reduction without electricity (e.g., Ni–P coatings).

Applications

- PCB manufacturing and corrosion-resistant coatings.

9.5.6 Other advanced techniques

- **Molecular Beam Epitaxy (MBE):** Ultrahigh vacuum and atomic-level control (for quantum dots, III–V semiconductors)[22].
- **Pulsed Electron Deposition (PED):** Similar to PLD but uses an electron beam.
- **Ion Beam Deposition (IBD):** Ionized atoms are deposited with high precision (optics and tribological coatings) (Table 9.5).

9.5.6.1 Conclusion

Different thin-film deposition techniques are chosen based on the following:

- **Material requirements** (metals, oxides, and polymers)
- **Precision needed** (atomic-level versus micron-scale)
- **Substrate limitations** (temperature sensitivity, size)
- **Cost and scalability**

Table 9.5 Comparison of key deposition techniques

Method	*Thickness control*	*Deposition rate*	*Uniformity*	*Temperature*	*Cost*
Thermal evaporation	Good	Medium	Moderate	Low–medium	Low
Sputtering	Excellent	Medium–high	Excellent	Low–medium	High
PECVD	Good	High	Good	Low	Medium
ALD	Atomic-level	Very slow	Excellent	Low–medium	Very high
Spin coating	Moderate	Fast	Poor (nonuniform edges)	Low	Very low

In summary, the choice of thin-film deposition technique depends on the desired film characteristics, substrate compatibility, material system, and application requirements. Understanding the fundamental mechanisms and advantages of PVD, CVD, and solution-based methods is essential for optimizing thin-film fabrication processes across various technological domains.

9.5.7 Self-assembly

Self-assembly[23] is the process by which molecules or nanoparticles spontaneously organize into predefined structures due to interatomic or intermolecular interactions. This technique leverages the intrinsic properties of the materials involved to build nanodevices without the need for external direction.

Self-assembly is particularly useful for creating complex structures at the nanoscale in an efficient and cost-effective manner.

Some other methods generally used in research are as follows:

9.6 ETCHING TECHNIQUES

Etching[24,25] is a fundamental process in nanofabrication, used to selectively remove materials from a substrate to define highly precise features. It directly determines the resolution, sidewall profile, and functional performances of the final device. Etching processes are broadly classified into wet (chemical) and dry (physical/plasma-based) techniques, each offering distinct advantages and limitations.

9.6.1 Wet etching

Wet etching involves the use of liquid-phase chemicals to dissolve target materials. It can be either isotropic or anisotropic, depending on whether its etching rate is independent of direction or dependent on crystal orientation.

9.6.1.1 *Isotropic etching*

Process: Etchant attacks material uniformly in all directions.
Example: HF solution for SiO_2 removal.
Advantages: Simple, cost-effective, and high selectivity.
Limitations: Poor anisotropy; undercutting is common.
Applications: Sacrificial layer removal and simple pattern transfer.

9.6.1.2 *Anisotropic etching*

Process: Etch rate varies with crystal orientation due to different bond densities.
Example: KOH etches Si(100) ~100× faster than Si(111).
Advantages: Produces well-defined, crystallographically determined features.
Limitations: Restricted to crystalline substrates.
Applications: MEMS fabrication, V-grooves, and pit formation in silicon wafers.

9.6.2 Dry etching

Dry etching uses gas-phase reactions, ion bombardment, or a combination of both to remove material. It offers higher anisotropies and smaller feature sizes compared to wet methods.

9.6.2.1 *Plasma etching/reactive ion etching (RIE)*

Process[25]**:** Reactive gases (e.g., CF_4 for Si and Cl_2 for metals)[26] are activated in plasma, forming volatile byproducts.
Advantages: Better control of feature profiles; scalable to sub-100 nm features.
Applications: CMOS gate etching and shallow trench isolation.

9.6.2.2 *Deep reactive ion etching (DRIE)*

Process: Alternates between SF_6 plasma etching and C_4F_8 polymer passivation (Bosch process).
Advantages: Very high aspect ratios (>30:1) are achievable.
Applications: MEMS structures,= and through-silicon vias (TSVs).

9.6.2.3 *Sputter etching (ion milling)*

Process: Inert ion bombardment (e.g., Ar^+) physically ejects atoms from the surface.
Advantages: Can etch nonvolatile or inert materials.

Limitations: Redeposition and surface damage.
Applications: Patterning magnetic thin films and hard mask removal.

9.6.2.4 Atomic layer etching (ALE)

Process[27]: Alternating self-limiting surface modification and low-energy ion removal.
Advantages: Atomic-scale precision and minimal surface damage.
Applications: Gate dielectric trimming and quantum device fabrication.

9.6.3 Advanced/specialized techniques

9.6.3.1 Focused ion beam (FIB) etching

Process: Direct ion-beam milling (commonly Ga^+), without the need for masks.
Advantages: Maskless and site-specific modifications.
Limitations: Slow, expensive, and potential Ga contamination.
Applications: Device prototyping and transmission electron microscope (TEM) lamella preparation.

9.6.3.2 Electrochemical etching

Process: Electric bias in the electrolyte promotes selective dissolution.
Advantages: Can produce porous structures with high surface area.
Applications: Porous silicon sensors and photonic crystals.

9.6.3.3 Cryogenic etching

Process: Plasma etching at cryogenic temperatures (e.g., −100°C) suppresses sidewall reactions.
Advantages: Smooth, vertical sidewalls without polymer passivation.
Applications: High-aspect-ratio silicon nanostructures (Table 9.6).

Table 9.6 Comparison table of different etching processes

Technique	*Anisotropy*	*Selectivity*	*Resolution*	*Example applications*
Wet isotropic	Low	High	~μm	SiO_2 sacrificial etch
Wet anisotropic	High	Moderate	~μm	MEMS V-grooves
RIE	Moderate	Moderate	<100 nm	CMOS gate etching
DRIE	Very high	Low	<100 nm	MEMS accelerometers
ALE	Ultrahigh	Ultrahigh	Atomic scale	Quantum devices
FIB/ion milling	High	Low	<10 nm	TEM sample prep

9.6.4 Decision tree: wet versus dry etching selection

```
Start
│
├── Is the substrate crystalline?
│   ├── Yes → Need crystallographic control? → Wet anisotropic (KOH, TMAH) .
│   └── No → Wet isotropic (HF, H3PO4).
│
├── Is feature size < 1 µm?
│   ├── Yes → Use dry etching.
│   │   ├── High aspect ratio? → DRIE.
│   │   ├── Atomic precision? → ALE.
│   │   └── General nanofabrication → RIE.
│   └── No → Wet etching preferred.
│
└── Special material or prototyping? → FIB, ion-millingor electrochemical.
```

9.6.5 Emerging trends

- **Directed Self-Assembly (DSA) + Etching:** Using block copolymer patterns as etch masks for sub-10 nm features.
- **Neutral Beam Etching:** Uses neutral radicals instead of charged ions to minimize plasma-induced damage, which is critical for sensitive quantum materials.
- **Cryo-ALE Hybrids:** Combining cryogenic plasma control with atomic layer precision for next-generation device architectures.

9.7 MATERIALS FOR NANODEVICE FABRICATION

The selection of materials for nanodevice fabrication is pivotal, as their performance at the nanoscale is dictated not only by intrinsic bulk properties but also by size-dependent quantum, surface, and interface effects. At reduced dimensions, phenomena such as **quantum confinement**, altered band structures, and enhanced surface reactivity become dominant, enabling functionalities unattainable in conventional microdevices.

9.7.1 Magnetic thin films

Magnetic nanostructures are central to spintronics[28], magnonics, and high-density data storage. Thin films of ferromagnetic materials (Fe, Co, Ni, and their alloys like $Ni_{80}Fe_{20}$—Permalloy) exhibit **tailored magnetic anisotropy, damping, and exchange stiffness at nanometer thicknesses.** Ferrimagnets (e.g., YIG—yttrium iron garnet) are valued for ultra-low magnetic damping, while antiferromagnets (e.g., NiO and Mn_2Au) **enable terahertz spin dynamics.** Rare-earth–transition-metal alloys (e.g., TbFeCo) provide high **perpendicular anisotropy and thermal stability,** essential for heat-assisted magnetic recording. Control over film microstructure, grain boundaries, and interfaces is critical for engineering **spin wave dispersion**, domain wall motion, and spin torque efficiency.

9.7.2 Semiconductors

Traditional semiconductors[29] such as Si, GaAs, and InP remain the foundation of nanoelectronics and optoelectronics. At sub-10 nm scales, quantum confinement alters carrier effective masses and density of states, enabling applications **in quantum dots, nanowires, and photonic nanostructures**. Wide-bandgap semiconductors (e.g., GaN and SiC) offer high breakdown voltages and thermal stability for power electronics, while 2D semiconductors like **MoS_2 exhibit direct bandgaps at monolayer** thickness, crucial for atomically thin transistors. Integration challenges include lattice mismatch, strain management, and surface passivation.

9.7.3 Carbon nanotubes (CNTs) and graphene

CNTs[30]—rolled graphene[31,32] cylinders—exhibit ballistic electron transport, high aspect ratios, and current-carrying capacities exceeding copper. **Single-walled CNTs (SWCNTs)** can behave as either semiconductors or metals depending on **chirality**, enabling transistor channel design at molecular scales. Graphene, with its gapless Dirac dispersion, shows ultrahigh **carrier mobility** (>200,000 $cm^2/V \cdot s$) and excellent **thermal conductivity** (~5000 $W/m \cdot K$), making it ideal for high-frequency electronics, transparent conductors, and thermal management layers. Key challenges include **bandgap engineering for graphene and chirality control for CNTs to achieve uniform device performance.**

9.7.4 Nanoparticles and nanocomposites

Metal nanoparticles (e.g., Au, Ag, and Pt) display localized surface plasmon resonances (LSPR) that can be tuned by their size and shape for sensing, photothermal therapy, and light-harvesting. **Polymer nanocomposites** combine mechanical flexibility with enhanced electrical/thermal conductivity by embedding nanofillers such as CNTs or graphene. **Core–shell** nanoparticles allow multifunctional behavior, for example, **magnetic–plasmonic** hybrids for combined imaging and therapy. In all cases, **surface functionalization** is critical for stability, biocompatibility, and integration into larger device architectures.

9.7.5 Two-dimensional (2D) materials beyond graphene

Transition metal dichalcogenides (TMDs)[33], such as MoS_2, WS_2, and WSe_2, possess strong **spin–orbit coupling, valley-selective optical transitions, and**

thickness-tunable bandgaps. Black phosphorus (BP) offers **high mobility** (~1000 $cm^2/V{\cdot}s$) and **anisotropic conduction,** making it attractive for polarization-sensitive optoelectronics. Hexagonal boron nitride (h-BN) serves as an atomically flat **dielectric with excellent thermal and chemical stability.** The van der Waals nature of 2D materials enables **heterostructures** with atomically sharp interfaces and novel quantum phenomena, free from lattice matching constraints.

9.7.6 Quantum materials

Quantum materials[34], where emergent properties arise from **strong electron correlations or topological band structures,** are redefining device possibilities. Topological insulators (e.g., Bi_2Se_3) support robust surface states immune to backscattering, enabling **dissipationless interconnects.** Weyl semimetals exhibit exotic transport such as **chiral anomaly-induced negative magnetoresistance.** Quantum dots (semiconductor nanocrystals) offer discrete energy levels for single-photon sources and quantum logic operations. These materials are integral to quantum computing, ultra-sensitive metrology, and spin–photon interfaces.

9.8 CHALLENGES IN NANODEVICE FABRICATION

Despite the remarkable progress in nanotechnology, several challenges[35,36] remain in nanodevice fabrication.

9.8.1 Challenges in bottom-up techniques

(e.g., self-assembly, chemical synthesis, DNA origami, vapor–liquid–solid growth)

Limited Spatial Control: While molecular self-assembly and chemical growth can produce nanoscale features with sub-10 nm resolution, the positioning and orientation of these features are often stochastic. Deterministic placement over a wafer-scale substrate is extremely difficult to achieve without hybrid pattern-guiding techniques.

Defect Density: Self-assembly tends to produce grain boundaries, vacancies, or dislocations that severely degrade device performance. Defect reduction over large areas remains an unsolved problem for commercial yields.

Integration with CMOS Processes: Many bottom-up methods involve chemistry, solvents, or processing temperatures that are incompatible with standard semiconductor workflows. This makes direct integration into existing fab lines challenging.

Throughput: Growth or assembly times can be long (hours to days), and batch-to-batch reproducibility is affected by environmental fluctuations (temperature, humidity, and precursor purity).

Material Constraints: Bottom-up growth recipes are often material-specific, limiting flexibility when multiple functional materials must be patterned in a single device.

9.8.2 Challenges in top-down techniques

9.8.2.1 Photolithography

Resolution Limits: Even with DUV and EUV systems, achieving a <20 nm half-pitch requires multiple patterning steps (LELE, SADP, and SAQP), which increases complexity, cost, and defect risk.

Cost: EUV lithography systems cost hundreds of millions of dollars; mask fabrication is slow and expensive. This is a major bottleneck for smaller fabs or research settings.

Mask and Alignment Tolerances: As feature sizes shrink, mask defects and overlay errors become increasingly significant, demanding more precise (and costly) metrology tools.

9.8.2.2 Electron beam lithography (EBL)

Throughput Limitations: EBL writes patterns serially, making wafer-scale patterning impractically slow (hours to days per wafer) and rendering it unsuitable for high-volume manufacturing unless used for mask writing.

Resist Sensitivity: High-resolution resists like HSQ or ZEP520A require very precise process control, and their exposure times increase with higher resolution, which further lowers throughput.

9.8.2.3 Scanning probe lithography (SPL)

Extremely Low Throughput: The tip-by-tip, pixel-by-pixel writing approach (e.g., thermal SPL and DPN) yields unparalleled resolution but is orders of magnitude too slow for wafer-scale production.

Tip Wear and Lifetime: Nanoscale tips degrade with use, altering resolution and pattern fidelity over time.

9.8.2.4 Cross-cutting scalability issues

Process Uniformity: Achieving uniform patterns over 300 mm wafers is challenging for any technique due to local heating, precursor depletion, and nonuniform exposure[37,38].

Yield Loss from Defects: A single particle, mask defect, or alignment error can kill an entire nanoscale device, making high-yield mass production demanding in cleanroom and metrology infrastructure. At the nanoscale, small defects can significantly affect the performance of devices[39]. The film should be as pure as possible. The impurities can create defects in the crystals, and it may cause uneven fabrications. In every step of the resist coating, the bubble formation or uneven thickness can cause unwanted defects in the etching process. Precise control over defect formation and mitigation is crucial to the development of reliable nanodevices.

9.8.2.5 *Device reliability*

With the decrease in the size of devices, their sensitivity to environmental factors, such as temperature, humidity, and radiation, increases. Ensuring long-term reliability and stability is critical for practical nanodevice applications.[40]

Multi-Material Integration: As devices become more complex (e.g., logic + memory + photonics), scalable patterning must handle dissimilar materials without damaging previously fabricated layers. Integrating different materials into functional nanodevices is challenging due to differences in physical properties such as thermal expansion, electrical conductivity, and chemical reactivity. Ensuring these materials work together efficiently requires careful design and fabrication, such as creating exchange bias by depositing antiferromagnetic materials inside PVD chambers.

Environmental and Economic Costs: High-resolution lithography requires ultra-pure chemicals, high-vacuum systems, and significant energy consumption, raising sustainability concerns.

For better clarity, a comparison between bottom-up and top-down nanofabrication processes is given in Table 9.7, considering the scalability issue.

Figure 9.13 presents a resolution–throughput trade-off map that contrasts top-down and bottom-up fabrication approaches. It shows that top-down techniques like photolithography offer **high throughput and uniform patterning over large areas but are fundamentally limited in resolution by optical diffraction and mask geometry**. In contrast, bottom-up methods such as **DNA origami can achieve near-atomic precision and highly complex architectures, but their throughput and scalable area** coverage are orders of magnitude lower due to the kinetics and yield of self-assembly processes. The plot highlights the broad spectrum of intermediate methods that lie between these extremes, underscoring the ongoing tension between **speed**

Table 9.7 A comparison table for bottom-up vs. top-down nanofabrication scalability

Parameter	*Bottom-up techniques (self-assembly, chemical synthesis, VLS growth, and DNA origami)*	*Top-down techniques (photolithography, EBL, and SPL)*
Typical resolution	1–10 nm (atomic precision possible locally)	Photolithography: ~20–50 nm (EUV <20 nm); EBL/SPL: <5 nm possible
Throughput	Low to moderate (hours to days per batch, depending on growth rate)	Photolithography: High (hundreds of wafers/hour); EBL/SPL: Very low (serial process)
Pattern placement control	Poor to moderate (requires templating for deterministic placement)	Excellent (mask-defined, deterministic positioning)
Defect density	High (grain boundaries, vacancies, and stochastic defects)	Low–moderate (mainly from particle contamination or mask defects)
Scalability to wafer level	Limited without hybrid approaches (guiding patterns and templating)	Mature for photolithography; EBL/SPL not scalable without parallelization
Cost per tool	Low to moderate (chemical reactors, ovens)	Photolithography, EBL, and SPL: very high ($1M–$5M)
CMOS compatibility	Often poor (solvent/temperature/material incompatibilities)	Excellent for photolithography; EBL/SPL adaptable
Multi-material capability	Material-specific; difficult to integrate multiple dissimilar materials	Flexible, but requires multistep processes
Environmental impact	Often lower chemical waste, but can have precursor toxicity issues	High-energy use and large chemical waste volumes (photoresists and developers)
Best use case	Ultra-small building blocks (nanowires, quantum dots, and molecular electronics)	High-volume IC production and precise device patterning

and precision in nanofabrication and illustrating why the choice of technique depends critically on the target application—be it large-area manufacturing or ultimate nanoscale control.

9.9 APPLICATIONS OF NANODEVICES

Nanodevices have a wide range of applications across various fields. Some of the most exciting applications include the following:

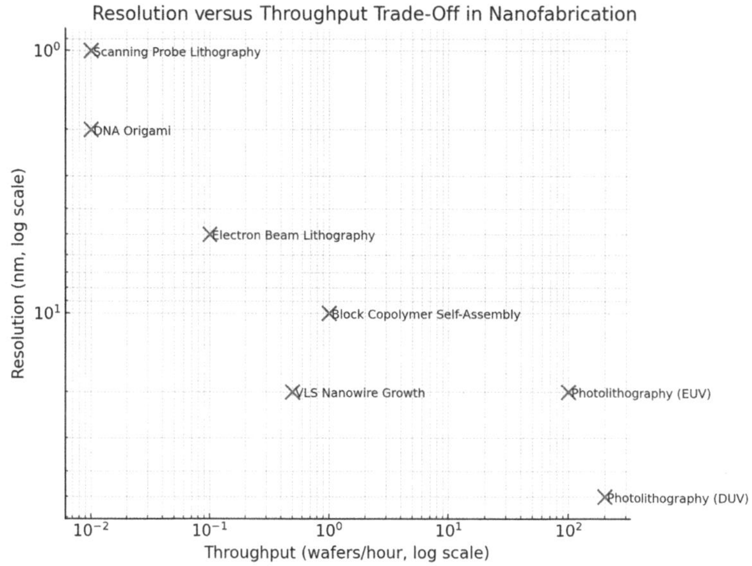

Figure 9.13 A resolution–throughput trade-off map: it shows that while top-down techniques like photolithography excel in throughput but are resolution-limited, bottom-up approaches like DNA origami achieve atomic precision but have orders-of-magnitude lower scalability.

9.9.1 Nanoelectronics

Miniaturization of transistors and other electronic components has led to faster, smaller, and more energy-efficient electronics.[41] Quantum devices and single-electron transistors, topological insulator-based quantum computing devices, are paving the way for future computing technologies.

9.9.2 Nanomedicine

Nanodevices are revolutionizing medicine with applications in drug delivery, diagnostic tools, and biosensors.[42] Nanoscale sensors can detect diseases at an early stage, while drug delivery systems can target specific cells or tissues.

9.9.3 Energy harvesting and storage

Nanodevices are being developed for more efficient energy harvesting and storage solutions,[43] such as high-capacity batteries, supercapacitors, and solar cells.

9.9.4 Sensors and actuators

Nanodevices are used in sensors[44] for detecting chemical, biological, and environmental parameters. Their high sensitivity and small size make them ideal for applications in environmental monitoring, security, and healthcare.

9.9.5 Conclusion

The fabrication of nanodevices is a rapidly evolving field that integrates advanced materials, novel fabrication techniques, and cutting-edge research. While there are numerous challenges in terms of scalability, defect control, and integration, the potential benefits in fields like electronics, medicine, and energy are immense. As technology continues to advance, we can expect nanodevices to play a pivotal role in shaping the future of various industries, leading to new discoveries and innovations that are currently beyond our imagination.[45–47]

9.10 EXAMPLE QUESTIONS AND ANSWERS

9.10.1 Fundamental concepts

Q1: Compare and contrast top-down and bottom-up nanofabrication approaches. Provide examples of techniques for each.

A1:

- **Top-down**: Starts with bulk material and etches/patterns it into nanostructures (e.g., photolithography and e-beam lithography).
- **Bottom-up**: Builds structures atom-by-atom (e.g., CVD, ALD, and self-assembly).
- **Advantages**: Top-down is scalable; bottom-up offers atomic precision.

9.10.2 Lithography techniques

Q2: Explain why EBL achieves higher resolution than photolithography.

A2: EBL uses electrons $(\lambda \approx 0.01\,\text{nm})$ instead of UV light $(\lambda \approx 193\,\text{nm})$, reducing diffraction limits. Electrons can be focused to nanometer-scale spots, enabling sub-10 nm patterning.

Q3: A photolithography system uses light with a wavelength of 193 nm. Calculate the theoretical minimum resolvable feature size (R) using the Rayleigh criterion $(R = k_1\lambda / \text{NA})$, where $k_1 = 0.61$ and $\text{NA} = 0.93$.

A3: $R = (0.61 \times 193\,\text{nm}) / 0.93 \approx \mathbf{126\,nm}$.

9.10.3 Deposition methods

Q4: Why is sputtering preferred over thermal evaporation for depositing alloys?

A4: Sputtering ejects target material stoichiometrically, preserving alloy composition. Thermal evaporation can cause segregation due to differing vapor pressures of components.

Q5: Calculate the deposition rate (D) in sputtering if a 100 nm film is grown in 5 minutes. Express D in nm/s.

A5: $D = 100\,\text{nm} / (5 \times 60\,\text{s}) \approx \mathbf{0.33\,nm/s}$.

9.10.4 Material challenges

Q6: How does ALD achieve atomic-level thickness control?

A6: ALD uses sequential, self-limiting surface reactions. Each cycle deposits one atomic layer by alternating between precursor and reactant pulses, ensuring exact thickness control.

Q7: A thin film has a defect density of $10^6/\text{cm}^2$. If each defect occupies $1\ \text{nm}^2$, what percentage of the film's area is defective?

A7: Defective area $= (10^6 \times 1\,\text{nm}^2) / (10^7\ \text{nm}^2 / \text{cm}^2) = 0.1\,\text{cm}^2$.

%defective $= (0.1\,\text{cm}^2 / 1\,\text{cm}^2) \times 100 = \mathbf{10\%}$.

9.10.5 Applications

Q8: Why are CNTs used in nanoelectronics?

A8: CNTs exhibit high electron mobility ($\sim 10^5\ \text{cm}^2/\text{Vs}$), ballistic transport, and tunable bandgaps, making them ideal for transistors and interconnects.

Q9: A quantum dot has a confinement energy (E) given by $E = \hbar^2\pi^2 / (2mL^2)$, where $m = 0.1m_e$ and $L = 5\,\text{nm}$. Calculate E (use $\hbar = 1.05 \times 10^{-34}\ \text{J}\cdot\text{s}$, $m_e = 9.11 \times 10^{-31}\ \text{kg}$).

A9: $E = (1.05 \times 10^{-34})^2 \times \pi^2 / \left(2 \times 0.1 \times 9.11 \times 10^{-31} \times (5 \times 10^{-9})^2\right)$

$\approx \mathbf{2.4 \times 10^{-20}\ J(0.15\,eV)}$.

9.10.6 Advanced topics

Q10: Describe how HiPIMS improves film density compared to DC sputtering.

A10: HiPIMS uses high-power pulses ($\sim\text{kW/cm}^2$) to generate highly ionized plasmas. Ions bombard the substrate, enhancing adatom mobility and producing denser, smoother films.

Q11: If a PECVD reactor deposits SiO_2 at 10 nm/min, how much time is needed for a 200 nm film? What if the rate drops by 20% due to plasma instability?

A11:

- **Normal:** 200 nm/10 nm/min = **20 minutes**.
- **Reduced Rate:** 200 nm/(0.8 × 10 nm/min) = **25 minutes**.

9.10.7 Practical scenarios

Q12: A resist spin-coated at 3000 RPM has a thickness (t) given by $t = k / \sqrt{\omega}$, where $k = 200\,\text{nm} \cdot \sqrt{\text{RPM}}$. Find t.
A12: $t = 200\,\text{nm} / \sqrt{3000} \approx$ **3.65 nm.**
Q13: Why is EUV lithography critical for sub-7 nm semiconductor nodes?
A13: EUV uses 13.5 nm light, enabling resolution below 10 nm. Traditional UV lithography cannot achieve this due to diffraction limits.

9.10.8 Co-sputtering

Q1: Explain the principle of co-sputtering and how it can be used to deposit alloy films with precise composition control. Provide an example.
A1:

- **Principle**: Co-sputtering uses multiple targets (e.g., Ti and Al) simultaneously in a single sputtering chamber. The composition of the deposited alloy (e.g., TiAl) is controlled by adjusting the power applied to each target, which governs their individual sputtering rates.
- **Example**: To deposit a Ti_3Al_2 film, apply higher power to the Ti target (faster sputtering rate) and lower power to the Al target. The ratio of sputtered atoms determines the final film stoichiometry.
- **Advantage**: Enables tunable alloy compositions without pre-alloyed targets.

MATH PROBLEMS ON CO-SPUTTERING

Problem

In a co-sputtering system, two targets—**Titanium (Ti)** and **Aluminum (Al)**—are used to deposit a **$TiAl_3$** thin film. The sputtering yield of Ti is **0.8 atoms/ion**, and that of Al is **1.2 atoms/ion**. The ion current density on the Ti target is **5 mA/cm²**, and on the Al target is **10 mA/cm²**.

1. **Calculate the deposition rate ratio (Ti:Al) in atoms/cm²·s.**
2. **What power ratio (Ti:Al) should be applied to achieve a stoichiometric $TiAl_3$ film?**
3. *(Assume power is proportional to ion current density and sputtering yield.)*

Given

- Charge of an ion (Ar^+), $q = 1.6 \times 10^{-19}\,\text{C}$
- Avogadro's number is not required (atomic fluxes are in relative terms).

Solution:

1. **Deposition Rate Ratio (Ti:Al)**
   The deposition flux (Γ) of each material is given by:

   $\Gamma = \text{Sputtering yield} \times \text{Ion flux}$

   where **Ion Flux** (J) = Ion Current Density / Charge per ion.
   - For Ti:
   - For Al:
   - Deposition Rate Ratio (Ti:Al):
2. **Power Adjustment for $TiAl_3$ Stoichiometry**
   The desired film is **$TiAl_3$**, meaning the atomic ratio should be **Ti:Al = 1:3**.
   - The **current deposition ratio is 1:3 (Ti:Al)**, which already matches the desired stoichiometry.
   - Since **power is proportional to ion current density × sputtering yield**, the power ratio should be:

Final Answer

1. The deposition rate ratio is **Ti:Al = 1:3**.
2. The power ratio should be **Ti:Al = 1:3** to achieve **$TiAl_3$** stoichiometry.

9.10.8.1 Key takeaways

- Co-sputtering allows **precise composition control** by adjusting target power based on sputtering yields and ion currents.
- **Stoichiometry depends on the following:**
  - Sputtering yield (material-dependent)
  - Ion current density (controlled by power)
- This principle is used for **alloy films** (e.g., TiN and TaAl) in semiconductors and coatings.

9.10.9 Magnetron sputtering

Q2: How does magnetron sputtering enhance deposition rates and film quality compared to conventional DC sputtering? Describe the role of the magnetic field.

A2:

- **Mechanism**: A magnetic field (from permanent magnets behind the target) traps electrons near the target surface, creating a high-density plasma. This increases ionization and sputtering efficiency.

- **Benefits**:
  1. **Higher Deposition Rate**: More Ar^+ ions bombard the target, ejecting material faster.
  2. **Lower Operating Pressure**: Reduced gas scattering improves film purity.
  3. **Better Film Uniformity**: Confined plasma minimizes substrate damage.
- **Example**: Magnetron sputtering is used for **ITO** coatings, where high rates and uniformity are critical for transparent conductive films.

REFERENCES

1. Allan, A. Semiconductor Technology ITRS Roadmap. In *ECWC 10 Conference at IPC Printed Circuit Expo, SMEMA Council APEX and Design Summit 05*; 2005; Vol. 3, pp. 1264–1268.
2. Collins, S. P.; Storrow, A.; Liu, D.; Jenkins, C. A.; Miller, K. F.; Kampe, C.; Butler, J. *Fundamentals of Microfabrication and Nanotechnology*, Third; Madou, M. J., Ed.; CRC Press, 2021.
3. Cui, Z. *Nanofabrication: Principles, Capabilities and Limits*, Second edition; Springer, 2017.
4. Gates, B. D.; Xu, Q.; Stewart, M.; Ryan, D.; Willson, C. G.; Whitesides, G. M. New Approaches to Nanofabrication: Molding, Printing, and Other Techniques. *Chem. Rev.* 2005, *105*, 1171.
5. Chou, S. Nanoimprint Lithography. *Technol. Rev.* 2003, *106*, 42.
6. Orloff, J.; Utlaut, M.; Swanson, L. *High Resolution Focused Ion Beams*; Springer New York, Science+Business Media, LLC, 1985; Vol. 17.
7. Levinson, H. J. Chapter 1: Overview of Lithography. *Principles of Lithography*; Levinson, H. J., Ed.; SPIE Press, 2010; pp. 1–6.
8. Mack, C. *Fundamental Principles of Optical Lithography*; John Wiley & Sons, Ltd., 2007.
9. Benschop, J.; Banine, V.; Lok, S.; Loopstra, E. Extreme Ultraviolet Lithography: Status and Prospects. *J. Vac. Sci. Technol. B Microelectron. Nanom. Struct. Process. Meas. Phenom.* 2008, *26*, 2204.
10. Pease, R. F.; Chou, S. Y. Lithography and Other Patterning Techniques for Future Electronics. *Proc. IEEE* 2008, *96*, 13.
11. Grigorescu, A. E.; Hagen, C. W. Resists for Sub-20-Nm Electron Beam Lithography with a Focus on HSQ: State of the Art. *Nanotechnology* 2009, *20*, 292001.
12. Zheng, X.; Calò, A.; Albisetti, E.; Liu, X.; Alharbi, A. S. M.; Arefe, G.; Liu, X.; Spieser, M.; Yoo, W. J.; Taniguchi, T.; Watanabe, K.; Aruta, C.; Ciarrocchi, A.; Kis, A.; Lee, B. S.; Lipson, M.; Hone, J.; Shahrjerdi, D.; Riedo, E. Patterning Metal Contacts on Monolayer MoS2 with Vanishing Schottky Barriers Using Thermal Nanolithography. *Nat. Electron.* 2019, *2*, 17.
13. Kaestner, M.; Rangelow, I. W. Scanning Probe Lithography on Calixarene towards Single-Digit Nanometer Fabrication. *Int. J. Extrem. Manuf.* 2020, 2, 032005.

14. Thornton, J. A. Influence of Substrate Temperature and Deposition Rate on Structure of Thick Sputtered Cu Coatings. *J. Vac. Sci. Technol.* 1975, *12*, 830.
15. Ohring, M. Chapter 6: Chemical Vapor Deposition. *Material Science of Thin Films*; Ohring, M., Ed.; Academic Press, 2002; pp. 277–355.
16. Kelly, P. J.; Arnell, R. D. Magnetron Sputtering: A Review of Recent Developments and Applications. *Vacuum* 2000, *56*, 159.
17. Eason, R. *Pulsed Laser Deposition of Thin Films*; Eason, R., Ed.; Wiley-Interscience, 2011.
18. Choy, K. L. Chemical Vapour Deposition of Coatings. *Prog. Mater. Sci.* 2003, *48*, 57.
19. Pierson, H. O. *Handbook of Chemical Vapor Deposition CVD. Principles, Technology, and Applications*, Second edition; ScienceDirect, 1999.
20. George, S. M. Atomic Layer Deposition: An Overview. *Chem. Rev.* 2010, *110*, 111.
21. Schlesinger, M.; Paunovic, M., Eds. *Modern Electroplating*, Fifth; John Wiley & Sons, Inc., 2010.
22. Cho, A. Y.; Arthur, J. R. *Molecular Beam Epitaxy*; ScienceDirect, 1975.
23. Scales, S. A.; Whitesides, G. M.; Grzybowski, B.; Griffin, C.; Zumer, S.; Whitesides, G. M.; Grzybowski, B. Linked References Are Available on JSTOR for This Article: Self-Assembly at All Scales. *Science* 2002, *295*, 2418.
24. Williams, K. R.; Muller, R. S. Etch Rates for Micromachining Processing. *J. Microelectromech. Syst.* 1996, *5*, 256.
25. You, A.; Be, M.; In, I. Plasma Etching—A Discussion of Mechanisms. *J. Vac. Sci. Technol.* 1980, *16*, 291.
26. Jansen, H.; Gardeniers, H.; De Boer, M.; Elwenspoek, M.; Fluitman, J. A Survey on the Reactive Ion Etching of Silicon in Microtechnology. *J. Micromech. Microeng.* 1996, *6*, 14.
27. Kanarik, K. J.; Lill, T.; Hudson, E. A.; Sriraman, S.; Tan, S.; Marks, J.; Vahedi, V.; Gottscho, R. A. Overview of Atomic Layer Etching in the Semiconductor Industry. *J. Vac. Sci. Technol.* 2015, *33*, 020802.
28. Parkin, S. S. P. Giant Magnetoresistance in Magnetic Nanostructures. *Annu. Rev. Mater. Sci.* 1995, *25*, 357.
29. Sze, S. M.; Ng, K. K. *Physics of Semiconductor Devices*, Third; Sze, S. M.; Ng, K. K., Ed.; Wiley-Interscience, 2006; Vol. 26.
30. Iijima, S. Helical Microtubules of Graphitic Carbon. *Nature* 1991, *354*, 737.
31. Novoselov, K. S.; Geim, A. K.; Morozov, S. V.; Jiang, D.; Zhang, Y.; Dubonos, S. V.; Grigorieva, I. V.; Firsov, A. A. Electric Field Effect in Atomically Thin Carbon Films. *Science* 2004, *306*, 666.
32. Geim, A. K.; Novoselov, K. S. The Rise of Graphene. *Nat. Mater.* 2007, *6*, 183.
33. Geim, A. K.; Grigorieva, I. V. Van Der Waals Heterostructures. *Nature* 2013, *499*, 419.
34. Hasan, M. Z.; Kane, C. L. Colloquium: Topological Insulators. *Rev. Mod. Phys.* 2010, *82*, 3045.
35. Gordon, E. M. Cramming More Components onto Integrated Circuits. *Scand. J. Stat.* 1998, *86*, 82.
36. The International Roadmap for Devices and Systems. The IRDSTM is Devised and Intended for Technology Assessment Only and Is Without Regard to Any Commercial Considerations Pertaining to Individual Products or Equipment. IRDS 2022: Executive Summary. IEEE. 2022.

37. Barman, A.; Gubbiotti, G.; Ladak, S.; Adeyeye, A. O.; Krawczyk, M.; Grafe, J.; Adelmann, C.; Cotofana, S.; Naeemi, A.; Vasyuchka, V. I.; Hillebrands, B.; Nikitov, S. A.; Yu, H.; Grundler, D.; Sadovnikov, A. V.; Grachev, A. A.; Sheshukova, S. E.; Duquesne, J. Y.; Marangolo, M.; Csaba, G.; Porod, W.; Demidov, V. E.; Urazhdin, S.; Demokritov, S. O.; Albisetti, E.; Petti, D.; Bertacco, R.; Schultheiss, H.; Kruglyak, V. V.; Poimanov, V. D.; Sahoo, S.; Sinha, J.; Yang, H.; Münzenberg, M.; Moriyama, T.; Mizukami, S.; Landeros, P.; Gallardo, R. A.; Carlotti, G.; Kim, J. V.; Stamps, R. L.; Camley, R. E.; Rana, B.; Otani, Y.; Yu, W.; Yu, T.; Bauer, G. E. W.; Back, C.; Uhrig, G. S.; Dobrovolskiy, O. V.; Budinska, B.; Qin, H.; Van Dijken, S.; Chumak, A. V.; Khitun, A.; Nikonov, D. E.; Young, I. A.; Zingsem, B. W.; Winklhofer, M. The 2021 Magnonics Roadmap. *J. Phys. Condens. Matter* 2021, *33*, 413001.
38. Haensch, W.; Nowak, E. J.; Dennard, R. H.; Solomon, P. M.; Bryant, A.; Dokumaci, O. H.; Kumar, A.; Wang, X.; Johnson, J. B.; Fischetti, M. V. Silicon CMOS Devices beyond Scaling. *IBM J. Res. Dev.* 2006, *50*, 339.
39. Ratnesh, R. K.; Goel, A.; Kaushik, G.; Garg, H.; Chandan; Singh, M.; Prasad, B. Advancement and Challenges in MOSFET Scaling. *Mater. Sci. Semicond. Process.* 2021, *134*, 106002.
40. Dubek, K.; Schneidhofer, C.; Dörr, N.; Rojacz, H.; Plank, B.; Schmid, U. Accelerated Lifetime Estimation and Failure Analysis of Micromachined Humidity Sensors under High Temperature and High Humidity Conditions. *Microelectron. Reliab.* 2025, *174*, 115885.
41. Frank, D. J.; Dennard, R. H.; Nowak, E.; Solomon, P. M.; Taur, Y.; Wong, H. S. P. Device Scaling Limits of Si MOSFETs and Their Application Dependencies. *Proc. IEEE* 2001, *89*, 259.
42. Ferrari, M. Cancer Nanotechnology: Opportunities and Challenges. *Nat. Rev. Cancer* 2005, *5*, 161.
43. Goodenough, J. B.; Kim, Y. Challenges for Rechargeable Li Batteries. *Chem. Mater.* 2010, *22*, 587.
44. Wang, Z. L. Nanopiezotronics. *Adv. Mater.* 2007, *19*, 889.
45. Waldrop, M. M. More Than Moore. *Nature* 2016, *530*, 144.
46. Pan, D. Z. Directed Self-Assembly for Advanced Chips. *Nat. Electron.* 2018, *1*, 530.
47. Bhushan, B. *Handbook of Nanotechnology*, Fourth; Bhushan, B. Ed.; Springer, 2017.

Chapter 10

Development in nanoscale device simulation and opportunities in electronic fabrication

B. Mounika and Sresta Valasa

10.1 INTRODUCTION

As semiconductor technology advances to the nanoscale, designing and manufacturing smaller, more efficient electronic devices has become increasingly complex. Traditional complementary metal-oxide semiconductor (CMOS) technology is reaching its physical limits, which means researchers must look for new materials, improved device structures, and advanced fabrication methods to meet performance demands. At such small scales, effects like quantum tunneling, increased leakage currents, and variability in manufacturing can negatively impact device performance [1–5]. Since building and testing physical prototypes is both expensive and time-consuming, researchers now depend on simulation tools to study and optimize device behavior before fabrication.

Over the years, nanoscale device simulation has evolved significantly. Earlier, researchers relied on drift-diffusion (DD) models, but these proved insufficient as device dimensions shrank [6–8]. Today, more advanced techniques, such as quantum transport models, atomistic simulations, and AI-based predictive modeling, allow for a deeper understanding of how electrons behave in nanoscale devices. For example, technology computer-aided design (TCAD) helps researchers simulate how different materials and structures affect transistor performance, simplifying the design of efficient devices before real-world production begins [9–11].

At the same time, fabrication technologies have also made great progress, enabling the production of next-generation semiconductor devices. Researchers are now using wide-bandgap materials like GaN and SiC to improve power and high-frequency applications [12–14]. Two-dimensional materials like MoS_2 and advanced dielectrics are also being explored to push the limits of what is possible in modern electronics [4]. Additionally, new fabrication techniques like extreme ultraviolet (EUV) lithography, atomic layer deposition (ALD), and molecular beam epitaxy (MBE) allow engineers to precisely control material thickness and structure, ensuring better performance and energy efficiency in transistors, memory devices, and even futuristic applications like neuromorphic and quantum computing [15–20].

DOI: 10.1201/9781003635123-10

This chapter will discuss the evolution of nanoscale device simulation, innovations in semiconductor materials, and the latest trends in fabrication technology. It will also highlight how predictive modeling helps researchers reduce costs, speed up development, and improve device performance. By combining simulation with advanced fabrication, scientists and engineers can develop next-generation electronic devices while overcoming the challenges of scaling technology to smaller and more efficient levels.

10.2 EVOLUTION OF NANOSCALE DEVICE SIMULATION TECHNIQUES

The continuous miniaturization of semiconductor devices has revolutionized modern electronics by, pushing transistor dimensions deep into the nanoscale regime. While this has enabled higher integration densities and enhanced device performance, it has also introduced significant challenges related to quantum effects, short-channel effects (SCEs), and process variability [21]. These complexities make traditional trial-and-error fabrication approaches impractical due to high costs and long development cycles. To overcome these challenges, researchers have turned to nanoscale device simulation, a field that has evolved significantly from classical modeling approaches to sophisticated quantum-based, multi-scale, and AI-driven simulation techniques.

10.2.1 Early models for semiconductor devices

In the early days of semiconductor design, device behavior was well understood through models based on classical physics, which treated electron transport as a macroscopic process, governed by drift and diffusion mechanisms. Some of the fundamental models developed during this period include.

10.2.1.1 Drift-diffusion model

The DD model is one of the earliest and most widely used approaches for simulating semiconductor devices [18–21], describing charge carrier transport based on drift and diffusion mechanisms. The drift mechanism refers to the movement of electrons and holes under the influence of an applied electric field, where carriers experience a force that directs their motion, contributing to current flow. In contrast, the diffusion mechanism occurs due to a concentration gradient, causing carriers to move from regions of higher concentration to lower concentration as they strive for equilibrium [22]. Both mechanisms play a crucial role in semiconductor device operation, influencing performance parameters such as conductivity and carrier mobility.

The DD model was effective for bulk metal-oxide semiconductor field effect transistors (MOSFETs) and early CMOS devices, but as device dimensions shrank below 100 nm, it failed to capture quantum mechanical effects and carrier velocity overshoot, leading to the development of more advanced models.

10.2.1.2 Hydrodynamic model

To enhance accuracy, the hydrodynamic model was developed as an extension of the DD model, incorporating carrier energy transport, velocity overshoot effects in short-channel devices, and nonequilibrium transport phenomena [23–25]. This model provided improved predictions for nanoscale devices by accounting for the dynamic behavior of charge carriers beyond the limitations of the DD model [24]. However, despite its advancements, it still relied on semiclassical approximations and could not fully capture purely quantum effects such as tunneling and quantum confinement, which are crucial in ultra-scaled semiconductor devices.

10.2.2 Transition to quantum transport models

With semiconductor devices scaling down to the sub-50 nm regime, classical transport models such as the DD and hydrodynamic models became increasingly inadequate [25, 26]. At these dimensions, quantum mechanical effects began to play a dominant role in carrier transport, necessitating more advanced modeling techniques. Among these effects, quantum tunneling allows electrons to penetrate energy barriers due to their wave-like nature, leading to significant leakage currents that classical models cannot capture [26]. Quantum confinement emerges in ultra-thin channels, where energy levels become discrete rather than continuous, altering carrier behavior and impacting threshold voltage and mobility [27]. Additionally, in ballistic transport, carriers traverse short-channel devices with minimal scattering, thereby making traditional transport assumptions less valid [28]. To accurately describe these phenomena and improve device modeling, researchers developed quantum transport models capable of addressing the limitations of classical approaches.

10.2.2.1 Schrödinger-Poisson solver

One of the fundamental quantum transport methods is the Schrödinger-Poisson (SP) solver, which self-consistently couples the Schrödinger equation with Poisson's equation to determine the electrostatic potential and carrier distribution within a device [29, 30]. The Schrödinger equation governs the quantum mechanical behavior of charge carriers, thereby defining their wavefunctions and discrete energy states, while Poisson's equation describes the electrostatic potential resulting from charge distributions. This approach is particularly useful for modeling advanced transistor architectures such as ultra-thin body (UTB) MOSFETs, Fin-field effect transistors (FinFETs), gate-all-around (GAA) FETs, and 2D material-based devices (e.g., MoS_2, WS_2, and graphene FETs), where quantum confinement effects significantly influence device operation [31–33]. By solving the Schrödinger equation alongside Poisson's equation, the SP method enables accurate calculation of sub-band energies, carrier distributions, and confinement effects in nanoscale devices. This is crucial for predicting threshold voltage shifts, mobility variations, and other performance parameters not captured by classical models.

10.2.2.2 Nonequilibrium Green's function (NEGF) method

The non-equilibrium Green's function (NEGF) method is widely employed for an even more comprehensive quantum transport analysis, particularly in transistors operating in the ballistic or near-ballistic regime. Unlike the SP approach, which primarily addresses quantum confinement, NEGF extends quantum transport modeling by incorporating coherence, tunneling, and scattering effects, thereby making it a powerful tool for analyzing carrier dynamics in ultra-scaled transistors [4–25]. This method is particularly valuable for studying the behavior of tunnel field-effect transistors (TFETs), nanowire transistors, and other emerging quantum devices where conventional transport models are no longer applicable. NEGF formulates charge transport using Green's functions, which describe quantum states and interactions within the device [34–36]. This approach enables precise modeling of nonequilibrium transport, including quantum tunneling, inelastic scattering, and quantum interference, which are critical for understanding the performance of modern nanoscale transistors.

By accurately capturing quantum mechanical effects, both the SP solver and the NEGF method have become indispensable in the design and optimization of next-generation semiconductor devices. As transistor dimensions continue to shrink and quantum effects become more pronounced, these models will play a vital role in advancing nanoscale electronics beyond the limits of classical transport theories.

10.2.3 Multi-scale and atomistic modeling

As semiconductor devices continue to scale down to the nanoscale, their behavior becomes increasingly complex due to quantum mechanical effects, interface interactions, and nanoscale transport phenomena. Traditional simulation approaches that rely on classical physics are no longer sufficient to capture these effects accurately. To address this challenge, researchers have developed multi-scale modeling techniques that integrate different levels of abstraction, ranging from atomic-scale simulations to full-device modeling [37, 38]. This approach ensures a more accurate representation of how materials and devices behave under various operating conditions, bridging the gap between fundamental material studies and practical semiconductor design.

10.2.3.1 Molecular dynamics (MD) simulation

Molecular dynamics (MD) is a powerful atomistic simulation method used to model the physical interactions of atoms and molecules over time [39, 40]. It is particularly valuable for studying material stability under varying temperature and pressure conditions, providing insights into phase transitions and structural integrity. Additionally, MD simulations are crucial for analyzing defect formation and migration in semiconductor materials, which is crucial for understanding reliability and degradation mechanisms. In heterostructures, MD enables the investigation of interfacial interactions between different material layers, shedding light on adhesion, strain effects, and electronic properties [40]. Furthermore, it plays a significant role in exploring thermal transport in nanoscale transistors and memory devices, aiding in the design of efficient heat dissipation strategies to improve device performance and longevity.

For example, in gallium nitride high electron mobility transistors (GaN HEMTs), MD simulations help understand how dislocations and interface defects impact electron transport. Similarly, in 2D materials like MoS_2, MD is used to analyze strain effects and mechanical stability when integrated into flexible electronic devices [41, 42]. By providing insight into atomic-level interactions, MD modeling plays a crucial role in optimizing material properties for better device performance.

10.2.3.2 Density functional theory (DFT) for material-level simulations

Density functional theory (DFT) is a quantum mechanical simulation technique used to predict the electronic properties of materials at the atomic scale. It provides valuable insights into the band structure and density of states of emerging semiconductor materials, aiding in the design of advanced electronic and optoelectronic devices [42]. Additionally, DFT is instrumental in analyzing the dielectric properties of high-k oxides such as HfO_2, ZrO_2, and Al_2O_3, which are critical for improving gate insulation and scaling in MOSFETs and MOSHEMTs [34–42]. It also allows researchers to study defect-induced changes in electronic properties, which directly impact carrier mobility and influence device reliability. Furthermore, DFT is crucial for understanding Schottky and Ohmic contact behavior at metal–semiconductor interfaces, thereby enabling the engineering of efficient contacts for high-performance nanoelectronic and power devices [42–45].

DFT has also been widely used to investigate emerging channel materials for advanced electronic applications. It plays a key role in analyzing the electronic properties of III-nitride alloys such as AlN, GaN, and InN, which are essential for high-power and high-frequency devices. Additionally, DFT enables the study of 2D materials like MoS_2, WSe_2, and graphene, which are promising candidates for next-generation FETs due to their superior carrier mobility and atomic-scale thickness [40–45]. By providing deep insights into band structure, carrier transport, and interfacial interactions, DFT continues to be an invaluable tool in the development of novel materials for future semiconductor technologies. Using DFT, researchers can predict material properties before experimental fabrication, thereby reducing the time and cost of developing new semiconductors.

10.2.3.3 Monte Carlo methods for carrier transport

Monte Carlo (MC) simulations are widely used to model carrier transport in semiconductor devices under realistic operating conditions. Unlike classical models that rely on simplified assumptions, MC methods provide a more accurate representation by accounting for various scattering mechanisms, including phonon, impurity, and interface scattering [46, 47]. These simulations are particularly useful for analyzing nonequilibrium transport in high-field conditions and understanding hot carrier effects in ultra-scaled transistors. By simulating the random motion of electrons based on probability distributions, MC techniques effectively capture complex transport phenomena [48–51]. They have been instrumental in optimizing nanoscale GaN transistors, where factors such as electron velocity overshoot and ballistic transport play a critical role in determining device performance.

10.2.3.4 TCAD simulations: bridging atomistic and device-level modeling

At the device level, TCAD has become the standard simulation framework for semiconductor design, integrating process simulation, device simulation, and circuit-level analysis. TCAD provides a comprehensive toolset for researchers and engineers to analyze the effects of material compositions, gate architectures, and fabrication processes on transistor performance. Popular TCAD platforms like Sentaurus TCAD and Silvaco Atlas are widely used to optimize advanced device architectures such as FinFETs, GAAFETs, and metal-oxide semiconductor high electron mobility transistors (MOSHEMTs). For instance, when designing next-generation MOSHEMTs with different oxide materials, TCAD simulations can predict threshold voltage shifts, leakage currents, and interface trap effects, enabling better optimization before fabrication [51–53]. Additionally, TCAD enables the study of oxide thickness effects and electrothermal performance in high-power radio frequency (RF) devices. TCAD accelerates prototyping by reducing reliance on costly and time-consuming trial-and-error fabrication processes, thereby contributing to more efficient semiconductor design and manufacturing.

10.3 ADVANCES IN NANOSCALE DEVICE SIMULATION

With the increasing complexity of semiconductor devices, simulation techniques have advanced beyond traditional physics-based models. The integration of artificial intelligence (AI), quantum computing, and high-performance computing (HPC) is reshaping how nanoscale device simulations are performed [42–54]. These new methodologies enhance predictive accuracy, accelerate simulations, and enable the exploration of novel device architectures that were previously difficult to model.

10.3.1 Machine learning and AI-driven simulation

One of the most transformative developments in recent years is the application of machine learning and AI-driven simulation. Instead of relying solely on physics-based models, researchers are now training AI algorithms on experimental and simulated datasets to make faster and more accurate predictions about material and device behavior. AI-assisted TCAD models can optimize semiconductor designs by predicting key parameters such as mobility, threshold voltage, and leakage current with high precision [55, 56]. Machine learning is also being used to accelerate MC simulations by reducing the number of computationally expensive iterations required to reach accurate results.

10.3.2 Quantum computing for nanoscale device simulation

Another significant breakthrough is the use of quantum computing for nanoscale device simulation. As semiconductor devices approach atomic-scale dimensions, quantum mechanical interactions become increasingly dominant. Conventional simulation methods struggle to model quantum effects such as electron tunneling and entanglement, which are critical in emerging device technologies like quantum dots and spintronic transistors [57–59]. Quantum computing platforms, such as those developed by IBM and Google, are being explored to perform complex quantum mechanical calculations more efficiently, enabling more accurate modeling of next-generation semiconductor devices.

10.3.3 High-performance computing (HPC) for multi-scale modeling

In addition, HPC has become essential for large-scale multi-physics simulations. With the increasing demand for complex simulations that integrate atomistic, quantum, and continuum models, traditional computing resources often fall short. The use of parallel processing, cloud-based simulation platforms, and GPU acceleration has enabled researchers to perform large-scale device simulations much faster than before [43–46]. For example, multi-scale simulations that would have previously taken weeks to complete can now be run in a matter of hours using HPC clusters. This capability is crucial for designing advanced transistors, memory devices, and optoelectronic components with unprecedented precision.

10.4 APPLICATIONS OF NANOSCALE DEVICE SIMULATION

Over the past decade, nanoscale device simulation has become an essential tool for the design and optimization of advanced semiconductor technologies. With continuous scaling of transistors and the emergence of novel materials, classical TCAD methods have evolved to incorporate quantum mechanical models such as the NEGF approach, multi-scale modeling techniques, and atomistic simulations [60–63]. These advanced methodologies enable accurate predictions of electronic transport, thermal effects, and reliability concerns in modern devices. As a result, nanoscale device simulations are widely applied in various domains, including logic and memory technology, high-frequency power electronics, and optoelectronic devices. One of the most critical applications of nanoscale simulations is in the development of next-generation transistors, such as FinFETs, GAA FETs, and 2D-material-based devices [64–66]. These simulations allow engineers to evaluate device performance under different material and structural configurations, optimizing critical parameters such as subthreshold slope, leakage current, and carrier mobility. In this view, V. R. Kotha et al. [4] reported the variation of substrate dielectric material (SDM) in Monolayer (ML) MoS_2 MOSFETs for digital, analog, RF, and circuit applications using NEGF approach in TCAD. The 2D schematic of the device is shown in Figure 10.1. The study evaluates SDMs ranging from low- to high-k values (SiO_2, Si_3N_4, Al_2O_3, ZrO_2, HfO_2, Ta_2O_5, and TiO_2). Results reveal that higher-k materials like Ta_2O_5 (k=32) enhance DC metrics: I_{on} improves by ~62.18% and I_{on}/I_{off} ratio by ~4× compared to SiO_2, attributed to reduced channel resistivity and improved dielectric screening, as shown in Figure 10.2(a). However, TiO_2(k=80) degrades analog/RF performance, showing a 44.72% reduction in cut-off frequency (f_T) and lower intrinsic gain (A_v), as visualized in Figure 10.2(d) due to increased gate capacitance (C_{gg}) (Figure 10.2(c)) and output conductance (g_{ds}). While higher-k SDMs boost transconductance (g_m), as shown in Figure 10.2(b), TiO_2's excessive C_{gg} and elevated channel resistivity offset these benefits. The analysis concludes that SDMs with k=18–32 strike an optimal balance by enhancing digital performance and sustaining analog metrics, whereas TiO_2 is unsuitable for RF applications. DIBL improvements and subthreshold swing (SS) are also noted for high-k SDMs up to k=32, but degrade with TiO_2 due to increased parasitic resistance.

Additionally, in power electronics, device simulations play a crucial role in designing high-efficiency wide-bandgap semiconductor devices like GaN and

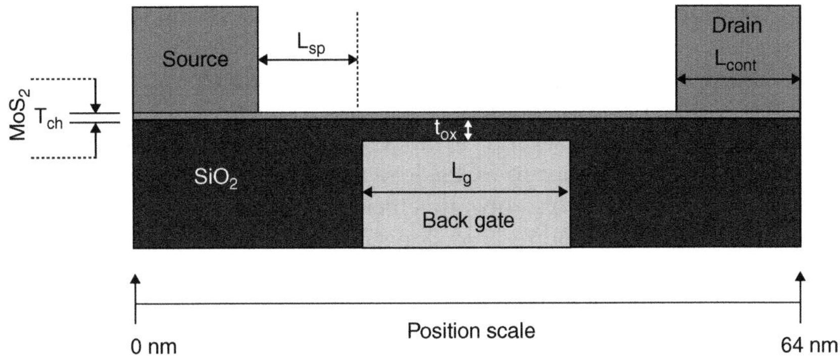

Figure 10.1 2D schematic of ML MoS_2 MOSFET.

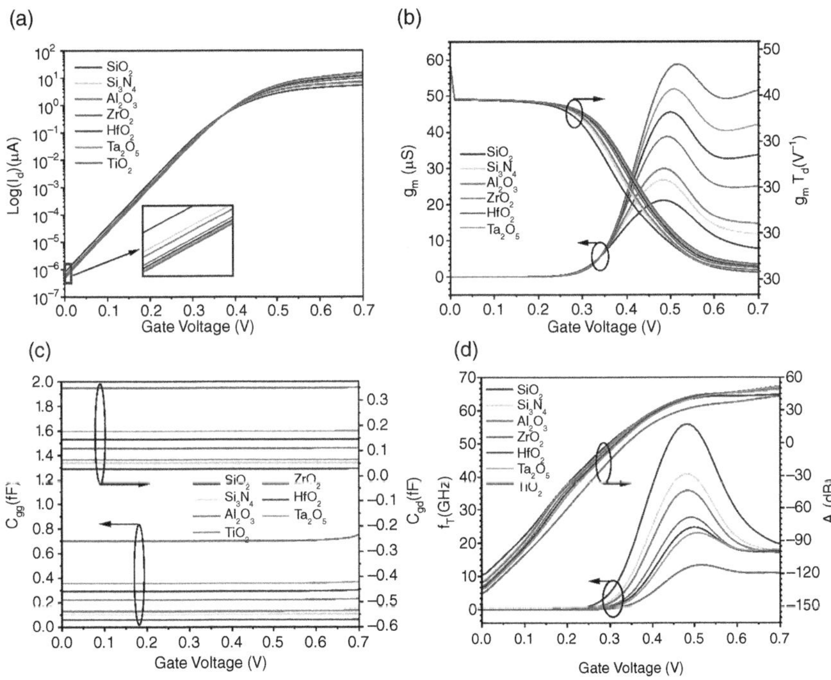

Figure 10.2 DC and analog/RF performance of ML MoS_2 MOSFET. (a) I_d vs V_g characteristics (b) g_m and g_m/I_d (c) C_{gg} and C_{gd} (d) f_T and A_V. (Kotha et al. [4])

SiC MOSFETs, where quantum mechanical effects influence breakdown voltage and thermal stability. Beyond traditional semiconductors, nanoscale simulations are also integral to emerging fields such as quantum computing, flexible electronics, and neuromorphic computing, where precise modeling of quantum interactions and novel materials is required for device functionality.

10.5 OPPORTUNITIES IN ELECTRONIC FABRICATION

Nanoscale devices are rapidly emerging as a strong alternative to traditional bulk transistors, especially in deep submicron technologies. These advanced device structures offer several advantages over conventional MOSFETs, including better resistance to SCEs, lower leakage current, and higher I_{on}/I_{off} ratios [63–66]. They also operate efficiently at lower power supply voltages while achieving subthreshold slopes below 60 mV/decade. One of the key factors driving their development is the availability of modern fabrication techniques, which allow for precise manufacturing with high accuracy. Additionally, the use of laser-based patterning has significantly simplified the creation of intricate nanoscale structures, complementing novel material growth processes to further enhance device performance.

As semiconductor technology progresses, fabrication techniques must achieve a high level of precision to ensure reliable performance. Among these, deposition and growth methods play a key role in forming the essential layers of electronic devices [59–63]. Chemical vapor deposition (CVD), including plasma-enhanced CVD (PECVD) and ALD, enables the controlled formation of ultra-thin films with excellent uniformity and precise thickness. These techniques are widely used to create high-k dielectrics and ultrathin gate oxides in advanced transistor technologies like MOSHEMTs and FinFETs. Physical vapor deposition (PVD), such as sputtering and thermal evaporation, is commonly used for depositing metal contacts and barrier layers [61–66]. Meanwhile, MBE and metal organic chemical vapor deposition (MOCVD) are essential for fabricating IIIV semiconductor devices, especially for high-frequency and power applications where precise layer stacking is required.

On the other hand, lithography techniques help define nanoscale features with high accuracy. Photolithography remains the most widely used method, with advancements like EUV lithography allowing feature sizes below 7 nm, which is crucial for scaling down transistors [57]. For even finer patterns, electron beam lithography (EBL) is used to directly write structures as small as a few nanometers, making it valuable for research and prototype development [61–67]. Nanoimprint lithography (NIL) offers a cost-effective approach for high-resolution patterning, which is useful for emerging technologies such as flexible electronics and photonic devices. Focused ion beam (FIB) lithography, although mainly used for defect repair, is also employed in nanoscale structuring where extreme precision is required [63].

With continuous advancements, new opportunities are emerging in electronic fabrication. The use of 2D materials like graphene and MoS_2 is enabling the development of ultra-thin, flexible, and low-power devices. Additionally, 3D integration and stacking of materials like GaN-on-Si and III-V/Si or SiC substrates are improving performance in high-power and RF applications. AI is also playing a crucial role in fabrication by optimizing processes, detecting defects, and improving manufacturing yield through machine learning. Moreover, the rise of quantum computing and neuromorphic engineering is driving the need for specialized fabrication techniques for spintronic, superconducting, and memristor-based devices, where atomic-scale precision is essential.

10.6 SUMMARY

Continuous advancements in semiconductor technology are enabled by the combined progress in nanoscale device simulation and fabrication techniques. As transistors and other electronic components reach atomic dimensions, traditional design methods are no longer sufficient, necessitating the use of advanced simulation tools to model carrier transport, material properties, and process variations with high accuracy. These simulation techniques, ranging from quantum transport models to multi-scale simulations, have significantly improved the ability to predict device behavior before fabrication, reducing development costs and accelerating innovation. At the same time, fabrication technologies have evolved to meet the demands of miniaturization and performance enhancement. The introduction of wide-bandgap semiconductors like GaN and SiC, two-dimensional materials like graphene and MoS_2, and advanced manufacturing techniques such as EUV lithography and ALD has enabled the production of highly efficient, ultra-scaled devices. These innovations are shaping the future of semiconductor technology, with applications in HPC, power electronics, AI, and quantum computing. By leveraging the synergy between simulation and fabrication, researchers and engineers can overcome the limitations of conventional semiconductor scaling and continue driving technological advancements in the field.

REFERENCES

1. M. R. Baklanov, J. F. De Marneffe, D. Shamiryan, A. M. Urbanowicz, H. Shi, T. V. Rakhimova, H. Huang, and P. S. Ho. Plasma processing of low-k dielectrics. *Journal of Applied Physics*, 113(4), 2013. https://doi.org/10.1063/1.4765297
2. J. Bardeen and W. H. Brattain. The transistor, a semi-conductor triode. *Physical Review*, 74(2), 1948, 230–231. https://doi.org/10.1103/PhysRev.74.230
3. V. M. Donnelly and A. Kornblit. Plasma etching: Yesterday, today, and tomorrow. *Journal of Vacuum Science & Technology, A: Vacuum, Surfaces, and Films*, 31(5), 2013. https://doi.org/10.1116/1.4819316
4. V. R. Kotha, S. Valasa, and N. Vadthiya. Insights into substrate dielectric engineering of monolayer MoS2 FET: Digital/analog/RF perspective to circuit implementation. *IEEE Transactions on Dielectrics and Electrical Insulation*, 31, 2024, C2.
5. C. Acikgoz, M. A. Hempenius, J. Huskens, and G. J. Vancso. Polymers in conventional and alternative lithography for the fabrication of nanostructures. *European Polymer Journal*, 47(11), 2011, 2033–2052. https://doi.org/10.1016/j.eurpolymj.2011.07.025

6. A. Agarwal, P. C. Pradhan, and B. P. Swain. Effects of the physical parameter on gate all around FET. *Sadhana—Academy Proceedings in Engineering Sciences*, 44(12), 2019. https://doi.org/10.1007/s12046-019-1232-8
7. J. A. del Alamo, X. Cai, J. Lin, W. Lu, A. Vardi, and X. Zhao. CMOS beyond Si: Nanometer-scale III–V MOSFETs. *IEEE Bipolar/BiCMOS Circuits and Technology Meeting*, 2017. https://doi.org/10.1109/BCTM.2017.8112904
8. K. Ariga, M. Nishikawa, T. Mori, J. Takeya, L. K. Shrestha, and J. P. Hill. Self-assembly as a key player for materials nanoarchitectonics. *Science and Technology of Advanced Materials*, 20(1), 2019, 51–95. https://doi.org/10.1080/14686996.2018.1553108
9. S. Selberherr, W. Fichtner, and H. Pötzl. MINIMOS – A program package to facilitate MOS device design and analysis. *Numerical Analysis of Semiconductor Devices and Integrated Circuits*, 1, 2016, 275–279.
10. W. R. Frensley. Wigner function model of a resonant tunneling semiconductor device. *Physical Review B*, 36, 1987, 1570–1580.
11. M. V. Fischetti. Master equation approach to the study of electronic transport in small semiconductor devices. *Physical Review B*, 59, 1999, 4901–4917. https://doi.org/10.1103/PhysRevB.59.4901
12. C. S. Lent and D. J. Kirkner. The quantum transmitting boundary method. *Journal of Applied Physics*, 67 (10), 1990, 6353–6359.
13. S. Datta. Nanoscale device modeling: The Green's function method. *Superlattices Microstructures*, 28, (4), 2000, 253–278.
14. L. Kadanoff and G. Baym, *Quantum Statistical Mechanics*. New York: W. A. Benjamin Inc., 1962.
15. J. M. Soler et al.. The SIESTA method for ab initio order N materials simulation. *Journal of Physics: Condensed Matter*, 14, (11), 2002, 2745. https://doi.org/10.1088/0953-8984/14/11/302
16. J. Taylor, H. Guo, and J. Wang. Ab initio modeling of quantum transport properties of molecular electronic devices. *Physical Review B*, 63, 2001, 245407. https://doi.org/10.1103/PhysRevB.63.245407
17. M. Brandbyge, J.L. Mozos, P. Ordejón, J. Taylor, and K. Stokbro. Density functional method for nonequilibrium electron transport. *Physical Review B*, 65, 2002, 165401. https://doi.org/10.1103/PhysRevB.65.165401
18. S. Smidstrup et al. Quantumatk: An integrated platform of electronic and atomic scale modelling tools. *Journal of Physics: Condensed Matter*, 32, (1), 2019, 015901. https://doi.org/10.1088/1361-648X/ab4007
19. M. G. Pala, P. Giannozzi, and D. Esseni. Unit cell restricted Bloch functions basis for first principle transport models: Theory and application. *Physical Review B*, 102, 2020, 045410. https://doi.org/10.1103/PhysRevB.102.045410
20. A. A. Mostofi, J. R. Yates, Y.S. Lee, I. Souza, D. Vanderbilt, and N. Marzari. wannier90: A tool for obtaining maximally localised Wannier functions. *Computer Physics Communications*, 178, (9), 2008, 685–699. [Online]. Available: https://www.sciencedirect.com/science/article/pii/S0010465507004936
21. M. Calderara, S. Brück, A. Pedersen, M. H. BaniHashemian, J. VandeVondele, and M. Luisier. Pushing back the limit of ab initio quantum transport simulations on hybrid supercomputers. *Proceedings of the International Conference for High Performance Computing, Networking, Storage and Analysis*, 2015, pp. 1–12.

22. N. Mingo and L. Yang. Phonon transport in nanowires coated with an amorphous material: An atomistic Green's function approach. *Physical Review B*, 68, 2003, 245406. https://doi.org/10.1103/PhysRevB.68.245406
23. J. Dai and Z. Tian. Rigorous formalism of anharmonic atomistic Green's function for three-dimensional interfaces. *Physical Review B*, 101, 2020, 041301. https://doi.org/10.1103/PhysRevB.101.041301
24. C.H. Wang and J. M. Taylor. Landauer formulation of photon transport in driven systems. *Physical Review B*, 94, 2016, 155437. https://doi.org/10.1103/PhysRevB.94.155437
25. Y. Lee, D. Logoteta, N. Cavassilas, M. Lannoo, M. Luisier, and M. Bescond. Quantum treatment of inelastic interactions for the modeling of nanowire field effect transistors. *Materials*, 13, (1), 2020, 60. [Online]. Available: https://www.mdpi.com/19961944/13/1/60
26. M. Luisier and G. Klimeck. Atomistic full band simulations of silicon nanowire transistors: Effects of electron phonon scattering. *Physical Review B*, 80, 2009, 155430. https://doi.org/10.1103/PhysRevB.80.155430
27. T. Frederiksen, M. Paulsson, M. Brandbyge, and A.P. Jauho. Inelastic transport theory from first principles: Methodology and application to nanoscale devices. *Physical Review*, 75, 2007, 205413. https://doi.org/10.1103/PhysRevB.75.205413
28. U. Aeberhard. Photovoltaics at the mesoscale: Insights from quantum kinetic simulation. *Journal of Physics D: Applied Physics*, 51, (32), 2018, 323002. https://doi.org/10.1088/1361-6463/aacf74
29. J. Cao, S. Fiore, C. Klinkert, N. Vetsch, and M. Luisier. Light matter interactions in Van Der Waals photodiodes from first principles. *Physical Review B*, 106, 2022, 035306. https://doi.org/10.1103/PhysRevB.106.035306
30. N. Mingo. Anharmonic phonon flow through molecular sized junctions. *Physical Review B*, 74, 2006, 125402. https://doi.org/10.1103/PhysRevB.74.125402
31. Y. Guo, M. Bescond, Z. Zhang, M. Luisier, M. Nomura, and S. Volz. Quantum mechanical modeling of anharmonic phonon-phonon scattering in nanostructures. *Physical Review B*, 102, 2020, 195412. https://doi.org/10.1103/PhysRevB.102.195412
32. Y. Asai. Nonequilibrium phonon effects on transport properties through atomic and molecular bridge junctions. *Physical Review B*, 78, 2008, 045434. https://doi.org/10.1103/PhysRevB.78.045434
33. F. Ducry, J. Aeschlimann, and M. Luisier. Electrothermal transport in disordered nanostructures: A modeling perspective. *Nanoscale Advances*, 2, 2020, 2648–2667. https://doi.org/10.1039/D0NA00168F
34. N. Mori, G. Mil'nikov, J. I. Iwata, and A. Oshiyama. RSDFTNEGF quantum transport simulation of ultrasmall field effect transistors. *Proceedings of the IEEE Electron Devices Technology and Manufacturing Conference*, 2020, pp.1–3.
35. C. Klinkert, A. Szabó, C. Stieger, D. Campi, N. Marzari, and M. Luisier. 2D materials for ultra-scaled field effect transistors: One hundred candidates under the ab initio microscope. *ACS Nano*, 14, (7), 2020, 8605–8615. https://doi.org/10.1021/acsnano.0c02983
36. Z. Ren, R. Venugopal, S. Goasguen, S. Datta, and M. Lundstrom. nanoMOS 2.5: A two dimensional simulator for quantum transport in double gate MOSFETs. *IEEE Transactions on Electron Devices*, 50, (9), 2003, 1914–1925.

37. H. R. Khan, D. Mamaluy, and D. Vasileska. Quantum transport simulation of experimentally fabricated nanoFinFET. *IEEE Transactions on Electron Devices*, 54, (4), 2007, 784–796.
38. M. Luisier, A. Schenk, W. Fichtner, and G. Klimeck. Atomistic simulation of nanowires in the sp d s 3 5* tight binding formal ism: From boundary conditions to strain calculations. *Physical Review B*, 74, 2006, 205323. https://doi.org/10.1103/PhysRevB.74.205323
39. J. Chang, L. F. Register, and S. K. Banerjee, Ballistic performance comparison of monolayer transition metal dichalcogenide MX2 (M = Mo, W; X = S, Se, Te) metaloxide semiconductor field effect transistors. *Journal of Applied Physics*, 115, (8), 2014, 084506.
40. Y. Lee, S. Fiore, and M. Luisier. Ab initio mobility of single layer MoS2 and WS2: Comparison to experiments and impact on the device characteristics. *Proceedings of International Electron Devices Meeting*, 2019, pp. 24.4.1–24.4.4.
41. A. A. Laturia, M. L. Van de Put, and W. G. Vandenberghe. Generation of empirical pseudopotentials for transport applications and their application to group IV materials. *Journal of Applied Physics*, 128, (3), 2020, 034306.
42. L. A. Agapito, S. Alkis, J. L. Krause, and H. P. Cheng. Atomistic origins of molecular memristors. *Journal of Physical Chemistry C*, 113, (48), 2009, 20713–20718. https://doi.org/10.1021/jp9060347
43. M. Palsgaard, T. Gunst, T. Markussen, K. S. Thygesen, and M. Brandbyge. Stacked Janus device concepts: Abrupt pn junctions and cross plane channels. *Nano Letters*, 18, (11), 2018, 7275–7281.
44. K. S. Thygesen and A. Rubio. Conserving GW scheme for nonequilibrium quantum transport in molecular contacts. *Physical Review B*, 77, 2008, 115333. https://doi.org/10.1103/PhysRevB.77.115333
45. A. N. Ziogas, T. BenNun, G. I. Fernández, T. Schneider, M. Luisier, and T. Hoefler. A datacentric approach to extreme scale ab initio dissipative quantum transport simulations. *Proceedings of the International Conference for High Performance Computing, Networking, Storage, and Analysis*, 2019, pp. 1–13.
46. P. Suveetha Dhanaselvam, and N. B. Balamurugan. Performance analysis of fully depleted triple material surrounding gate (TMSG) SOI MOSFET. *Journal of Computational Electronics*, 13(2), 2014, 449–455. https://doi.org/10.1007/s10825-014-0554-3
47. A. Talukdar, A. K. Raibaruah, and K. C. Deva Sarma. Dependence of electrical characteristics of Junctionless FET on body material. In *Procedia Computer Science*, (Vol. 171, pp. 1046–1053), eds. S. M. Thampi, S. Madria, X. Fernando, R. Doss, S. Mehta, and D. Ciuonzo, Elsevier B.V., 2020. https://doi.org/10.1016/j.procs.2020.04.112
48. L. S. Tan, and M. Hong. Laser nano-patterning for large area nanostructure fabrication. *International Journal of Optomechatronics*, 2(4), 2008, 382–389. https://doi.org/10.1080/15599610802438425
49. Y. Wang, T. Li, and H. Yang. Nanofabrication, effects and sensors based on micro-electro mechanical systems technology. *Philosophical Transactions of the Royal Society A: Mathematical, Physical and Engineering Sciences*, 371 (2000), 2013. https://doi.org/10.1098/rsta.2012.0315
50. Y. R. Wang, S. M. Olaizola, I.S. Han, C. Y. Jin, and M. Hopkinson. Direct patterning of peri odic semiconductor nanostructures using single-pulse

nanosecond laser interference. *Optics Express*, 28(22), 2020, 32529–32539. https://doi.org/10.1364/OE.397709

51. J. Widiez, J. Lolivier, M. Vinet, T. Poiroux, B. Previtali, F. Dauge, M. Mouis, and S. Deleonibus. Experimental evaluation of gate architecture influence on DG SOI MOSFETs performance. *IEEE Transactions on Electron Devices*, 52(8), 2005, 1772–1779. https://doi.org/10.1109/TED.2005.851824
52. J. L. Wilbur and G. M. Whitesides. Self-assembly and self-assembled monolayers in micro and nanofabrication. In *Nanotechnology* (pp.331–369). Springer, 1999. https://doi.org/10.1007/978-1-4612-0531-9_8
53. Q. Xie, M. H. Hong, L. H. Van, and T. C. Chong. Nano-patterning by laser interference lithography. *5th Pacific RIM Conference on Lasers and Electro-Optics*, 2003. https://doi.org/10.1109/CLEOPR.2003.1277242
54. X. Yan, S. Huang, Y. Wang, Y. Tang, and Y. Tian. Bottom-up self-assembly based on DNA nanotechnology. *Nanomaterials*, 10(10), 2020, 1–14. https://doi.org/10.3390/nano10102047
55. J. Yang, K. Kim, Y. Lee, K. Kim, W. C. Lee, and J. Park. Self-organized growth and self assembly of nanostructures on 2D materials. *FlatChem*, 5, 2017, 50–68. https://doi.org/10.1016/j.flatc.2017.07.004
56. L. Yang, J. Wei, Z. Ma, P. Song, J. Ma, Y. Zhao, Z. Huang, M. Zhang, F. Yang, and X. Wang. The fabrication of micro/nano structures by laser machining. *Nanomaterials*, 9(12). 2019. https://doi.org/10.3390/nano9121789
57. P. D. Ye. Main determinants for III V metal-oxide-semiconductor field-effect transistors (invited). *Journal of Vacuum Science & Technology, A: Vacuum, Surfaces, and Films*, 26(4), 2008, 697–704. https://doi.org/10.1116/1.2905246
58. Y. Yoshida, K. Ohnishi, Y. Matsuo, and S. Watanabe. Design of a patterned nanostructure array using a nanosecond pulsed laser. *AIP Advances*, 8(4), 2018. https://doi.org/10.1063/1.5027060
59. D. H. Yun, B. S. Shin, J. H. Park, Y.-W. Ma, C. Y. Gwak, D. B. You, and B. Kim. 360 nm Continuous wave laser-based contact or non-contact laser interference nano lithography. *Journal of Nanoscience and Nanotechnology*, 20, 2020, 128–134. https://doi.org/10.1166/jnn.2020.17261
60. N. F. Za'bah, K. S. K. Kwa, L. Bowen, B. Mendis, and A. O'Neill. Top-down fabrication of single crystal silicon nanowire using optical lithography. *Journal of Applied Physics*, 2012, 024309. https://doi.org/10.1063/1.4737463
61. F. Zahoor, F. A. Hussin, F. A. Khanday, M. R. Ahmad, I. M. Nawi, C. Y. Ooi, and F. Z. Rokhani. Carbon nanotube field effect transistor (Cntfet) and resistive random access memory (RRAM) based ternary combinational logic circuits. *Electronics (Switzerland)*, 10(1), 2021, 1–20. https://doi.org/10.3390/electronics10010079
62. M. Zhang, Y. Guo, J. Zhang, J. Yao, and J. Chen. Simulation study of the double-gate tunnel field-effect transistor with step channel thickness. *Nanoscale Research Letters*, 15(1), 2020. https://doi.org/10.1186/s11671-020-03360-7
63. S. Zhang. Review of modern field effect transistor technologies for scaling. *Journal of Physics: Conference Series*, 1617(1), 2020. https://doi.org/10.1088/1742-6596/1617/1/012054
64. S. Zhang, X. Ma, J. Zhang, R. Chen, Y. Pan, C. Yu, L. Dong, Y. Wei, and G. R. Arce. Fast optical proximity correction based on graph convolution network. *Proceedings of SPIE—The International Society for Optical Engineering* (Vol. 11613), 2021. https://doi.org/10.1117/12.2583773

65. Y. Zhang, A. Zubair, Z. Liu, M. Xiao, J. Perozek, Y. Ma, and T. Palacios. GaN FinFETs and trigate devices for power and RF applications: Review and perspective. *Semiconductor Science and Technology*, 36(5), 2021, 054001. https://doi.org/10.1088/1361-6641/abde17
66. F. Zhou and H. Liu. Direct nanofabrication using DNA nanostructure. *Methods in Molecular Biology* (Vol. 1500, pp. 217–235), eds. Ke, Yonggang, Humana Press Inc., 2017. https://doi.org/10.1007/978-1-4939-6454-3_15
67. T. Zou, B. Zhao, W. Xin, Y. Wang, B. Wang, X. Zheng, H. Xie, Z. Zhang, J. Yang, and C. Guo. High-speed femtosecond laser plasmonic lithography and reduction of graphene oxide for anisotropic photoresponse. *Light: Science & Applications*, 9(1), 2020. https://doi.org/10.1038/s41377-020-0311-2

Chapter 11

An overview of nanoscale device modeling and requirements in the semiconductor industry

E. Ramola and J. Charles Pravin

11.1 INTRODUCTION

The semiconductor industry's relentless pursuit of device miniaturization has pushed conventional complementary metal-oxide-semiconductor (CMOS) technologies to their physical and performance limits, especially as feature sizes approach and surpass the 10 nm threshold. This transition has catalyzed a shift in modeling paradigms from classical to quantum-aware and multiscale device modeling approaches. Early works on drift-diffusion (DD) and hydrodynamic (HD) models provided reliable approximations for micrometer-scale transistors, which were primarily governed by classical transport and bulk effects. However, as channel lengths decreased, these models lost accuracy due to quantum confinement, tunneling, ballistic transport, and short-channel effects (SCEs) [1].

One of the most significant breakthroughs in nanoscale device modeling is the adoption of the nonequilibrium Green's function (NEGF) formalism, which allows a quantum-mechanical description of carrier transport under nonequilibrium conditions. Datta's foundational work [2] demonstrated the effectiveness of NEGF in modeling nanotransistors by accurately capturing tunneling and confinement phenomena in 1D and 2D systems. NEGF has since become a standard tool in quantum transport simulations for devices like fin field-effect transistors (FinFETs), tunneling FETs (TFETs), and heterojunction devices. However, its computational expense has driven research toward simplified approximations, hybrid models, and hardware-accelerated solutions.

Another critical modeling approach is Monte Carlo (MC) simulation, which offers a stochastic solution to the Boltzmann transport equation (BTE). It effectively accounts for scattering events and nonlocal transport phenomena, making it suitable for high-field and ultra-scaled devices. Jungemann et al. [3] extended this technique to incorporate full-band-structure effects and multi-valley transport in III–V and Si-based materials. However, MC simulations are also computationally intensive and require sophisticated statistical handling to achieve convergence.

DOI: 10.1201/9781003635123-11

To bridge quantum and classical models, researchers have developed quantum-corrected DD and HD models by integrating effective potential or density-gradient (DG) terms to mimic quantum behavior without requiring full quantum solutions. These models offer a trade-off between computational efficiency and physical accuracy, making them ideal for large-scale circuit simulations where full NEGF solutions are impractical [4].

The emergence of novel device architectures like FinFETs, gate-all-around (GAA) FETs, nanosheet FETs, and vertical nanowire transistors has further complicated the modeling landscape. These 3D architectures require robust mesh generation, anisotropic mobility modeling, and surface scattering effects, which are not easily captured in traditional 2D modeling frameworks. Device simulators such as Synopsys Sentaurus and Silvaco Atlas have incorporated enhanced 3D capabilities and quantum corrections to support these needs.

Material innovation has also played a transformative role. The use of high-k dielectrics (e.g., HfO_2), metal gates, and strained silicon improved electrostatics and mobility in short-channel devices. More recently, two-dimensional (2D) semiconductors such as MoS_2, WS_2, and black phosphorus have emerged, offering superior electrostatics, atomically thin channels, and potential for heterogeneous integration. Modeling these materials requires atomistic simulation methods like tight-binding, density functional theory (DFT), and molecular dynamics, often interfaced with transport solvers to study their device performance [5].

In parallel, variability and reliability modeling have gained prominence due to the stochastic nature of nanoscale fabrication and long-term degradation mechanisms. Fluctuations in line-edge roughness, oxide thickness, and dopant placement can drastically affect threshold voltage, drive current, and subthreshold swing. Advanced statistical modeling methods, including principal component analysis, Gaussian process regression, and MC variability extraction, are used to capture these effects. Moreover, reliability concerns such as bias temperature instability (BTI), random telegraph noise (RTN), hot-carrier injection (HCI), and time-dependent dielectric breakdown (TDDB) are modeled using time-domain stress simulations and aging-aware compact models [6].

As the demand for rapid simulation grows, the integration of machine learning (ML) and data-driven modeling has begun to revolutionize device modeling. ML models can approximate device behavior based on large-scale simulation or measurement data, which enables real-time prediction of device characteristics, yield estimation, and design optimization. Recent studies have demonstrated the use of neural networks, support vector machines, and Gaussian processes in learning compact models from physical simulation data [7]. However, these models often lack physical interpretability and require careful training and validation to ensure generalizability.

Compact modeling, critical for circuit-level simulation, has also evolved to reflect nanoscale physics. Models such as BSIM-CMG (for multi-gate FETs) and ASM-HEMT (for GaN HEMTs) are widely adopted, offering parameterized, physics-based representations of modern devices. These models are calibrated using Technology Computer-Aided Design (TCAD) simulations and experimental data, ensuring both accuracy and performance in analog, radio frequency (RF), and digital design flows [8].

Finally, the semiconductor industry's roadmap toward sub-5 nm nodes, heterogeneous integration, and 3D packaging underscores the increasing need for multiscale and multi-domain modeling that combines quantum, electrical, thermal, and mechanical analyses. Cross-disciplinary efforts between academia, industry, and electric design automation (EDA) tool providers are enabling new platforms capable of capturing the full device-to-system spectrum, from atomic-scale effects to chip-level performance.

11.2 LIMITATIONS OF CONVENTIONAL MODELS IN NANOSCALE DEVICE MODELING

The aggressive downscaling of semiconductor devices, particularly beyond the 20 nm technology node, has revealed the inadequacies of classical device modeling techniques, once considered robust and reliable for micrometer-scale technologies. The foundational models in this category—namely the DD and HD approaches—are built upon assumptions that no longer hold true at the nanoscale, where quantum effects dominate, and classical physics alone fails to describe carrier dynamics accurately.

11.2.1 Drift-diffusion model: scope and shortcomings

The DD model has long been the cornerstone of semiconductor device simulation due to its simplicity and computational efficiency. It couples the Poisson equation with the continuity equations for electrons and holes, and describes current densities using classical drift and diffusion terms. These models assume local quasi-equilibrium, isotropic mobility, and frequent carrier scattering, thereby ensuring diffusive transport characteristics.

However, as device dimensions shrink below 50 nm, these assumptions break down. In particular, the notion of local equilibrium is challenged by nonlocal field effects and carrier velocity overshoot. Surface and interface scattering, which intensifies in thin-body and ultra-scaled devices, reduces effective mobility and introduces phenomena not accounted for in the DD model. Moreover, the DD formulation is unable to incorporate quantum-mechanical effects such as wavefunction confinement, source-to-drain tunneling, or gate leakage currents through ultrathin dielectrics [1].

11.2.2 Hydrodynamic models: improvements and limitations

To address some of the limitations of DD models, HD models introduce energy transport equations and consider carrier temperature and energy relaxation. These models offer better accuracy in high-field regions and in describing hot-carrier effects. Nonetheless, they are still derived from the semiclassical BTE and fail to capture the inherently quantum nature of electron transport in nanoscale devices.

For example, in devices where the channel length is comparable to the electron mean free path, the transport regime becomes quasi-ballistic. Under such conditions, carriers do not fully thermalize within the device, and the HD approximation of carrier temperature becomes physically invalid. Moreover, HD models are unable to address phase-coherent transport or quantum confinement, which are increasingly relevant in nanowire, FinFET, and GAA architectures [9].

11.2.3 Quantum confinement and tunneling effects

One of the most significant limitations of classical models is their inability to capture quantum confinement effects. In ultrathin body devices or materials with atomically thin channels (e.g., 2D semiconductors), charge carriers are confined in one or more dimensions, altering the density of states and subband structure. This impacts threshold voltage, mobility, and capacitance—parameters critical to device performance and reliability.

Additionally, classical models neglect quantum tunneling, which becomes particularly important in short-channel devices. Direct tunneling, from source to drain or through the gate dielectric leads to increased off-state leakage currents, thereby affecting both static power consumption and device reliability. These phenomena require quantum transport models, such as the NEGF formalism, to be accurately described [2].

11.2.4 Short-channel effects and ballistic transport

With continued scaling, SCEs such as drain-induced barrier lowering (DIBL), threshold voltage roll-off, and increased subthreshold swing become more pronounced. These effects arise due to the inability of the gate to maintain full electrostatic control over the channel, a phenomenon poorly captured in conventional models.

Moreover, in sub-20 nm nodes, a substantial portion of electrons traverse the channel without undergoing scattering, marking the onset of ballistic or quasi-ballistic transport. Conventional models that rely on mobility and scattering-limited drift currents are unable to capture this nonequilibrium behavior. Ballistic transport models or full-band MC simulations are required for an accurate description of carrier dynamics in this regime [10].

11.2.5 The need for nonlocal and quantum-aware models

Classical models are inherently local, meaning they assume carrier behavior at a point depends only on the properties at that point. In nanoscale devices, this assumption fails due to quantum-mechanical effects such as wavefunction penetration into barriers, coherence, and interference. These effects are nonlocal and require spatial correlations and wavefunction-based modeling for accurate prediction.

Quantum-corrected models—such as DG approaches or quantum DD equations—attempt to introduce these effects in a simplified manner. However, for rigorous analysis, quantum transport solvers based on NEGF or Wigner functions become indispensable [11]. These frameworks inherently account for phase coherence, quantum tunneling, and energy quantization, making them essential tools for modeling modern transistor technologies.

The transition from classical to quantum regimes in device operation necessitates a fundamental shift in modeling paradigms. While DD and HD models provided the foundation for decades of semiconductor development, their limitations in the nanoscale era have become evident. Addressing these requires the integration of quantum transport theories, nonlocal phenomena, and multiscale frameworks, setting the stage for more accurate, predictive, and physically consistent device simulations.

11.3 QUANTUM-MECHANICAL MODELING APPROACHES

As device dimensions continue to shrink below 10 nm, the role of quantum-mechanical phenomena in determining carrier transport behavior becomes increasingly dominant. Classical models, which assume particle-like behavior of electrons, cannot adequately represent effects such as tunneling, quantum confinement, and phase coherence. To overcome these limitations, various quantum-mechanical modeling approaches have been developed to explicitly consider the wave nature of electrons, the discrete nature of energy levels, and the nonequilibrium conditions in nanoscale devices. This section explores foundational and advanced methods used in quantum-mechanical device modeling, emphasizing the NEGF formalism and Schrödinger–Poisson (SP) solvers.

11.3.1 The need for quantum transport models

In nanoscale transistors, where channel lengths are comparable to or shorter than the mean free path of carriers, quantum effects dominate transport behavior. Phenomena such as:

- Quantum confinement, where carrier motion is restricted in one or more dimensions.
- Direct source-to-drain tunneling, which increases subthreshold leakage currents.
- Wave interference, leading to conductance oscillations or resonances.
- Quantum capacitance, impacting device charging behavior, is either missed or inaccurately predicted by classical models.

To faithfully model these effects, a quantum transport formalism capable of describing the nonlocal, phase-coherent, and statistical behavior of carriers is required. This necessity is addressed primarily by two modeling paradigms: (i) self-consistent Schrödinger–Poisson (SP) solvers, and (ii) the nonequilibrium Green's function (NEGF) formalism.

11.3.2 Schrödinger–Poisson solvers

The SP approach is one of the earliest quantum corrections implemented in device simulation. It involves the following:

- Solving the time-independent Schrödinger equation to obtain discrete energy levels and wavefunctions in confined structures (such as quantum wells or inversion layers).
- Coupling these results self-consistently with the Poisson equation to determine the electrostatic potential distribution.

This method precisely models quantum confinement and its corresponding quantized energy levels, which are essential for characterizing advanced architectures like FinFETs, SOI MOSFETs, and GAA FETs. Nonetheless, the SP approach lacks an inherent mechanism for transport modeling. By assuming an equilibrium state or relying on indirect current approximations, its predictive accuracy is compromised in high-bias or far-from-equilibrium scenarios.

Despite these limitations, SP solvers are widely used for calibrating threshold voltage shifts, capacitance–voltage characteristics, and effective mass models in compact device simulations [12].

11.3.3 Nonequilibrium Green's function (NEGF) formalism

To overcome the limitations of SP solvers and model nonequilibrium, quantum-coherent transport, the NEGF formalism has emerged as the gold standard in nanoscale device simulation. Introduced into semiconductor modeling by Datta and others [2], NEGF provides a rigorous, atomistic, and fully quantum-mechanical framework for computing current, charge, and potential distribution under an applied bias.

The NEGF formalism involves the following:

- Solving Green's functions for the device Hamiltonian, which describes the probability amplitude of an electron propagating from one point to another.
- Incorporating contact self-energies that represent the effect of source and drain terminals.

- Modeling scattering via appropriate self-energy terms, allowing inclusion of phonons, impurities, or interface roughness.
- Computing observable quantities such as current and charge density directly from the Green's functions.

NEGF facilitates the modeling of various phenomena, including:

- Ballistic and quasi-ballistic transport.
- Tunneling (both direct and band-to-band).
- Quantum interference and resonant tunneling.
- Sub-band occupation and quantum capacitance.

Furthermore, NEGF's compatibility with tight-binding and effective mass Hamiltonians highlights its versatility for both material and device-level modeling. It has been widely used to simulate diverse architectures, such as silicon nanowires, TFETs, FinFETs, carbon nanotube FETs (CNTFETs), and 2D semiconductor transistors [13].

Despite its accuracy, NEGF is computationally intensive, especially in 2D and 3D device geometries. This has motivated research into mode-space reduction, recursive algorithms, hardware acceleration, and hybrid approaches that combine NEGF with semiclassical models to achieve system-level scalability.

11.3.4 Quantum transport in emerging devices

NEGF and similar quantum transport solvers are also essential for evaluating the performance limits and switching characteristics of emerging devices such as the following:

- TFETs, which exploit band-to-band tunneling.
- Negative capacitance FETs (NC-FETs), where quantum charge behavior interacts with ferroelectric materials.
- Spintronic and topological devices, where quantum spin and surface states play dominant roles.
- 2D material transistors, where monolayer thickness and low dielectric screening require atomistic modeling.

Quantum-mechanical modeling approaches represent a fundamental shift in how nanoscale devices are analyzed and designed. By accurately capturing phenomena such as quantum confinement, tunneling, and phase-coherent transport, techniques like the SP method and NEGF formalism enable predictive and physically accurate simulations for modern and emerging transistor architectures. While computational demands remain a challenge, ongoing research in numerical optimization and hardware acceleration

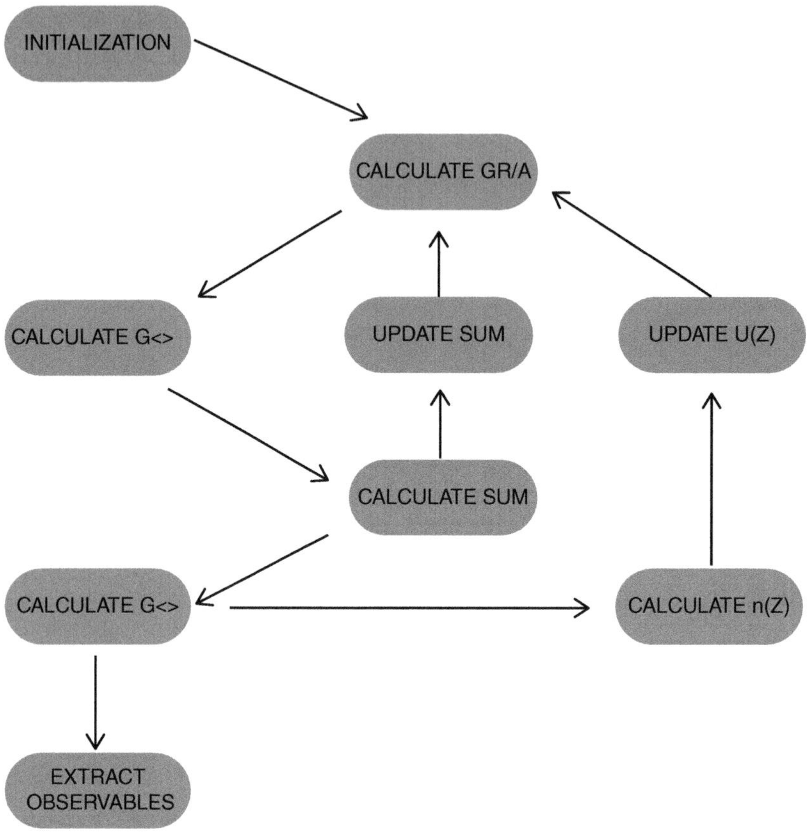

Figure 11.1 Self-consistent NEGF algorithm for Green's function and observable calculation.

continues to extend their applicability to larger, more complex device structures (Figure 11.1).

11.4 MONTE CARLO SIMULATIONS IN NANOSCALE DEVICE MODELING

11.4.1 Basics of solving the Boltzmann transport equation (BTE)

The BTE is a fundamental equation describing carrier transport in semiconductors, accounting for spatial and momentum evolution of the carrier distribution function $f(r,k,t)$ under external forces and scattering events [14, 15]. In its general form, the BTE is expressed as follows:

$$\partial f / \partial t + V_k \cdot \Delta r f + F / \Delta k f = (\partial f / \partial t)_{coll}$$

where V_k is the group velocity, F is the external force (e.g., due to electric fields), and the right-hand term represents changes due to collisions/scattering.

Since direct analytical solutions of the BTE are generally impractical for realistic semiconductor devices, the MC method is used as a statistical, particle-based numerical solution technique [1, 14]. This method tracks a large ensemble of carriers in phase space, where deterministic free-flight motion is interspersed by stochastic scattering events. The free-flight time is determined from the total scattering rate, and the type of scattering event is chosen based on relative probabilities. For nanoscale devices, MC simulations accurately capture nonequilibrium transport effects such as velocity overshoot, ballistic transport, and hot-carrier dynamics, which cannot be fully addressed by DD or HD models [15, 16].

11.4.2 Full-band and multi-valley Monte Carlo simulations

11.4.2.1 Full-band Monte Carlo

The full-band MC approach incorporates the complete energy–momentum $(E-k)$ relationship obtained from numerical band-structure calculations using empirical pseudopotential or ab-initio methods [1, 17]. This ensures accurate modeling of high-field transport where carriers occupy non-parabolic regions of the band structure. Such modeling is essential in compound semiconductors (e.g., GaAs and GaN) and emerging wide-bandgap materials (e.g., SiC), where parabolic approximations are insufficient.

11.4.2.2 Multi-valley Monte Carlo

In materials such as silicon and GaAs, the conduction band features multiple valleys (local minima) in momentum space. The multi-valley MC approach explicitly accounts for inter-valley scattering, enabling accurate modeling of high-field transport phenomena. For example, in silicon, carriers may scatter from the central Γ\Gamma valley to higher energy side valleys (Δ\Delta, L), which affects mobility and saturation velocity [16, 17].

Hybrid approaches combining full-band and multi-valley MC methods use realistic band structures to determine valley positions, effective masses, and scattering rates, improving predictive capability for device performance [15, 17].

11.4.3 Strengths and limitations

MC simulations offer a physically rigorous framework capable of capturing microscopic transport phenomena, including ballistic motion, velocity overshoot, and hot-carrier effects, which are inadequately described by DD or HD models [14, 15]. The method can be applied to materials with complex or highly non-parabolic band structures without relying solely on simplified effective mass approximations, thereby making it highly versatile for compound semiconductors and emerging materials [1]. MC simulations are particularly valuable for investigating high-field and short-channel transport, because they can explicitly incorporate various scattering mechanisms such as phonon, impurity, alloy disorder, and surface roughness scattering [16].

However, these advantages come at the cost of high computational demands, as large particle ensembles and long simulation times are required to achieve statistically converged results [14]. The inherent statistical noise in the particle-based method often necessitates ensemble averaging to improve accuracy [16]. Additionally, classical MC methods do not inherently capture quantum-mechanical effects like tunneling and quantum confinement, which must be introduced via quantum-corrected approaches [17]. Furthermore, the accuracy of MC simulations is strongly dependent on the quality and availability of detailed material parameters, including scattering rates and full-band-structure data [1].

11.5 QUANTUM-CORRECTED SEMICLASSICAL MODELS IN NANOSCALE DEVICE SIMULATION

As device dimensions scale down to the nanometer regime, purely classical transport models fail to capture essential quantum-mechanical phenomena such as confinement, tunneling, and quantum interference [18, 19]. To bridge the gap between quantum transport methods (e.g., NEGF formalism) and computationally efficient classical approaches (e.g., DD), quantum-corrected semiclassical models have been developed. These models enhance traditional Boltzmann-based frameworks by incorporating quantum corrections into the potential or carrier transport equations, thereby enabling the simulation of nanoscale devices within a manageable computational cost.

11.5.1 Density-gradient and effective potential methods

The DG method introduces quantum corrections into classical transport equations by adding terms proportional to the gradient of the carrier density, derived from the Bohm quantum potential [20, 21]. For electrons, the quantum potential term Q takes the following form:

$$Q = -\left(\hbar^2 / 2m^*\right)\nabla^2 \sqrt{n} / \sqrt{n}$$

where "n" is the carrier density and m^* is the effective mass. This term accounts for quantum confinement and tunneling by modifying the electrostatic potential experienced by carriers. The DG approach can be integrated into DD or HD transport models with minimal computational overhead, making it attractive for device simulators such as Sentaurus or ATLAS [21, 22].

The effective potential method is another approach where the classical potential ϕ is replaced with a quantum-corrected potential ϕ_{eff} obtained from solutions to the Schrödinger equation or its approximations [23]. This method directly incorporates sub-band energies and wavefunction penetration into the transport equations, improving accuracy for ultrathin-body MOSFETs, quantum wells, and nanowires.

11.5.2 Quantum drift-diffusion and quantum hydrodynamic models

Quantum DD (QDD) models extend the conventional DD equations by adding quantum potential terms from the DG method or effective potential approach [20]. The continuity and current density equations remain structurally similar to the classical case, but are solved using the quantum-corrected potential. This allows the model to capture size quantization effects in ultra-scaled devices while preserving the computational efficiency of DD solvers [24].

Quantum HD (QHD) models are derived from the moments of the Wigner transport equation or from the Madelung formulation of the Schrödinger equation [25, 26]. These models incorporate quantum pressure and quantum potential terms into the momentum and energy balance equations, enabling the simulation of phase coherence, interference effects, and nonlocal transport phenomena. Compared with QDD models, QHD models provide a more accurate description of transient quantum effects but are also more computationally demanding.

11.5.3 Applicability in circuit-level simulation

In large-scale circuit simulation environments, fully quantum mechanical device models are computationally impractical due to their complexity and high memory requirements. Quantum-corrected semiclassical models, such as DG, effective potential, QDD, and QHD, offer a practical compromise [18, 21]. By embedding quantum corrections into semiclassical frameworks, these models enable the inclusion of quantum confinement and tunneling effects in **compact models** used within Simulation Program with Integrated Circuit Emphasis (SPICE)-like simulators [27]. This facilitates accurate prediction of circuit performance in nanoscale CMOS technologies, tunnel FETs, and multi-gate devices without resorting to prohibitively expensive full quantum simulations (Figure 11.2).

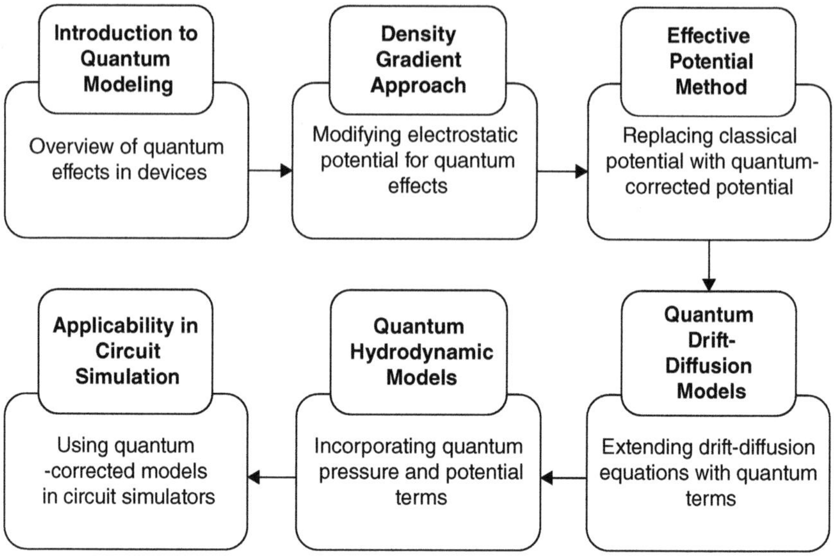

Figure 11.2 Quantum modeling in device simulation.

11.6 DEVICE SIMULATION TOOLS

The continuous scaling of semiconductor devices to the nanoscale regime necessitates accurate and predictive simulation frameworks to bridge the gap between theoretical modeling and experimental fabrication. TCAD platforms serve as indispensable tools for device engineers, enabling the detailed analysis of electrical, thermal, and optical behaviors before committing to costly prototyping [1, 28].

11.6.1 Overview of common TCAD platforms

Among the widely used TCAD tools, Synopsys Sentaurus and Silvaco Atlas have established themselves as industry standards. Synopsys Sentaurus offers a comprehensive suite for both 2D and 3D (three-dimensional) device simulations. Its physics-based models include DD, HD, and quantum-corrected transport equations, which are essential for analyzing SCEs, leakage mechanisms, and carrier mobility degradation in advanced CMOS and high-electron-mobility transistors (HEMTs) [29].

Silvaco Atlas provides a modular framework for semiconductor device simulation, supporting a broad spectrum of materials, including silicon (Si), gallium nitride (GaN), silicon carbide (SiC), and emerging 2D materials like MoS_2. It also offers multi-physics capabilities, enabling coupled electrothermal, optoelectronic, and radiation-effect simulations, which are crucial in designing devices for power electronics, photovoltaics, and space applications [30].

11.6.2 TCAD capabilities and dimensional modeling

Modern TCAD platforms offer 2D and 3D modeling environments, allowing designers to select the dimensional complexity best suited to their application. 2D simulations are computationally less intensive and are often sufficient for planar device structures, providing fast turnaround for initial design iterations. 3D simulations, on the other hand, are indispensable for modern architectures such as FinFETs, GAA FETs, and vertical transistors, where spatial nonuniformities, fringing fields, and localized self-heating effects significantly influence device performance [31].

Moreover, 3D TCAD modeling enables the inclusion of layout-aware effects, allowing engineers to study parasitic coupling, interconnect delays, and process-induced variations. This is increasingly important as device performance at sub-5 nm nodes becomes strongly dependent on manufacturing tolerances and 3D geometry [32].

11.6.3 Integration with layout and process tools

An important evolution in TCAD environments is their integration with layout design tools and process simulators. For instance, Sentaurus Process can simulate key fabrication steps such as oxidation, ion implantation, lithography, and etching, thereby generating realistic doping profiles and physical geometries for use in Sentaurus Device simulations [29]. Similarly, Silvaco's Victory Process and DeckBuild interfaces provide seamless process-to-device modeling workflows [30].

This integration also supports compact model extraction, enabling the generation of SPICE-compatible parameters directly from TCAD results. Such workflows bridge the gap between device-level physics and circuit/system-level simulation, ensuring that circuit designers work with models that accurately capture nanoscale effects [27].

11.6.4 Applications in emerging device technologies

TCAD tools are increasingly employed in the development of novel semiconductor devices, such as normally off GaN HEMTs, TFETs, and 2D-material-based FETs. These platforms allow researchers to assess band-to-band tunneling rates, interface trap densities, and thermal dissipation before fabrication. Additionally, simulation-driven optimization reduces experimental iterations, thereby lowering both time-to-market and development costs [33].

In conclusion, advanced TCAD platforms like Synopsys Sentaurus and Silvaco Atlas are essential to modern semiconductor research and manufacturing. Their ability to perform 2D/3D physics-based simulations, integrate with fabrication and layout workflows, and provide compact model generation ensures that device engineers can optimize designs with high accuracy before entering the fabrication stage. This integration of simulation into the semiconductor design cycle is critical in addressing the challenges of advanced technology nodes and emerging device concepts.

11.7 INDUSTRIAL REQUIREMENTS AND EDA INTEGRATION

The semiconductor industry operates in a highly competitive environment where rapid product development, performance optimization, and cost efficiency are critical to market success. To meet these challenges, device modeling serves as a fundamental bridge between materials research, fabrication technology, and electronic system design. By accurately capturing the physical behavior of devices, modeling enables predictive design, accelerates time-to-market, and reduces the number of costly fabrication cycles required for product validation [28, 34].

11.7.1 Role of modeling in semiconductor R&D and fabrication

In semiconductor research and development (R&D), modeling helps evaluate novel device structures, material systems, and fabrication processes before large-scale investment in manufacturing. This involves simulating electrical, thermal, optical, and reliability characteristics under various operating conditions [35]. For instance, predictive modeling can determine the feasibility of implementing high-mobility channel materials such as germanium (Ge) or indium gallium arsenide (InGaAs) in advanced CMOS technology nodes [36].

In fabrication, modeling plays a critical role in process optimization. TCAD tools are routinely integrated into the Design-Technology Co-Optimization (DTCO) workflow, where device physics, process parameters, and layout considerations are co-optimized to meet power, performance, and area targets [36]. Additionally, modeling supports yield enhancement by identifying process sensitivities, enabling more robust manufacturing flows.

11.7.2 Interfacing with electronic design automation (EDA) tools

While TCAD provides an accurate physics-based representation of devices, EDA tools enable full-chip verification, system-level simulation, and manufacturing sign-off. The interface between TCAD and EDA environments is essential for applying device-level insights to higher abstraction levels within the design flow [37].

This integration typically involves the extraction of compact models from TCAD simulations. Compact models, such as BSIM for MOSFETs or ASM-HEMT for GaN devices, translate complex device physics into computationally efficient forms suitable for circuit simulation. Once extracted, these models are incorporated into SPICE-based simulators (e.g., HSPICE and Spectre) for analog, digital, and mixed-signal circuit analysis [38].

11.7.3 Support for PDKs, SPICE, and compact model libraries

A key element in enabling TCAD–EDA interoperability is the Process Design Kit (PDK), provided by semiconductor foundries and containing device models, layout and design rules, parasitic extraction data, and verification scripts tailored to a specific fabrication process. Compact models obtained from TCAD simulations are calibrated and incorporated into PDKs to ensure that circuit designers work with accurate, process-dependent parameters. These kits typically include SPICE-compatible model libraries, facilitating compatibility with diverse EDA platforms, and provide multiple process corners (e.g., typical, slow, and fast) along with environmental variations in temperature and supply voltage to support worst-case verification. Compact model libraries are regularly updated to reflect process refinements and technology scaling, updates that are essential for maintaining strong correlation between simulated and fabricated device performance, especially at advanced nodes of 3 nm and below.

Industrial semiconductor development relies heavily on accurate device modeling and seamless TCAD–EDA integration. By enabling predictive design, ensuring compatibility with PDKs, and supporting SPICE-based compact model libraries, the modeling workflow bridges the gap between R&D and high-volume manufacturing. This integration is a key driver in achieving shorter design cycles, higher yield, and improved product reliability.

11.8 CONCLUSION

The continuous scaling of semiconductor devices into the nanoscale regime has transformed device modeling from a supportive design activity into a central enabler of innovation across research, development, and manufacturing. Accurate physics-based models integrated with TCAD environments provide essential insight into the electrical, thermal, and reliability characteristics of emerging device architectures before physical prototyping. Such predictive capabilities reduce design iterations, lower development costs, and accelerate time-to-market, all of which are critical factors in a highly competitive industry.

Moreover, the effective integration of TCAD outputs with EDA frameworks ensures that detailed device physics can be leveraged at the circuit and system levels through compact models, SPICE-compatible libraries, and PDKs. This TCAD–EDA synergy supports DTCO, enabling simultaneous advancements in device structure, fabrication processes, and system performance.

As technology nodes approach and surpass the 3 nm threshold, the importance of modeling extends beyond traditional CMOS scaling, encompassing novel materials, heterogeneous integration, and advanced packaging. The modeling ecosystem must evolve to address increased variability, reliability challenges, and multi-physics interactions inherent in nanoscale devices. Maintaining a close correlation between simulated predictions and measured results will remain a decisive factor for ensuring design accuracy and optimizing yield in next-generation semiconductor products.

Nanoscale device modeling, supported by robust industrial requirements and advanced EDA integration, is not merely a technical necessity but a strategic capability for sustaining progress in the semiconductor industry. Future developments will rely on enhanced physical accuracy, scalable computational methods, and closer collaboration between academia, foundries, and design houses to meet the demands of emerging electronic applications.

REFERENCES

1. M. Lundstrom, *Fundamentals of Carrier Transport*, 2nd ed. Cambridge: Cambridge Univ. Press, 2000.
2. S. Datta, *Quantum Transport: Atom to Transistor*. Cambridge: Cambridge Univ. Press, 2005.
3. C. Jungemann, B. Neinhüs, and B. Meinerzhagen, *Hierarchical Device Simulation: The Monte Carlo Perspective*. New York: Springer, 2003.
4. M. V. Fischetti and S. E. Laux, "Monte Carlo simulation of transport in technologically significant semiconductors of the diamond and zinc-blende structures—Part I: Homogeneous transport," *IEEE Trans. Electron Devices*, vol. 38, no. 3, pp. 634–649, Mar. 1991.
5. Q. H. Wang, K. Kalantar-Zadeh, A. Kis, J. N. Coleman, and M. S. Strano, "Electronics and optoelectronics of two-dimensional transition metal dichalcogenides," *Nat. Nanotechnol.*, vol. 7, no. 11, pp. 699–712, Nov. 2012, doi:10.1038/nnano.2012.193
6. T. Grasser, Ed., *Bias Temperature Instability for Devices and Circuits*. New York: Springer, 2014.
7. A. Zhan and A. Asenov, "Data-driven compact modelling of nanoscale devices: Recent advances and outlook," *IEEE Trans. Electron Devices*, vol. 67, no. 10, pp. 4137–4143, Oct. 2020.
8. Y. Cheng, C. Hu, P. K. Ko, and S. C. Tam, "BSIM3 model for analog and digital circuit design," *IEEE J. Solid-State Circuits*, vol. 29, no. 3, pp. 538–547, Mar. 1994.
9. T. Grasser, "Advanced device modelling and simulation," in R. A. Meyers (Ed.), *Encyclopedia of Sustainability Science and Technology* (pp. 57–90), Springer, 2012.
10. E. Pop, "Energy dissipation and transport in nanoscale devices," *Nano Res.*, vol. 3, no. 3, pp. 147–169, 2010.

11. M. V. Fischetti and S. E. Laux, "Monte Carlo study of electron transport in silicon inversion layers," *Phys. Rev. B*, vol. 48, no. 4, pp. 2244–2274, 1993.
12. M. V. Fischetti, "Effective electron mobility in Si inversion layers in metal–oxide–semiconductor systems with a high-κ insulator: The role of remote phonon scattering," *J. Appl. Phys.*, vol. 90, no. 9, pp. 4587–4608, 2001.
13. M. Luisier and G. Klimeck, "Atomistic full-band simulations of silicon nanowire transistors: Effects of electron–phonon scattering," *Phys. Rev. B*, vol. 80, no. 15, 155430, 2009.
14. C. Jacoboni and P. Lugli, *The Monte Carlo Method for Semiconductor Device Simulation*. New York: Springer, 1989.
15. M. V. Fischetti and S. E. Laux, "Monte Carlo analysis of electron transport in small semiconductor devices including band-structure and space-charge effects," *Phys. Rev. B*, vol. 38, no. 14, pp. 9721–9745, Oct. 1988.
16. D. K. Ferry, *Semiconductors*, 2nd ed. New York: Macmillan, 1991.
17. A. Asenov, S. Roy, and M. J. Kennedy, "Monte Carlo simulation of transport in high electron mobility transistors," *IEEE Trans. Electron Devices*, vol. 45, no. 12, pp. 2505–2513, Dec. 1998.
18. M. V. Fischetti and S. E. Laux, "Monte Carlo simulation of electron transport in small semiconductor devices including band-structure and space-charge effects," *Phys. Rev. B*, vol. 38, no. 14, pp. 9721–9745, Oct. 1988.
19. M. Lundstrom, *Fundamentals of Nano Transistors*. Singapore: World Scientific, 2017.
20. C. Gardner, "The quantum hydrodynamic model for semiconductor devices," *SIAM J. Appl. Math.*, vol. 54, no. 2, pp. 409–427, 1994.
21. R. Brunetti, C. Jacoboni, and F. Nava, "Quantum corrections to the drift-diffusion equation for high-field, short-channel MOSFETs," *Semicond. Sci. Technol.*, vol. 7, no. 12, pp. 171–178, Dec. 1992.
22. Synopsys, *Sentaurus Device User Guide, Version O-2018.06*. Mountain View, CA: Synopsys Inc., 2018.
23. T. Ando, A. B. Fowler, and F. Stern, "Electronic properties of two-dimensional systems," *Rev. Mod. Phys.*, vol. 54, no. 2, pp. 437–672, Apr. 1982.
24. W. Hänsch, *The Drift Diffusion Equation and Its Applications in MOSFET Modeling*. New York: Springer, 1991.
25. F. A. Buot, "Nonequilibrium quantum transport physics in nanosystems," *J. Comput. Electron.*, vol. 1, no. 1–2, pp. 57–123, 2002.
26. D. K. Ferry and J. R. Barker, "The physics and modeling of small semiconductor devices," *J. Comput. Electron.*, vol. 2, pp. 1–5, 2003.
27. Y. Taur and T. H. Ning, *Fundamentals of Modern VLSI Devices*, 2nd ed. Cambridge: Cambridge Univ. Press, 2009.
28. C.S. Rafferty, "Technology computer-aided design (TCAD) in semiconductor processing," *IEEE Trans. Electron Devices*, vol. 43, no. 7, pp. 1045–1053, Jul. 1996.
29. Synopsys, *Sentaurus Device User Guide*. Mountain View, CA: Synopsys Inc., 2023.
30. Silvaco, *Atlas Device Simulation Software Manual*. Santa Clara, CA: Silvaco Inc., 2023.
31. J.-P. Colinge, *FinFETs and Other Multi-Gate Transistors*. New York: Springer, 2008.

32. H.-S. P. Wong et al., "Device design considerations for nanoscale CMOS," *Proc. IEEE*, vol. 87, no. 4, pp. 537–570, Apr. 1999.
33. S. Mishra et al., "Performance analysis of normally-off GaN HEMTs using TCAD simulations," *IEEE Trans. Electron Devices*, vol. 67, no. 5, pp. 1871–1878, May 2020.
34. H. Iwai, "Roadmap for 22 nm and beyond," *Microelectron. Eng.*, vol. 86, no. 7–9, pp. 1520–1528, Jul.–Sep. 2009.
35. S. Selberherr, *Analysis and Simulation of Semiconductor Devices*. New York: Springer-Verlag, 1984.
36. S. Natarajan et al., "A 14 nm logic technology featuring 2nd-generation FinFET, air-gapped interconnects, self-aligned double patterning and a 0.0588 μm^2 SRAM cell size," in *IEEE International Electron Devices Meeting (IEDM)*, Dec. 2014, pp. 71–73.
37. Y. Cao, *Predictive Technology Model for Nano-CMOS Design Exploration*. Berkeley, CA: Springer, 2011.
38. W. Liu et al., "BSIM4.6.0 MOSFET model user's manual," Univ. California, Berkeley, CA, Tech. Rep., 2013.

Chapter 12

Molecular electronics

Madhuparna Chakraborty, Weiqiang Yuan, and Flavia P.N. Inbanathan

12.1 INTRODUCTION

Nanoscale electronic devices and circuits play a critical role in the nanotechnology era, paving the way for the miniature but highly efficient devices. However, with traditional silicon-based transistors reaching around 2 nm, these miniature devices are subjected to quantum tunneling effects and pose tremendous challenges for further miniaturization. Nevertheless, to uphold Moore's law, researchers are building new device designs by using molecules and atoms and exploiting the unique electronic properties of different materials, a process that has resulted in molecular electronics. It is a rapidly growing field, built on the concepts of biochemistry, physics, and electrical engineering. It specifically focuses on exploiting the properties of individual molecules and molecular-scale structures, using a bottom-up approach for electronic devices and their applications. Molecular electronic devices can perform complex functions at the nanoscale, and the research works have led to significant breakthroughs, including electron transport, molecular self-assembly [1], and the development of flexible electronics.

Molecular electronics focuses on the electronic functions of molecules, thereby offering potential alternatives to replace conventional Si- and Ge-based electronic materials and devices [2]. The optical, chemical, and electrical properties of the molecules make them suitable for designing miniaturized, energy-efficient, and highly functional electronic devices. Researchers have made significant progress in developing various types of molecular switches [3], which could be used in nano processors, data storage, and beyond. Molecular electronics include molecular switches, molecular wires, molecular diodes, molecular transistors, and so on.

Molecular switches are single molecules that transition between two stable states in response to external stimuli such as electrical current, temperature changes, chemical environment, or light. The reversible switching of charge states in 2D-layered molecular conductors highlights the tunable

DOI: 10.1201/9781003635123-12

nature of molecular switches under multiple stimuli, paving the way for applications in molecular circuitry. As the computing industry demands miniaturized and more efficient processors, molecular switches could become fundamental components in developing such nanodevices. Molecular switches can change their electronic state in response to external stimuli (like light, voltage, or pH). The focus on molecular switches and other nanoscale components is driving the evolution of electronic, optoelectronic, and sensing applications within nanotechnology.

Molecular wires [4] can transport electrical signals with minimal or no loss. Recent works in photoswitchable devices and molecular transistors showed enhanced conductivity [5], signaling a new era in electronic systems, covering the areas from quantum computing [6] to high-frequency applications. Several studies have provided insight into the molecular electron transport and conductance [7].

Researchers have demonstrated significant advancements in molecular electronics, such as a rotaxane [8] molecular structure that switches between different electronic states in response to an applied voltage; alignment of molecular orbitals with electrode states, which impacts electron flow across molecular junctions; single-molecule transistors using a benzene molecule; electron transport [9, 10] in molecular systems, which can be both ballistic and tunneling; conjugated organic molecules exhibiting bistable behavior, similar to that of logic gates; reproducible molecular junctions using self-assembly process; and molecular diodes exhibiting rectifier behavior [11]. Other significant research works exhibited enhanced conductance properties [12] in molecular structures by quantum interference [13–15] and tunneling. With the prevalent use of organic semiconductors and carbon nanotubes (CNTs), the scope of molecular electronics has been expanded because of their unique electronic properties. Studies have also demonstrated progress in light-sensitive molecular devices based on photochromatic molecules [16], the construction of qubit gates [17] using electron scattering in molecules, promising field-effect characteristics for organic electronics, molecular THz switches [18] operating at room temperature, and molecular junctions exhibiting electroluminescence for advanced photonic applications [19]. Scientific works have suggested that, by replacing silicon with molecular systems, remarkable progress can be accomplished in flexible and wearable electronics [20].

While significant progress has been made in demonstrating the functionality of molecular devices, integrating them into scalable systems remains a major challenge. The difficulties lie in achieving reproducibility and stability in molecular-scale devices. Additionally, issues related to the interface between molecules and metal electrodes, such as contact resistance and instability, remain unresolved. With this brief introduction, the next section discusses in detail the evolution of molecular electronics.

12.2 EVOLUTION OF MOLECULAR ELECTRONICS: FROM CONDUCTANCE TO INTELLIGENCE

12.2.1 Early theoretical foundations

The origin of molecular electronics can be traced back to the mid-20th century, when scientists first explored the electrical properties of organic molecules. The concept of "molecular engineering," introduced by Von Hippel, laid the groundwork for a bottom-up approach to construct electronic systems at the molecular level [21]. In 1959, physicist Richard Feynman delivered his famous lecture, "There's Plenty of Room at the Bottom," at the American Physical Society meeting at Caltech [22]. His visionary speech emphasized the potential of electronic miniaturization and encouraged researchers to explore molecular-scale fabrication. Although chemical synthesis was already well-developed, the ability to establish direct electrical connections to individual molecules remained a challenge. The studies on intramolecular electron transfer, which focused on how electrons move within molecules in solution, set the stage for the eventual development of molecular electronics [23].

12.2.2 Conceptual breakthrough in the 1970s

The first experimental evidence supporting electron transport through single molecules emerged in 1971. Scientists Mann and Kuhn used the Langmuir-Blodgett (LB) technique to create structured fatty acid monolayers between metal electrodes and measured their electrical conductivity. Their findings revealed an exponential decrease in conductivity with increasing molecular length, providing crucial insights into charge transport at the nanoscale [24]. Furthermore, molecular electronics gained significant attention in 1974 when Mark Ratner and Ari Aviram proposed the idea of a molecular rectifier [25, 26]. Their theoretical model demonstrated that organic molecules could function as electronic components by allowing electrical current to flow preferentially in one direction, like a diode. This groundbreaking proposal introduced the possibility of molecular-scale circuits, although practical implementation was still beyond the reach of existing technology.

12.2.3 Experimental advancements in the 1980s and 1990s

The 1980s and 1990s marked a pivotal era for molecular electronics, as significant experimental advancements allowed researchers to investigate charge transport at the molecular scale. This period was characterized by the development of advanced microscopy techniques, advanced lithographic techniques for molecular structures, and the first direct measurements of electrical conductance through single molecules. The rise in scanning probe microscopes has allowed more and more refined experiments that provide better control for trapping a single molecule between two electrodes [27]. The introduction of scanning tunneling microscopy (STM) and atomic force microscopy enabled scientists to visualize, manipulate, and measure the electronic properties of individual molecules with atomic precision. STM provided direct evidence of electron transport through molecular junctions, confirming theoretical predictions from the 1970s [27]. By applying voltage biases, researchers could study the conductance of molecular layers and gain insights into charge transfer mechanisms at the nanoscale.

Another key development was the use of self-assembled monolayers (SAMs) to create highly organized molecular structures on surfaces, allowing researchers to systematically arrange molecules with controlled orientation and thereby facilitating better studies of their electronic behavior [27].

These methods provided a way to fabricate molecular junctions, enabling detailed investigations of their conductive properties. A pioneering effort to measure single-molecule charge transport was carried out by Mark Reed's research group in the late 1990s and early 2000s. Their work significantly contributed to the advancement of techniques for molecular-scale measurements, providing valuable insights into the transport properties of various molecules and of junctions [28]. When a molecular junction is formed, the molecule interacts with the metallic electrode, resulting in molecular orbitals overlapping with the electronic states of the electrode, forming a new hybrid electronic wave function. The strength of this coupling can vary significantly, ranging from conjugated states that extend across the entire junction to the formation of orbital nodes that act as barriers, restricting electron transport between the electrodes. Scientists utilized mechanically controllable break junctions (MCBJs) and STM-based break junctions to investigate electron flow through single molecules. These experiments revealed how molecular length, electronic structure, and environmental factors influenced charge transport, laying the groundwork for future molecular electronic components [29].

Figure 12.1 depicts an example of the molecular junction formation and conductance measurement. Figure 12.1(a–e) demonstrates the formation of molecular junctions utilizing STM. Figure 12.1(f) depicts conductance measurements where molecular monolayers are deposited on a flat Au substrate [30]. Figure 12.1(g) shows a conductance measurement of a 90-mer DNA sample. One strand of the DNA was modified at its 3′ end with a thiol linker, while the complementary strand had the same linker attached at the 5′ end. These thiol-functionalized ends were anchored to either the STM tip or the Au(111) surface. The STM tip was repeatedly brought into contact and withdrawn from the DNA-modified Au(111) surface, while the conductance was measured as a function of tip displacement during the retraction process [31].

Various theoretical models have been proposed to explain electron transport through molecular junctions, including both temperature-independent

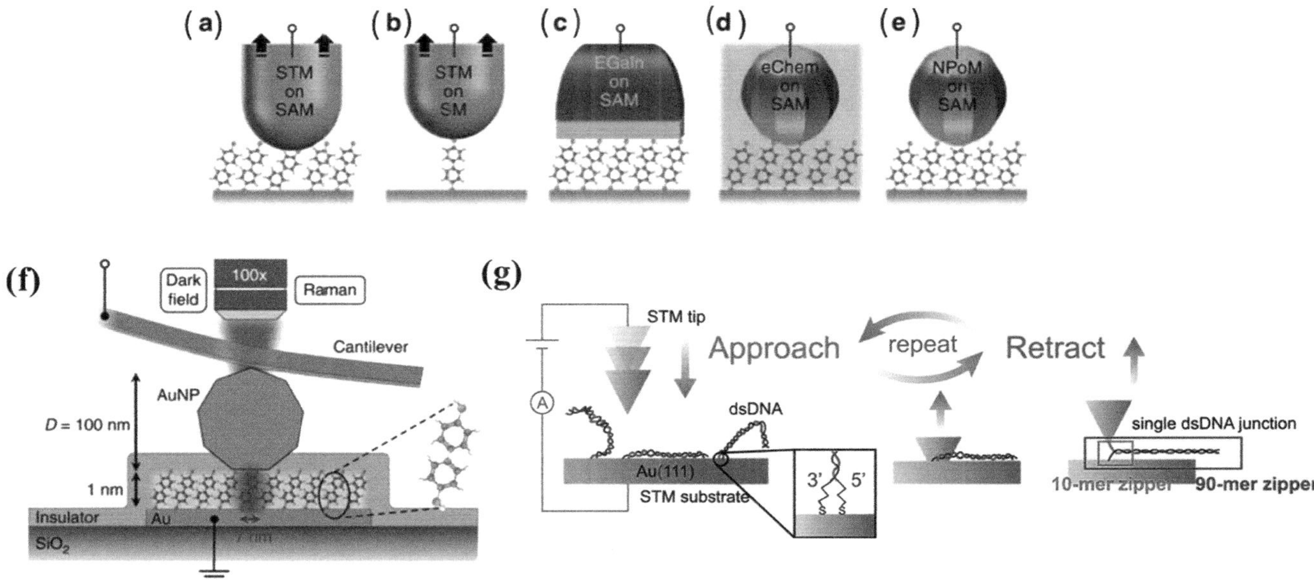

Figure 12.1 (a, b) Junctions formed by scanning tunneling microscope (STM) tip with pull-off from (a) SAMs. (b) Single molecules. (c) Eutectic GaIn alloy contacts. (d) Large-area electrochemical cells making indirect contact with AuNPs on SAMs. (e) The scheme here uses a single plasmonic nanoparticle contact. (f) Au nanoparticle on SAM, which is electrically contacted by a conductive transparent cantilever, giving real-time optical access to the junction. (g) Schematic of scanning tunneling microscopy–break-junction (STM-BJ) measurements.

(f: Adapted from Kos et al. [30]; g: Adapted from Harashima et al. [31].)

mechanisms like coherent tunneling (superexchange) and Fowler-Nordheim tunneling, as well as temperature-dependent processes such as thermionic (Schottky) emission and hopping conduction. While these models differ primarily in their dependence on applied bias, they are broadly categorized into two groups: tunneling, which is independent of temperature, and hopping, which relies on temperature variations [29].

By the late 1990s, researchers had started demonstrating key molecular electronic components that mimicked the behavior of traditional semiconductor devices at the molecular scale. These experiments confirmed that organic molecules could serve as functional electronic components, validating earlier theoretical models. By developing the ability to construct and study molecular-scale circuits, researchers laid the foundation for the breakthroughs of the 2000s, in which molecular electronic devices became increasingly practical and sophisticated.

12.2.4 Development of molecular devices

The early 2000s saw rapid progress in the experimental realization of molecular devices, with researchers successfully demonstrating molecular devices such as diodes, transistors, and memory elements. Single-molecule transistors showcased the potential for molecular-scale switching, further validating the viability of molecular electronics.

12.2.4.1 Molecular diode

Molecular diodes can be categorized into two types: rectifying diodes and resonant tunneling diodes, as shown in Figure 12.2. Figure 12.2(a) illustrates a rectifying diode with a structure consisting of two sections, S1 and S2, separated by an insulating group, R, where the alignment of the lowest unoccupied molecular orbital (LUMO) and the highest occupied molecular orbital (HOMO) enables conduction in only one direction. The structure of a resonant tunneling diode based on molecular technology is shown in Figure 12.2(b) [32]. As depicted, an electron-donating group and an electron-accepting group are connected by a CH_2 unit, which serves as the barrier. A traditional diode consists of a silicon p – n junction situated between

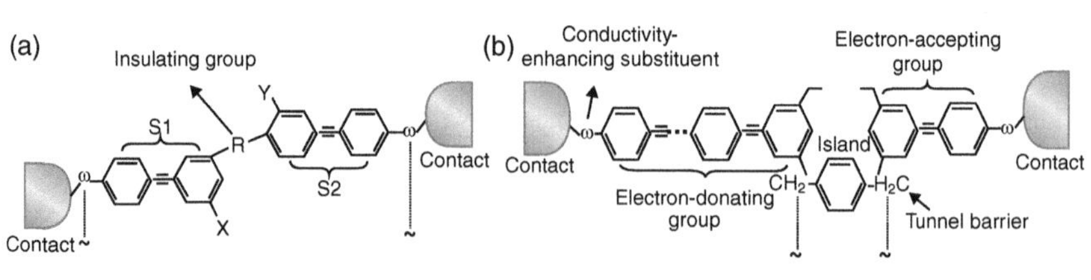

Figure 12.2 (a) Structure of a rectifying diode. (b) Structure of a tunneling diode. **(Adapted from Ghasemi et al. [32].)**

two electrodes, an anode and a cathode, allowing current to flow in only one direction. The asymmetry of the p – n junction creates an ON/OFF environment for electron transport. Single-molecule diode design initially involved asymmetric molecules that function like a p – n junction or use symmetric molecules with different metal electrodes [32]. However, these methods resulted in low rectification ratios and weak forward current. To overcome these limitations, Capozzi et al. [33] developed a highly efficient molecular rectifier. They used symmetric oxidized thiophene derivatives, whose molecular energy levels were nearly aligned with the Fermi levels of gold electrodes [33]. The key to achieving rectification in the scheme lies in breaking symmetry by varying the surface area of each electrode exposed to an ionic solution. This asymmetry, combined with precise energy level alignment, caused a positive voltage to significantly increase current while a negative voltage suppressed it effectively. In this method, electron transport is primarily governed by quantum tunneling, with its tunneling efficiency depending on the alignment of the molecule's discrete energy levels with the continuous spectrum of the electrode. Over the years, different types of molecular diodes have been developed, each leveraging distinct transport mechanisms and structural optimizations to enhance rectification efficiency and stability. Traditional molecular diodes operating in the tunneling regime were limited to rectification ratios of approximately 10^3. Advancements in fabrication techniques have enabled the creation of single-molecule diodes with high rectification ratios. For instance, researchers have developed diodes based on metal-oxide clusters (polyoxometalates) that exhibit rectification ratios greater than 100, sustaining current densities exceeding 10^5 A/cm^2. [34]. Researchers have developed molecular diodes capable of operating at frequencies up to 17.8 GHz. These diodes utilize gold nanocrystal electrodes and ferrocenyl undecanethiol molecules, demonstrating significant potential for high-speed electronic applications [35]. Tunable rectification, achieved via electrochemical control, utilizes the redox-active molecules for real-time switching of diode behavior, thereby offering dynamic electronic properties [36]. In recent years, supramolecular diodes have emerged as a new class of molecular rectifiers, leveraging non-covalent interactions such as $\pi - \pi$ stacking and hydrogen bonding to enhance charge transport and rectification ratios. These self-assembled systems offer improved stability and tunability, making them a promising alternative for nanoscale electronics [37].

12.2.4.2 Molecular transistor

The first single-molecule field-effect transistor (FET), developed in 2000, utilized a single C_{60} molecule to bridge nanogap gold (Au) electrodes formed via the electromigration break junction technique. When a voltage is applied to the silicon dioxide (SiO_2) insulating layer through the gate electrode, the current flowing through the C_{60} molecular junction can be modulated. To enhance gate control, electron beam lithography is employed to create nanogap Au source/drain electrodes on an aluminum gate electrode pad coated with a 3 nm oxide dielectric layer. Subsequently, a single divanadium (V_2) molecule is integrated into the nanogap Au electrodes to construct a single-molecule FET [38]. The design of single-molecule FETs must take several key factors into account. First, the gate design involves selecting the appropriate gate type, determining the dielectric layer material, and positioning the gate relative to the source and drain electrodes. Second, the molecular structure must be carefully engineered to optimize performance. Third, the molecule–electrode interface plays a crucial role, particularly in ensuring proper connection and electronic coupling between the molecule and electrodes. Through innovative designs, significant advancements have been made in single-molecule FETs, leading to discoveries in fundamental scientific principles and potential applications [38].

Extensive research is underway to enhance electrical conduction across the metal–molecule junction in molecular transistors. Tersoff et al. [39] demonstrated that contact resistance at the nanotube–metal junction has a greater impact on charge transfer than the intrinsic channel conductance. The presence of a Schottky barrier at this junction restricts current flow, thereby limiting conductance. To mitigate this barrier and enable smoother electron flow, Javey et al. [40] proposed the use of palladium (Pd) contacts on semiconducting single-walled carbon nanotubes (SWCNTs). Palladium is particularly effective because of its high work function and strong wetting interaction with CNTs, which improves charge injections. Additionally, as noted by Ebbesen et al. [41], modifying the geometry of CNTs allows for control over their metallic or semiconducting properties. This geometric tunability enables precise manipulation of their electronic characteristics, offering promising advancements in molecular transistor performance.

Single-molecule FETs operate on quantum mechanical principles. In addition to simply switching states, the gate also modulates molecular orbitals, thereby influencing both the electronic and spin states of the molecule. In cases of strong coupling, electron interactions can be tuned, greatly enhancing the flexibility of device control. At low temperatures, high-resolution measurements have revealed intricate gate-dependent physical phenomena such as the Coulomb blockade and the Kondo effect [38].

Figure 12.3 illustrates the two-stage Kondo effect observed using a single-molecule transistor platform. Guo et al. [42] showed that the Kondo temperature $\left(T^*\right)$ is influenced by the effective exchange of energy between two molecular levels. The transistor in Ref. [42] was fabricated using electromigration techniques, with a schematic diagram shown in Figure 12.3(a). Figure 12.3(b) provides a scanning electron microscopy image of the junction, where the core component is a single manganese phthalocyanine (MnPc) molecule. Transition metal phthalocyanine molecules exhibit strong coupling to metal substrates, and their HOMOs consist mainly of the metal ion's d orbital electrons. By varying the transition metal ion, the orbital structure and spin ground state can be easily adjusted, making these

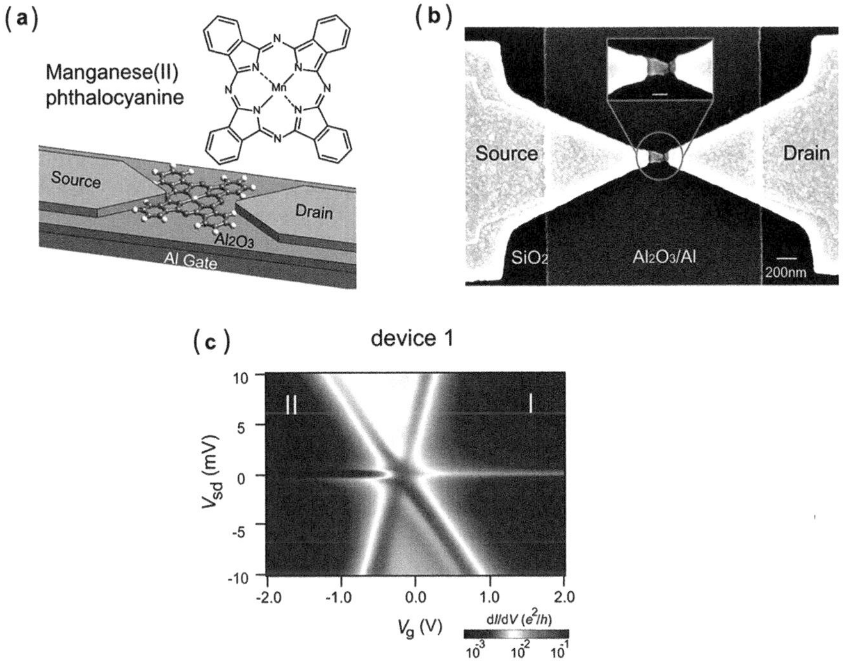

Figure 12.3 (a) Structure of Manganese (II) phthalocyanine molecule transistor device. (b) SEM image of the device after breaking, a gold nanowire over an Al_2O_3 / Al gate, a nanogap formed in the gold nanowire by electromigration. (c) Color plots of differential conductance (dI / dV) as a function of bias voltage (V_{sd}) and gate voltage (V_g) at 280 mK to 20 K.

(Adapted from Guo et al. [42].)

molecules ideal for studying exotic Kondo phenomena. Electrical measurements were performed from 280 mK to 20 K, as the device becomes unstable above 20 K. A bias voltage (V_{sd}) applied across the source and drain electrodes was used to control the energy of electrons passing through the device, while a gate voltage (V_g) applied to an aluminum electrode adjusted the molecule's chemical potential. Excited states arising from the internal vibrations of the molecules were used to confirm that the measured signals came from individual MnPc molecules. Figure 12.3 shows differential conductance (dI / dV) plots for two representative molecular devices, presented as a function of bias voltage (V_{sd}) and gate voltage (V_g). Coulomb blockade behavior was observed, with differential conductance peaks identifying two-electron blocking regions, I and II, corresponding to adjacent charge states of the molecule. A sharp zero-bias peak was detected in region I, while a gate-tunable zero-bias split peak appeared in region II, reflecting distinct Kondo processes linked to varying conditions in the charge states of the MnPc molecules.

12.2.4.3 Molecular switches

A molecular switch is a nanoscale system capable of undergoing controlled and reversible changes in its physical or chemical properties in response to external stimuli such as light, temperature, or chemical signals. These switches provide precise molecular-level control, attracting significant research interest due to their potential applications in nanotechnology, data storage, and molecular electronics. By leveraging binary states (0 and 1), molecular switches function similarly to conventional digital devices, enabling data storage and processing. The dithienylethene (DTE) molecule, first synthesized in the late 1980s, is one of the earliest examples of molecular switches. Experimental studies demonstrated that an aqueous solution of DTE transitions from transparent to yellow upon exposure to ultraviolet (UV) light [43]. This color change results from structural rearrangements at the molecular level, where the central ring of the molecule closes to form a new isomer with distinct optical characteristics. The process is fully reversible—irradiation with visible light restores the molecule to its original open-ring structure. This phenomenon, known as photoswitching, allows molecular systems to store information, with the open and closed-ring configurations corresponding to ON and OFF states, respectively. Another class of molecular switches, ruthenium nitrosyl complexes, emerged in the 1990s [43]. These compounds exhibit unique photochromic behavior but involve more intricate switching mechanisms. Over the years, researchers have developed various types of molecular switches, each featuring distinct mechanisms and applications. These switches can be broadly categorized as follows:

- Crown Ether Switches: These cyclic molecules bind with positive ions, undergoing conformational changes that alter ion conductivity (Figure 12.4(a)). External triggers, such as light exposure or pH shifts, can modify their ion affinity, significantly impacting ion transport properties [44].
- Rotaxanes: These molecules feature a "thread-and-ring" structure, where a ring moves along a linear chain (Figure 12.4(b)). Switching occurs when the ring shifts between different binding sites, driven by external stimuli such as electrical signals or pH changes [45].
- Photochromic Switches: These systems rearrange structurally between isomers with different optical properties upon light irradiation (Figure 12.4(c)). Essential performance factors include high stability and minimal thermal isomerization, making them suitable for molecular-scale electronic applications [46].
- Nanoparticle Switches: These switches involve nanoparticles that exhibit tunable optical properties due to structural phase transitions (Figure 12.4(d)). Their applications include high-density optical memory storage and long-term data retention, making them valuable for advanced nanotechnology solutions [47].

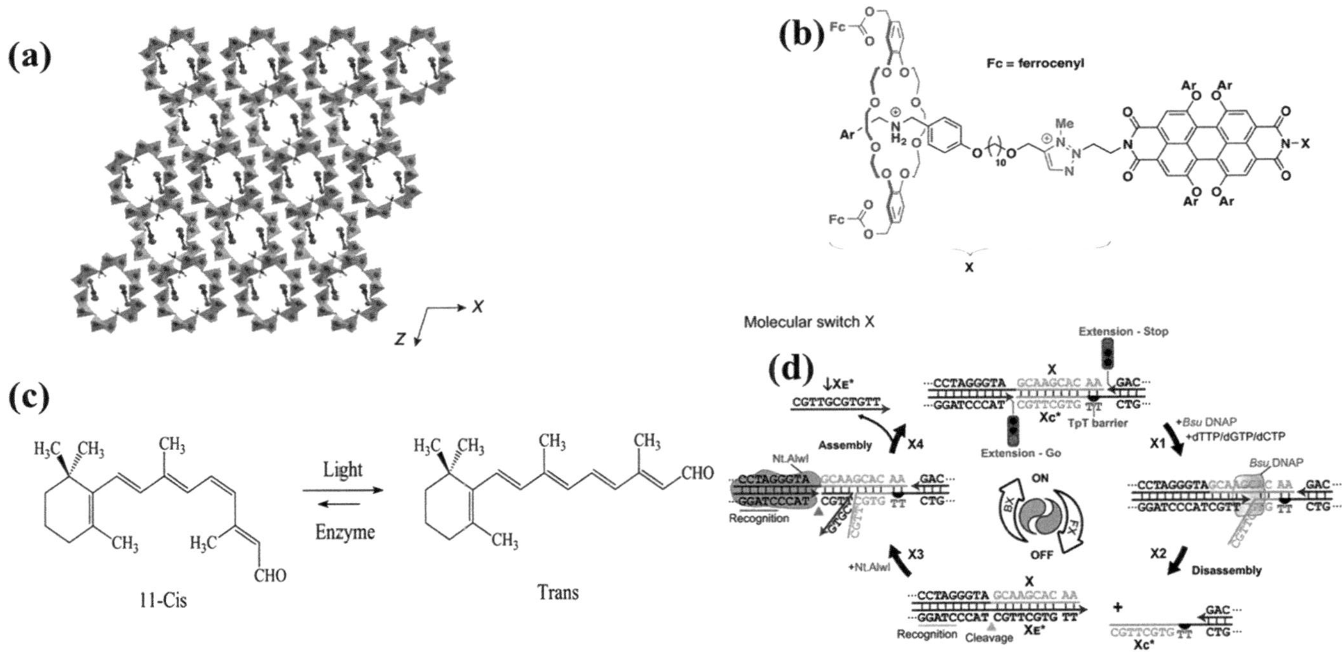

Figure 12.4 (a) Structure of crown ether, 1,10-diaza-18-crown-6-ether. (b) Molecules with tunable photophysical properties showing a rotaxane structure. (c) Chemical structures of 11-*cis* and *trans* retinal and switching between them, illustrating photochromic switches. (d) In type x switches, the initial coupled state (ON state) consists of two structural partners linked by complementary sticky-end segments x / xC*. Segment x is encoded using A, C, and G nucleotides (V nucleotides), while segment xC* is composed of C, G, and T nucleotides (B nucleotides), specifically designed to define the enzymatic working sites of the nanoparticle switch.

(a: Adapted from Ye [44]; Adapted from Riebe [45]; c: Adapted from Hamedian et al. [46]; d: Adapted from Kang et al. [47].)

The continuous advancement of molecular switch technology is driving innovations in nanodevices, smart materials, and next-generation computing. With ongoing research, these nanoscale systems are expected to play a pivotal role in the future of molecular electronics.

12.2.4.4 Molecular memory

Molecular memory is an advanced data storage technology that harnesses the unique electronic, magnetic, and structural properties of individual molecules to encode information with unparalleled density and efficiency. Unlike conventional semiconductor-based memory which relies on bulk materials, molecular memory operates at the nanoscale, utilizing mechanisms such as redox switching, charge trapping, and conformational changes to store binary states (0 and 1) with exceptional stability. These molecular states can be manipulated using external stimuli like electric fields, light, or chemical reactions, enabling fast and energy-efficient data processing.

Ref. [48] presents a molecular memory system integrated with a downstream processor, enabling an all-chemical RS flip-flop. This system allows for logically controlled, reversible capture and release of molecules in both aqueous solutions and solid-state environments. By extending the scope of molecular-logic-based computation, this approach impacts previously unexplored areas of chemistry, such as pollutant removal, chemical signal generation, and the programmed release of functional species from discrete supermolecules. In Ref. [48], digital control over the capture and release of molecular guests via molecular inputs enables the creation of sequential logic devices with input–output homogeneity, a key feature in biological information processing. In this design, the memory component is sequentially integrated with an INHIBIT gate, where fluorescence emission serves as the final output. Notably, all output states of the device are optically self-indicating [48]. The ketone/alcohol system functions as a bistable switch, as shown in Figure 12.5(a), with its state determined by input conditions. Applying a high concentration of reductant (set S input) converts the ketone to alcohol, setting the output to (Out1 = 1, Out2 = 0). Conversely, applying a high concentration of oxidant (reset R input) converts alcohol back to ketone and switches the output to (Out1 = 0, Out2 = 1). In the absence of input, the system retains its previous state. This behavior mirrors an Reset Set (RS) flip-flop, a fundamental memory element in computing. Here, the entire system—including inputs, outputs, and processing—is chemically driven. While molecular RS flip-flops exist in photochromic, enzyme, and DNA-based systems, this work demonstrates a redox-driven version., whose corresponding truth table is shown in Figure 12.5(b) [48].

12.2.5 Integration in the 2000s

Hybrid approaches that integrate molecular components with existing semiconductor technology emerged as a practical solution for interfacing

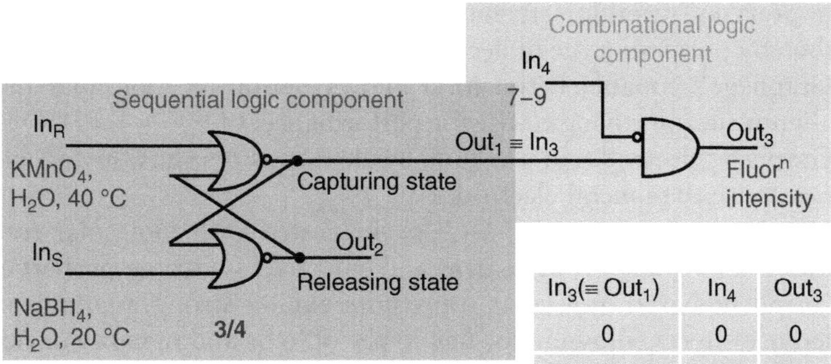

$In_3(\equiv Out_1)$	In_4	Out_3
0	0	0
1	0	1
0	1	0
1	1	0

Figure 12.5 (a) Integrated sequential-combinational logic demonstrating redox-induced reversible guest capture and release, where Out1 represents the capturing state and Out2 represents the releasing state. (b) Truth table for guest-induced fluorescence quenching of the device, with the output digitized using a threshold value.

(Adapted from Daly et al. [48].)

molecular electronics with traditional circuits. Incorporating carbon-based materials such as CNTs and graphene further expanded the possibilities, offering improved conductivity and structural stability [49].

Graphene's atomic thinness, high conductivity, and tunable electronic structure make it an ideal platform for stabilizing molecular junctions and improving charge transport efficiency.

Molecular diodes enable rectification at the nanoscale, allowing current to flow preferentially in one direction. Graphene's unique properties enhance molecular diode performance in several ways:

- Graphene forms strong $\pi-\pi$ interactions with organic molecules, improving molecular alignment and stability.
- Unlike metal electrodes, graphene reduces metal-induced molecular energy shifts, preserving molecular orbital integrity.
- Graphene's high carrier mobility facilitates efficient charge injection into molecular junctions.
- By tuning graphene's work function through doping or functionalization, researchers can optimize rectification behavior.

Graphene plays a crucial role in molecular transistors, where charge transport through a single molecule is controlled via an external gate. Some advantages are as follows:

- Graphene's smooth surface provides a defect-free interface, minimizing scattering and enhancing molecular device reproducibility.

- It prevents metallic screening effects seen in traditional electrodes, thereby preserving the molecule's gate response.
- Graphene's tunable Fermi level allows optimized molecular orbital alignment, improving transistor performance.
- Graphene electrodes in single-molecule transistors show lower contact resistance than metal electrodes.
- Graphene quantum dots (GQDs) integrated with molecular transistors enable charge confinement and enhanced switching properties.
- Graphene-based molecular transistors exhibit strong quantum interference effects, allowing for new types of logic and memory devices.

Researchers study nanoscale structures where quantum mechanical effects dominate, causing electrons to behave as waves, rather than particles, as shown in Figure 12.6. As transistors continue to shrink in size, they become less efficient and more prone to errors due to quantum tunneling, where electrons leak through the device even when it is supposed

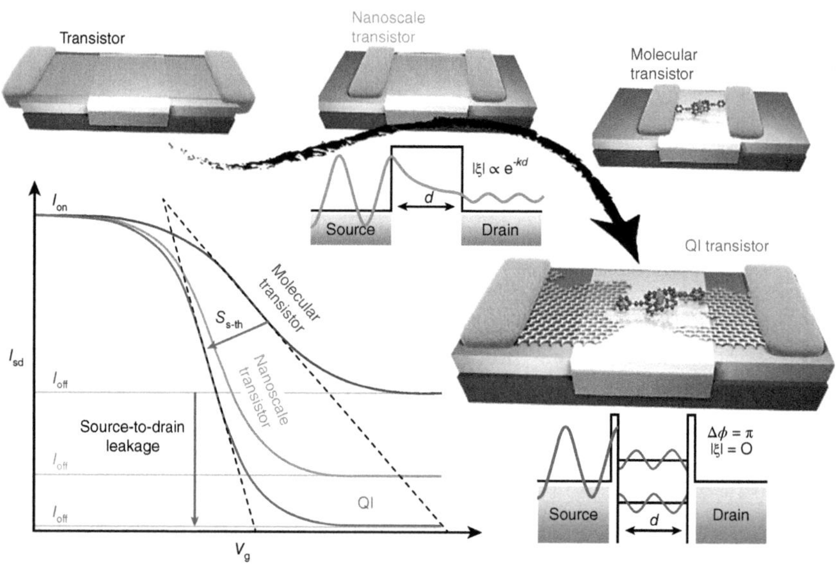

Figure 12.6 As the source-to-drain distance (d) in a transistor shrinks to the nanometer scale, quantum tunneling through the potential barrier defining the OFF state increases exponentially, leading to high leakage current and degraded subthreshold swing. At molecular-scale dimensions (<5 nm), source-drain leakage becomes a critical challenge unless quantum interference between two coherent conduction channels suppresses transmission. In such systems, interference effects play a key role in reducing leakage and improving device performance, as illustrated in the I–V curve.

(Adapted from Chen et al. [50].)

to be off. To address this issue, researchers are investigating alternative switching mechanisms and materials to mitigate this effect. Leveraging these quantum effects, they developed a novel transistor that utilizes a single zinc porphyrin molecule as its conductive channel, positioned between two graphene electrodes [50]. By applying a voltage to the electrodes, the researchers were able to regulate electron flow through the molecule using quantum interference. Quantum interference occurs when electron waves interact, either reinforcing each other (constructive interference) or canceling each other out (destructive interference). In this new transistor, switching is achieved by controlling the nature of electron interference—constructive interference turns the device on, whereas destructive interference switches it off. A key advantage of this design is its high ON/OFF ratio, allowing precise control over the transistor's state. The suppression of quantum tunneling through destructive interference minimizes electron leakage, thereby significantly improving efficiency. Another breakthrough is the transistor's remarkable stability. While previous single-molecule transistors have only lasted a few switching cycles, this device has demonstrated durability over hundreds of thousands of cycles without degradation. Additionally, the researchers discovered that quantum interference enhances the transistor's subthreshold swing, which determines how efficiently the device responds to changes in gate voltage. Their transistor achieved a subthreshold swing of 140 mV/dec, outperforming many other single-molecule transistors [50].

The integration of graphene with molecular diodes and transistors represents a major step toward scalable and high-performance molecular electronics. By leveraging the outstanding electrical and mechanical properties of graphene, researchers are advancing the development of next-generation nanoelectronic devices, including flexible electronics, ultra-fast logic circuits, and energy-efficient computing architectures.

12.2.6 Molecular intelligence

The future of molecular electronics is increasingly shaped by the integration of artificial intelligence (AI). As AI continues to advance, its fusion with molecular electronics opens new avenues for progress in the field. AI-driven methods facilitate the identification of intricate molecular behavior patterns, optimization of experimental conditions, and precise prediction of molecular junction properties. Techniques such as machine learning (ML) and deep learning (DL) offer robust tools for processing and analyzing the vast and complex datasets generated in single-molecule experiments [51].

ML is generally classified into supervised and unsupervised learning. Supervised learning involves training a model using both input data and corresponding correct outputs, allowing it to predict target variables for new data. This includes regression, which forecasts continuous values, and classification, which assigns categorical labels, such as identification of chemical species. In contrast, unsupervised learning helps interpret data without explicit labels or values. In single-molecule measurements, two primary unsupervised learning techniques are feature extraction, which reduces the dimensionality of multidimensional data, and clustering, which groups data into distinct categories [52].

Lemmer et al. [53] were the first to apply clustering techniques (as seen in Figure 12.7(a)) to conductance traces in single-molecule measurements, using a multiparameter vector-based classification (MPVC) method. In this approach, conductance traces are represented as vectors (as illustrated in Figure 12.7(b)), and essential clustering features are extracted by transforming each trace into characteristic parameters (as given in Figure 12.7(c)) [53]. In the chapter, Shang et al. [54] developed a multi-prototype clustering (MPC) algorithm to detect low-probability single-molecule events in three primary scenarios: evaluating junction formation probability (JFP), identifying rare stretching configuration changes through mechanical regulation, and examining structural modifications driven by electric fields. Their results demonstrate that the algorithm successfully distinguishes molecular junction events from pure tunneling in JFP analysis. ML is playing an increasingly vital role in theoretical calculations by reducing computational expenses. Convolutional neural networks (CNNs), trained on molecular dynamics (MD) and nonequilibrium Green's function (NEGF) simulations, have been employed to predict experimental conductance more efficiently than conventional first-principles calculations. Additionally, ML-based regression models have been developed to estimate single-molecule conductance using experimental data [54].

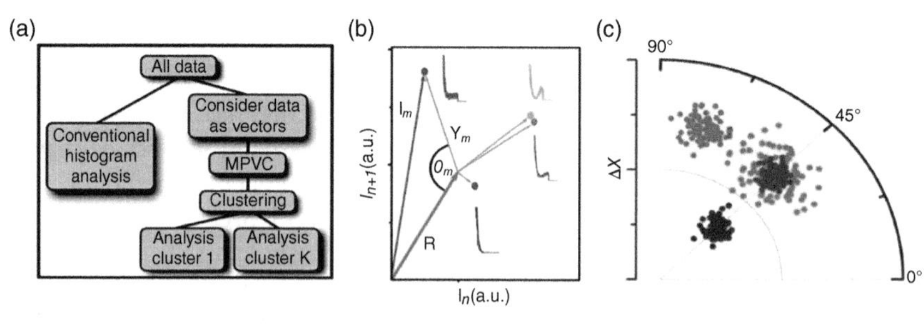

Figure 12.7 (a) Flowchart of the MPVC process. (b) Conductance trace converted to feature vectors. (c) Conductance cluster.

(Adapted from Lemmer et al. [53].)

12.3 KEY COMPONENTS OF MOLECULAR ELECTRONICS

As discussed earlier in the evolution section, molecular electronics is a new research field focused on creating electronic components and circuits from single molecules or groups of molecules. This approach offers the potential to overcome the limitations of conventional silicon-based electronics by shrinking the device sizes, enhancing operation speeds, and achieving novel functionalities [27, 55]. In general, molecular switches, wires, diodes, memory, and sensors are significant components in molecular electronics. This section will briefly explain and expand on some typical device structures and operating principles.

12.3.1 Molecular switches

Molecular switches refer to molecules that can reversibly switch between two or more stable states when external stimulation or excitation is applied, such as heat, light, pH, or voltage [27, 56]. They are critical components for controlling and manipulating molecular functions in nanotechnology, synthetic biology, and smart materials. Molecular switches operate through various mechanisms, including conformational changes, photoisomerization, or redox reactions [27].

12.3.1.1 Crown ether switches

Defined by a cyclic structure featuring ether linkages (–O–) interspersed with repeating carbon units, their general formula is $(C_2H_4O)n$, with "n" indicating the number of ether groups and, consequently, the size of the ring. These crown ethers can be categorized based on their cavity sizes, allowing for selective binding with specific cations. For instance, 18-crown-6 [57] is particularly adept at binding with potassium ions (K^+) due to its cavity dimensions, as illustrated in Figure 12.8 [58].

On the other hand, there are also: 14-crown-4, which preferentially binds to Li^+; 15-crown-5, which preferentially binds to Na^+; 18-crown-6, which preferentially binds to K^+; and 21-crown-7, which preferentially binds to Cs^+ [59, 60]. The special binding properties make the crown ether systems suitable for selective sensing purposes for certain ions, which will be demonstrated more in the application section. Besides their selective binding abilities, they can also serve as a foundation for collaboration with other functional groups to enable switching mechanisms.

Crown ethers paired with functional groups like azobenzene can act as photoswitches [58]. Figure 12.9(a) illustrates how the aromatic rings of two benzo-15-crown-5 molecules are linked by an azo connection [58]. When subjected to the correct light wavelength, the N=N double bond shifts from trans to cis, drawing the two macro rings closer. Additionally, it can operate

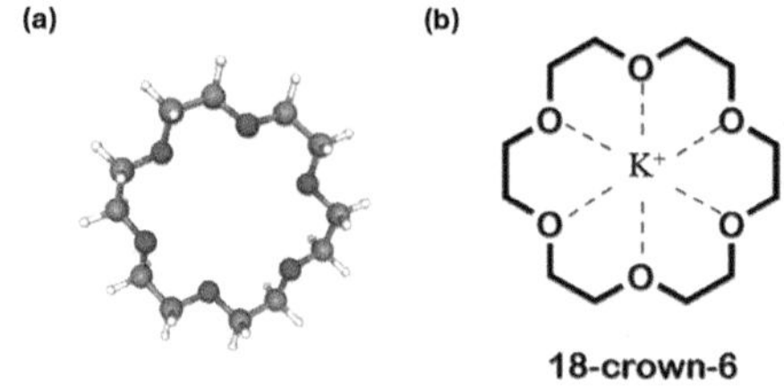

Figure 12.8 (a) Molecular structure of the crown ether of 18-crown-6. (b) Potassium complex of the crown ether 18-crown-6.

(a: Adapted from PubChem [57]; b: Modified from Gokel et al. [58].)

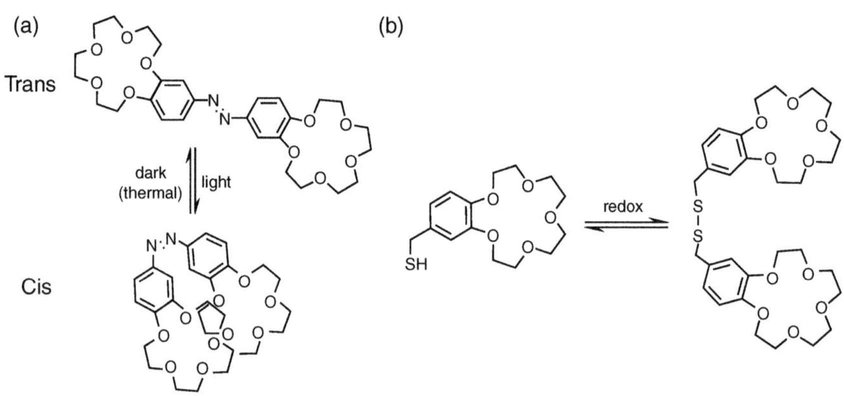

Figure 12.9 (a) Photochemical switching of two azo-linked benzo-15-crown-5 molecules. (b) Crown ether-based redox switch.

(Adapted from Gokel et al. [58].)

as a redox switch; for instance, a benzo 19-crown containing five oxygen atoms can create a disulfide bridge through the redox reaction depicted in Figure 12.9(b) [58].

12.3.1.2 Rotaxane

Rotaxanes are mechanically interlocked molecular structures, generally featuring a linear axle surrounded by a macrocyclic ring. This configuration allows for distinctive molecular movement, with the ring able to slide along the axle when triggered by particular stimuli, including temperature, pH, and electrochemical potential [61]. As a form of molecular machinery, rotaxanes demonstrate significant promise for use in molecular electronics, memory storage, and signal transduction [61, 62].

Figure 12.10(a) illustrates the typical switching behavior of a rotaxane. The figure depicts a blue molecular chain with large molecular groups

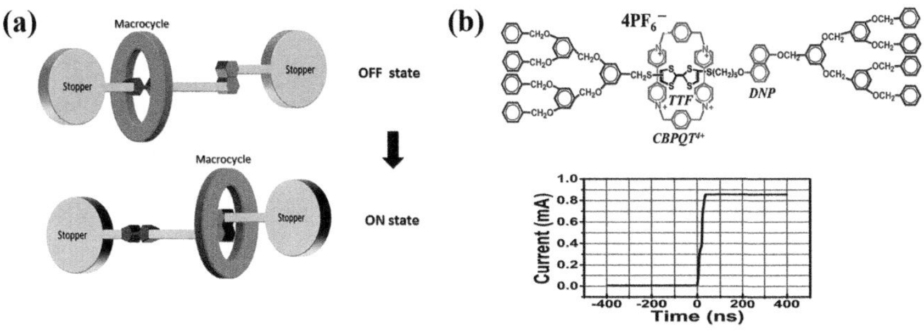

Figure 12.10 (a) Rotaxane switch working mechanism. (b) Rotaxane switch. **(a: Adapted from Wu et al. [61]; b: Adapted from Feng et al. [62].)**

serving as stoppers on each end. Along the chain, there are two distinct anchor points: point 1, marked in red, and point 2, marked in green. Under normal conditions, the macrocycle resides at anchor point 1, representing the "OFF" state. When an external stimulus, such as voltage, is applied, an oxidation state change at anchor point 1 (loss of electrons) causes the release of the macrocycle ring, which carries a positive charge, pushing it toward anchor point 2 through electrostatic force [61]. This results in a metastable state, shifting to the "ON" state. The entire process can be reversed when a reverse bias is applied. Figure 12.10(b) shows an example of a typical molecular structure of a rotaxane used in solid-state rotaxane-based Langmuir–Blodgett (LB) thin films on Indium Tin Oxide (ITO) glass. This rotaxane features CBPQT4+ as the macrocycle, and tetrathiafulvalene (TTF) along with 1,5-dioxynaphthalene (DNP) as its two anchor points. When an external bias is applied, the rotaxane transitions from the OFF state (low conductivity) to the ON state (high conductivity). This external bias triggers the loss of two electrons from the TTF site, causing the CBPQT4+ to move toward the DNP site. The switching process occurs quickly, with a transition time of about 60 ns [62].

12.3.1.3 Photochromic switches

Photochromic switches are a kind of molecular switch that utilize reversible photoisomerization reactions. The key mechanism is a ring-opening and closing process, which causes shifts in electrical properties, enabling the molecule to function like a photoswitch. Dihydroazulene (DHA) and its derivatives are considered promising materials for developing a photoswitch that is sensitive to light irradiation. The major cycle of the photo/thermo reaction between DHA and vinylheptafulvene (VHF) is shown in the following Figure 12.11 [63].

Under light irradiation, DHA is converted to VHF via the ring-opening process. Conversely, applying heat induces a ring-closing that effectively

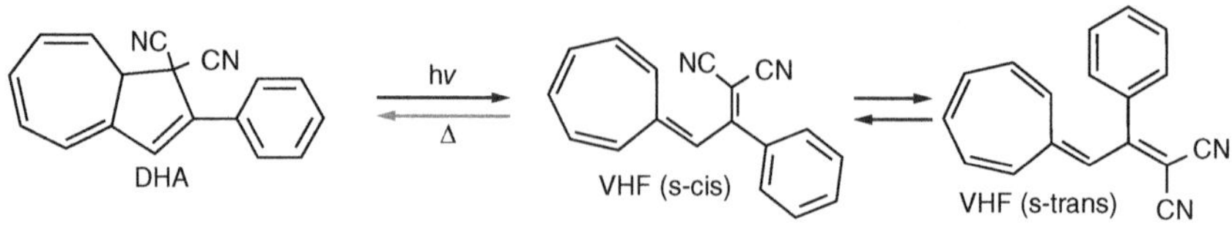

Figure 12.11 The principle of photo/thermoreaction between DHA and VHF. **(Modified from Broman and Nielsen [63].)**

converts VHF back to DHA [63, 64]. Research has indicated a significant difference in electrical properties between the two isomers. Specifically, DHA exhibits high resistivity, contrasting with the low resistivity observed in VHF [63]. Based on the distinct electrical behaviors of DHA and VHF, Broman et al. developed the photochromic switch on the silver nanogap, as shown in Figure 12.12 [63, 64].

Due to the high resistivity of DHA, the switch is initially "off." It is switched "on" when visual light is shone on it, as DHA is converted to VHF, enabling carriers to tunnel through the gap. This entire process can be reversibly performed by applying heat to the switch [63, 64]. In addition to DHA, diarylethene [65, 66] and azobenzene [67, 68] are also considered to construct the photochromic switch, as these molecules shift between isomers with different electrical and optical properties when exposed to light of various wavelengths. For example, diarylethenes demonstrate remarkable thermal stability, resistance to fatigue, and photosensitivity. They undergo a shift between open-ring and closed-ring forms upon exposure to various light wavelengths [69]. This switching effect is quantifiable in situ by regulating light radiation and analyzing the conductivity of the molecular

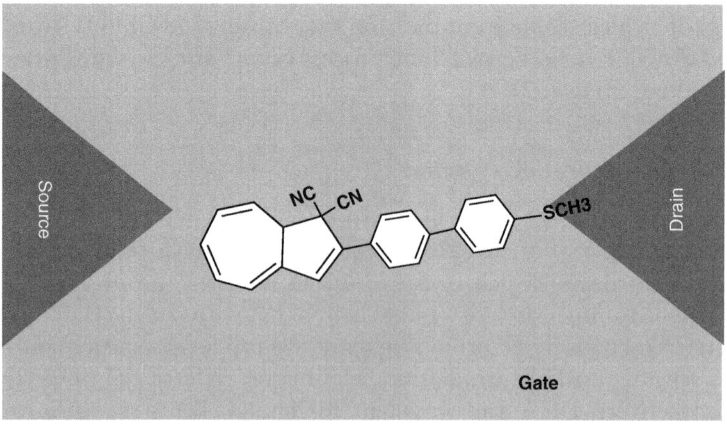

Figure 12.12 Schematic of DHA/VHF-based photochromic switch. **(Modified from Broman et al. [64].)**

junction in both the open and closed states [69]. Recent studies indicate that employing graphene or CNTs as electrodes bridged by diarylethene, which features three methylene (CH_2) groups on each side of its molecular backbone, can significantly enhance the effectiveness of photonic switches. These switches activate with monochromatic visible light (~240 μW/cm^2, $\lambda = 540$ nm) and deactivate under UV light (~100 μW/cm^2, $\lambda = 365$ nm) [70]. Azobenzene derivatives like the one mentioned in Figure 12.11, consisting of two benzene rings linked by a nitrogen–nitrogen double bond, can also be reversibly isomerized between trans and cis conformations upon light irradiation, leading to changes in electronic properties [71].

12.3.1.4 Thermo-responsive switches

A crucial switching mechanism depends on the molecule's conformational change, which alters its state in response to temperature variations. For example, the bianthrone-based molecular switch is the most representative switch based on the conformational change principle; the typical device schematic is shown in Figure 12.13.

The device is constructed on a nanogap with two separated metallic contacts. The bianthrone molecule is directly connected to one side of the metallic contact [72]. External conditions like temperature change can effectively induce the isomers' transition. The conformational change will narrow the tunneling gap between the bianthrone molecule and the metallic contact not connected to the bianthrone molecule, so the tunneling current is increased correspondingly while the external bias is present across the junction [72].

12.3.1.5 Molecular wires

Molecular wires are another critical component in building molecular electronics. They are supposed to work as conduits or interconnections on a nanometer scale to transport carriers through individual molecules or small collections of molecules within molecular circuits [73]. Therefore, molecular structure, energy level alignment, and electrode contact properties significantly affect electron transport efficiency in molecular wires [73, 74].

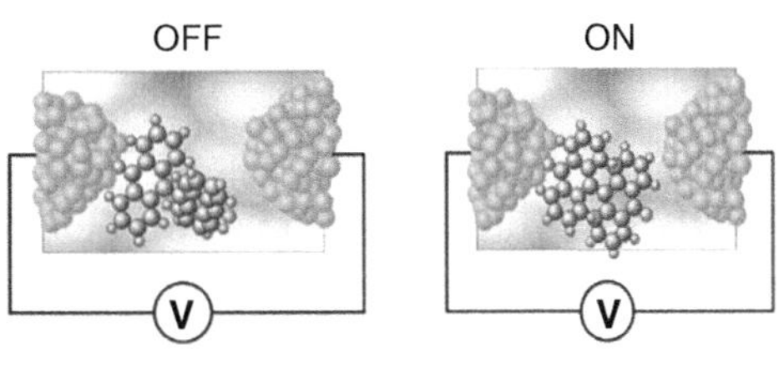

Figure 12.13 Schematic of bianthrone-based molecular switch.

(Adapted from Lara-Avila et al. [72].)

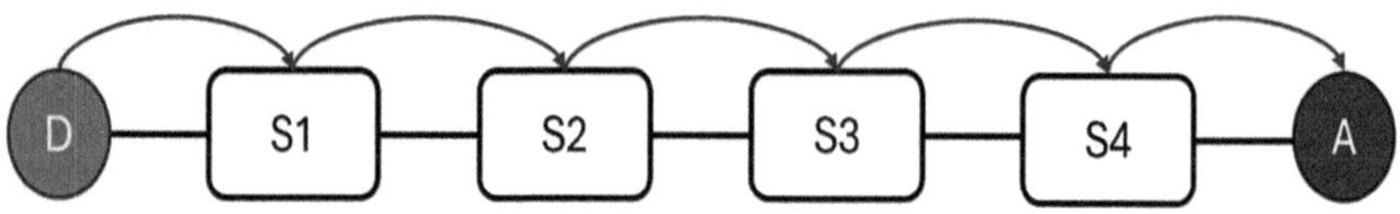

Figure 12.14 Hopping electron transport.

(Modified from Low and Marqués-González [74].)

One mechanism for carrier transport in molecular wires involves π-conjugation, which consists of alternating single and double bonds, to create a pathway for electron delocalization [74]. The simplified electron hopping mechanism is shown in Figure 12.14.

The hopping process can be considered as transferring carriers from the donor (D) to the acceptor (A) across a series of identical sites [74]. To achieve efficient hopping transport, reorganization energy (λ) and energy differences between hopping sites (ΔEhs) should be minimized, thereby reducing the activation energy for charge transfer [75]. Ryo Asakawa et al. [75] have recently successfully engineered a series of nanometer-scale π -extended molecules featuring a fused π-conjugated unit based on fused thienobenzo[b]indacenothiophene (TBID) that acts as a hopping site to reduce λ. The conformational twists between fused units effectively localize the π-conjugation to reduce ΔE_{hs}, thus facilitating better charge transport. The molecular structure and development of twisted molecular wires are shown in Figure 12.15.

Spectroscopic and electrochemical measurements confirmed the anticipated electronic structures of the synthesized oligomers. Conductance measurements indicated that these novel structures exhibited higher conductance and lower activation energy than non-fused oligothiophenes [75]. These samples demonstrated enhanced conductivity compared to non-fused oligothiophenes. The precise control over the sizes of these pi-conjugated areas enabled substantial single-molecule conductance in the oligomers, owing to their strong molecular structures [75–77].

Variants of porphyrin-based molecules represent another area of research in molecular wires, notable for their unique shapes and electronic properties. These wires consist of porphyrin units connected by various bridges that provide specific pathways for electronic transport [78]. Figure 12.16 displays the schematic structure of the single-molecule conductance measurements conducted through STM break junctions [79]. The porphyrin wires include meso–meso ethyne bridges, specifically the α,ω-di[(4′-thiophenyl)ethynyl]-terminated PZnn compounds, such as dithiol-PZn1, dithiol-PZn2, and dithiol-PZn3.

These porphyrin wires incorporate meso–meso ethyne bridges, demonstrating quasi-ohmic behavior. This indicates that the measured resistance rises linearly with the length of the molecules [79]. Moreover, recent studies have determined that the conductance of fused oligo-porphyrin nanowires

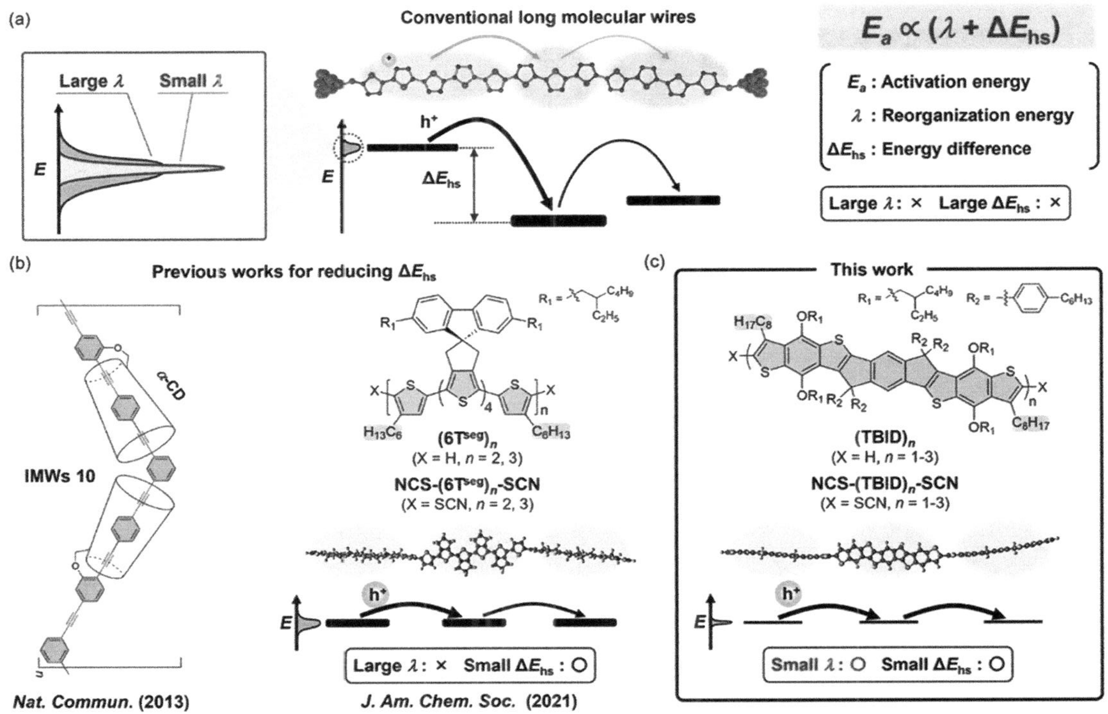

Figure 12.15 (a) Schematic of hopping transport and relationship between E_a, λ, and ΔE_{hs}; h^+ stands for holes. (b) Molecular structures of IMWs10, $(6T^{seg})_n$, and NCS$-(6T^{seg})_n-$SCN. (c) Molecular structures of $(TBID)_n$, NCS$-(TBID)_n-$SCN along with their schematic structure models and energy diagrams. The difference in the line thickness in the energy diagrams represents fluctuations due to thermal motion.

(a: Adapted from Asakawa et al. [75]; b: Adapted from Terao et al. [76] and Ie et al. [77]; c: Adapted from Asakawa et al. [75].)

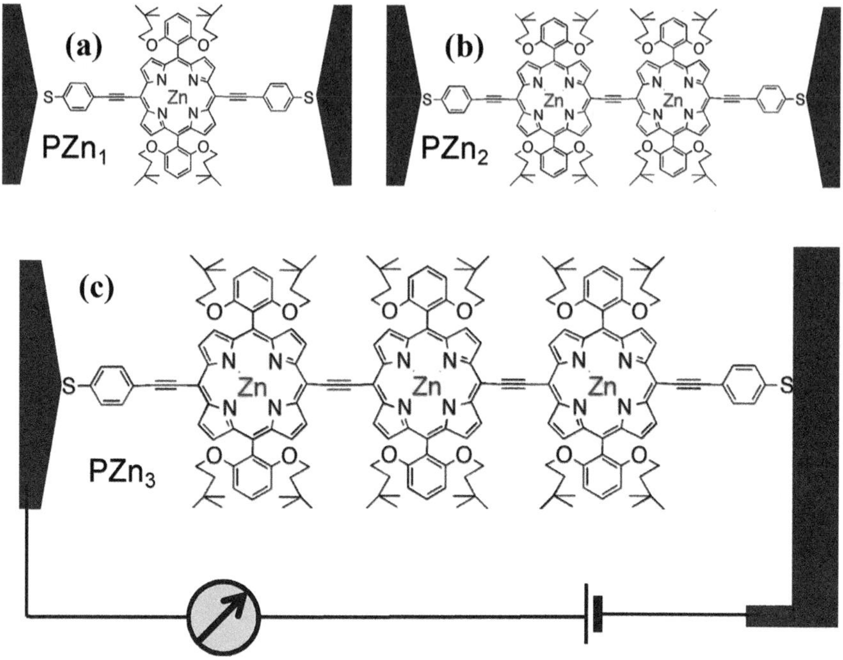

Figure 12.16 Single-molecule conductance measurements using STM break junctions with (a) dithiol-PZn1, (b) dithiol-PZn2, and (c) dithiol-PZn3.

(Adapted from Li et al. [79].)

does not adhere to the typical trend of decreasing conductance with longer lengths. Figure 12.17 shows a structure testing platform that utilizes graphene electrodes along with different oligo-porphyrin nanowires.

Instead of quasi-ohmic behavior, researchers have demonstrated that the conductance of fused porphyrin can either increase or remain independent of the length of these nanowires. Moreover, it reveals a critical correlation between the length of the nanowires and their electrical conductance, attributed to the unique electronic properties of the porphyrins [80]. The decline in the HOMO–LUMO gap with increasing length counteracts the intrinsic resistance associated with longer molecular chains, leading to a notable increase in conductance [80]. As research progresses, gaining a deeper insight into the electronic characteristics of porphyrin-based systems showcases their potential in nanodevices and materials science. Investigating the structural arrangements and electronic properties of these molecular wires is crucial for future advancements in organic and molecular electronics.

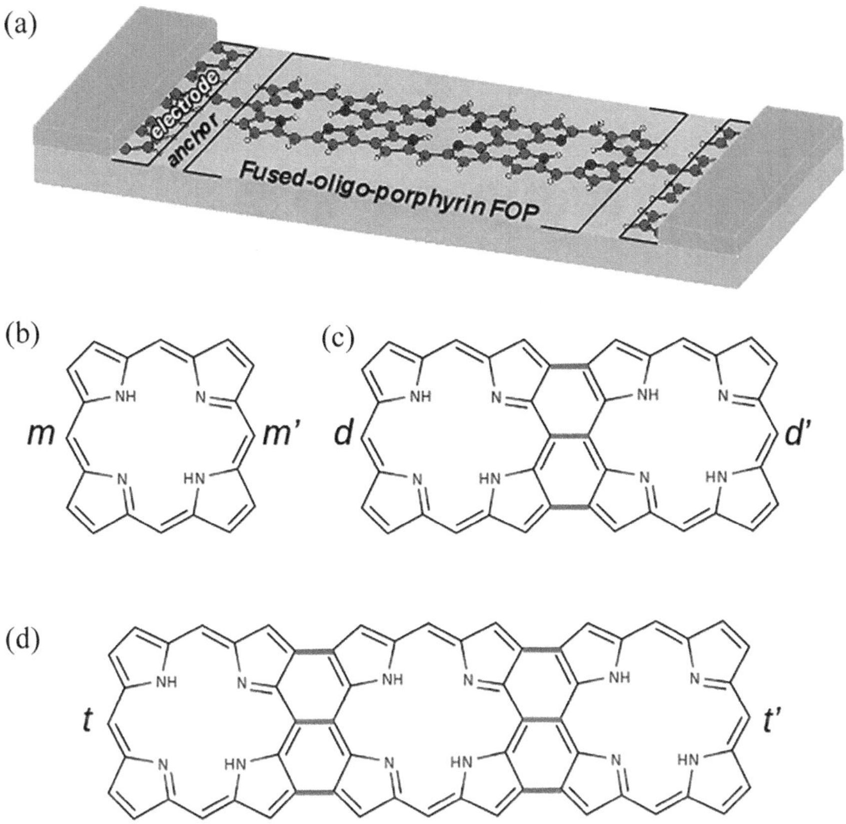

Figure 12.17 (a) The molecular junction contains a fused porphyrin trimer. (b) A porphyrin monomer is connected to electrodes from connection points m and m′. (c) A fused porphyrin dimer comprises two monomers linked by three single bonds (red) and connects to electrodes from connection points d and d′. (d) A fused porphyrin trimer is connected to electrodes from connection points t and t′.

(Adapted from Algethami et al. [80].)

12.3.1.6 Molecular transistor

The conventional semiconductor-based MOSFET uses an electric field directed at the gate to regulate the current flow between the source and drain electrodes. An early design proposing the use of molecular transistors is known as the molecular relay, as shown in Figure 12.18, which depends on conformational changes to modulate the current [81]. This device is designed with conductive molecular wires, a switching gate, and a switching atom. When the electric field is applied, the red switching atom is displaced, causing the entire conductive molecular wire to become non-conductive and

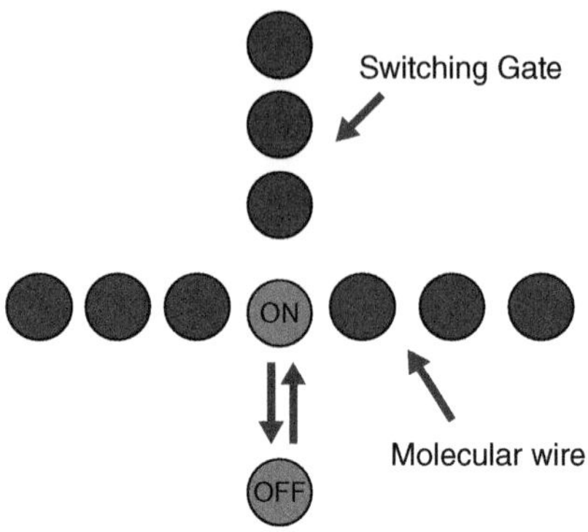

Figure 12.18 The schematic of the early molecular relay.

(Modified from Wada et al. [81].)

halt the current flow. Thus, current conduction is governed by the presence or absence of the switching atom within the molecular wire [81].

Rotaxanes have recently emerged as promising options for molecular transistors due to their unique bistable properties via the conformational changes. These properties enable them to switch between two distinct states that represent binary data critical for digital electronics, as shown in the molecular switch section. The use of nanoscale molecular switches with rotaxanes as the medium, along with crossbar Pt/Ti nanowires as electrodes, has proven their potential for high-density memory applications [82]. Single-electron tunneling (SET) molecular switches operate on the principle of quantum tunneling, where individual electrons are also allowed to move through barriers at the molecular scale, enabling the control of electrical conductivity within the device. The basic structure of the SET switch is shown in Figure 12.19.

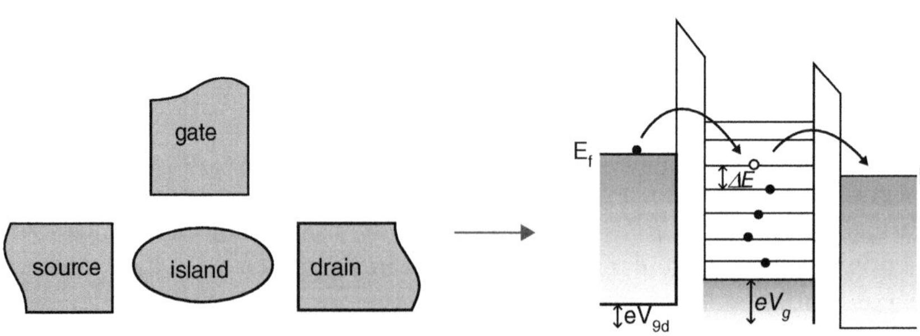

Figure 12.19 The schematic of the device structure of the SET switch and its working principles.

(Adapted from Jia Yen et al. [83].)

A tunneling electron behaves like a discrete charge due to the stochasticity of tunneling events. Figure 12.19 illustrates the SET switch, which comprises the source electrode, the island, the drain electrode, and a gate with tunnel junctions linking them. When no bias voltage is applied, the Fermi levels of the source and drain remain in equilibrium until a bias is introduced. In this scenario, an electron can tunnel through the junctions from the source to the drain via the island, as long as there is an available energy state at the island that lies between the Fermi levels of the electrodes. By manipulating the gate voltage, one can shift energy levels, thus controlling the addition and removal of electrons and effectively switching the device ON and OFF [83]. This process involves a molecular bridge connecting two electrodes, through which electrons can tunnel from one electrode to the other through the molecular switch, thereby toggling between conductive and non-conductive states [84, 85]. The effectiveness of this switching mechanism largely relies on the unique molecular structures used in these systems, which often consist of organic molecules that can exist in multiple stable states. This configuration and electron dynamics act as a key principle for creating advanced molecular electronic devices. The research described a three-terminal setup featuring a central "island" made from a single thiophene ring, which is linked to three conjugated arms via saturated connecting units, which serve as tunneling barriers. As a result, by changing the voltage on the "gate" electrode, the tunneling current between the source and drain can be effectively managed while achieving high switching speed [84].

12.4 APPLICATIONS OF MOLECULAR ELECTRONICS

The potential applications of molecular electronics are vast and impactful, covering a wide range of fields that highlight the versatility and transformative nature of this emerging technology. From consumer electronics to advanced medical devices, molecular electronics provides innovative solutions that promise to revolutionize existing systems and establish new paradigms across various industries. In this section, we will explore some key application areas for molecular electronics already in the market.

12.4.1 Flexible molecular devices

A key common use of molecular devices is for lighting purposes. Organic light-emitting diodes (OLEDs) have become a groundbreaking technology in molecular electronics. OLEDs have drastically transformed the display industry, offering rollable, foldable, and wearable screen solutions. The molecule-based flexibility of OLEDs allows them to be fabricated on thin substrates such as plastic films or ultrathin glass, enabling the creation of displays that are not only lightweight and compact but also capable of contorting into unconventional shapes. This adaptability is exemplified in rollable television systems and foldable smartphones, which capitalize on the deep blacks and vibrant colors inherent to OLED technology. For example, a typical flexible OLED panel is shown by Lei Zhou et al. [86], who have demonstrated a highly efficient flexible light-emitting diode using plastic embedded with silver (Ag) networks (PEANs) as the transparent electrodes, achieving an external quantum efficiency of approximately 50%. They created white OLEDs using this flexible electrode structure, and the device structure, along with the actual device in operation, is illustrated in Figure 12.20.

This flexible OLED comprises an organic molecular layer consisting of two emissive layers. The first layer is N,N′-dicarbazolyl-3,5-benzene (mCP) combined with 8 wt% of bis(3,5-difluoro-2-(2-pyridyl)phenyl-(2-carboxypyridyl)iridium-(III) (FIrpic) for blue emission, while the second layer features mCP doped with 6 wt% of iridium(III) bis(4-phenylthieno [3,2-c]pyridinato-N,C20) acetylacetonate (PO-01) to produce yellow emission [86]. The blend of blue and yellow yields a white emission. Such flexible OLEDs are capable of being mass-produced using the roll-to-roll process, paving the way for high-performance wearable optoelectronics with reduced manufacturing and material costs [86].

Another promising area for flexible molecular devices is in biomedical applications, where adaptability and biocompatibility are essential. A recent study demonstrates that biodegradable molecular electronic devices exhibit a novel capability for controlled drug delivery [87]. Figure 12.21 illustrates the device along with its operating principle.

In these biodegradable molecular devices, the biodegradable molecular poly (lactic acid-trimethyl carbonate) (P(LA-TMC)), preloaded with a drug, incorporates Zn wires as an RF coil to capture energy [87]. Zinc also functions as a heating element, utilizing an MgO layer for insulation to prevent

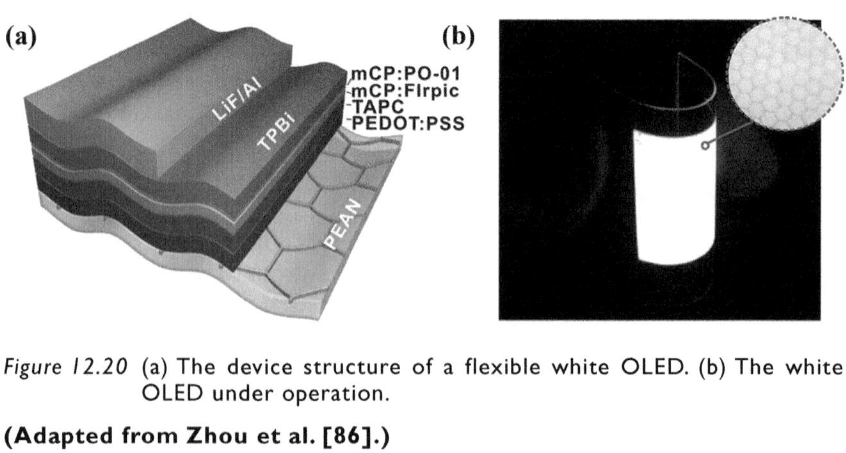

Figure 12.20 (a) The device structure of a flexible white OLED. (b) The white OLED under operation.

(Adapted from Zhou et al. [86].)

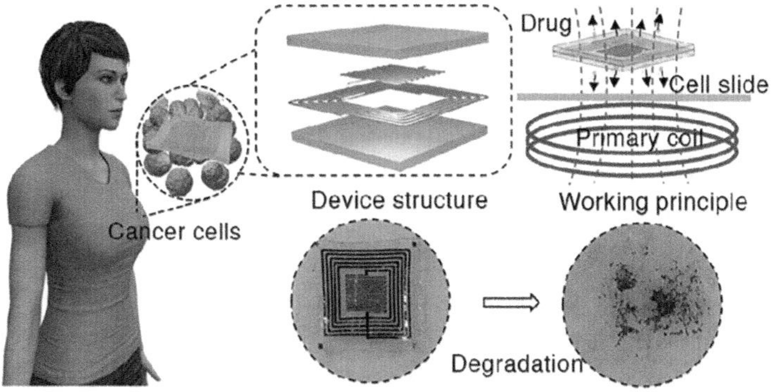

Figure 12.21 Biodegradable flexible molecular devices for drug delivery.

(Adapted from Li et al. [87].

short circuits between the Zn wire and the heating element. When an external alternating magnetic field is applied, the Zn heating element warms the drug-loaded P(LA-TMC)[87]. This study focuses on a biodegradable flexible electronic device incorporating the cancer drug paclitaxel. The researchers demonstrated their ability to release the drug in a controlled way while maintaining the mechanical integrity of the device, an essential factor for sustained biomedical applications [87]. These advancements could pave the way for innovative implantable medical devices capable of both monitoring and delivering therapies.

12.4.2 Energy harvesting and storage

Molecular materials could play a key role in energy-collecting devices such as organic solar cells (OSC), providing sustainable alternatives to conventional energy generation technologies. Recent research has reported that chemically welded silver nanowire (AgNW)-based transparent electrodes serve as critical components in flexible OSC (Figure 12.22) architectures [88]. By using ionic liquid (IL) type reducing agents to achieve atomic-level contacts between nanowires, researchers have engineered highly stable electrodes. The OSC has been constructed on such electrodes using molecular PM6:Y6 and M6:BTP-eC9:PC71BM as the active layer. The recorded power conversion efficiencies of using M6:BTP-eC9:PC71BM are up to 17.52%. The device structure is shown in Figure 12.23.

Research indicates that specific molecular compounds function as highly effective electrodes, greatly enhancing the efficiency of rechargeable lithium-ion batteries (LIBs). Among these, perylene polyimide emerges as a noteworthy option for organic electrodes in LIBs [89]. The configuration of LIBs utilizing polyimide as the upper electrode is illustrated in Figure 12.23.

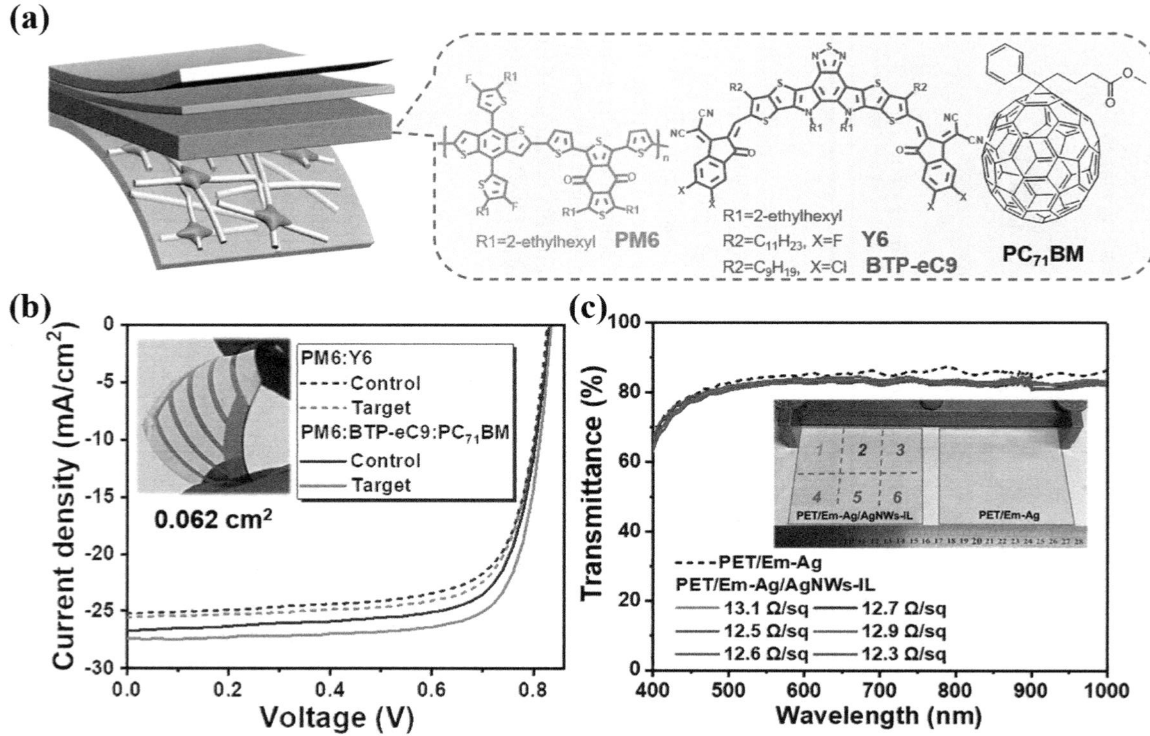

Figure 12.22 (a) Schematic representation of a flexible OSC, showing the molecular structures of the donor PM6 and the acceptors Y6, BTP-eC9, and PC71 BM. (b) J – V characteristics of the 0.062-cm^2 control and targeted flexible OSCs when illuminated by AM1.5G at 100 mW/cm^2. (c) Transmittance and resistance data from six distinct regions on a large-area FTE, with the configuration of PET/Embedded AgNW/AgNWs-IL.

(Adapted from Zeng et al. [88].)

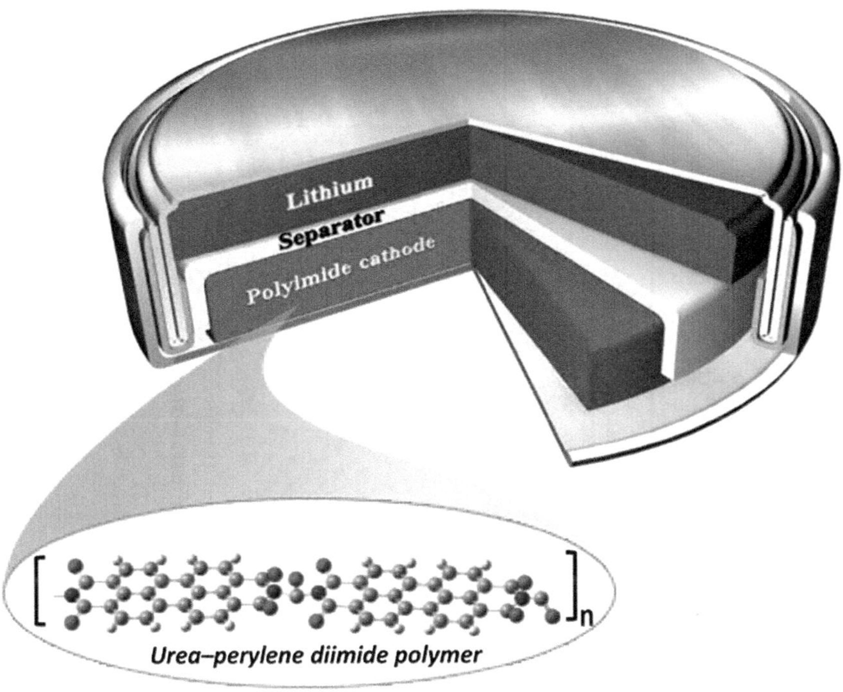

Figure 12.23 The LIB cell structure using polyimide as electrode.

(Adapted from Sharma et al. [89].)

Perylene-based polyimide LIBs are noted for their excellent electrochemical properties, including high specific capacity and impressive cycling stability, making them attractive for energy storage applications [89, 90]. The charge and discharge process of this kind of metal–organic battery has been shown in Figure 12.24.

12.4.3 Sensors

Molecular electronics can facilitate the creation of extremely sensitive wearable sensors to monitor individuals' health conditions. Recent research highlights the wearable sensors based on polymictic composition Dupond 5036 [91]. Wearable technologies have ushered in groundbreaking noninvasive techniques for tracking health biomarkers, as illustrated in a study on a dual biofluid biosensor embedded in a flexible tattoo platform. The device is displayed in Figure 12.25.

This innovative device can simultaneously extract and analyze sweat and interstitial fluid (ISF) using iontophoresis, a method that improves biomarker detection by accurately reflecting blood glucose levels without invasive procedures. Employing electrochemical biosensors, it measures glucose

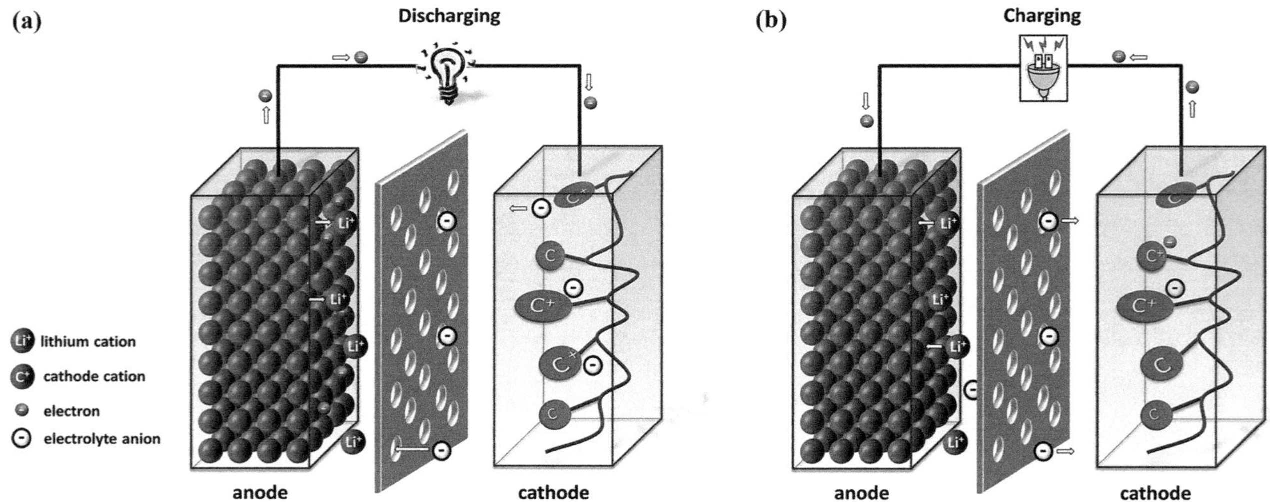

Figure 12.24 Diagram illustrating the processes of (a) discharging and (b) charging in a metal–organic battery. **(Adapted from Muench et al. [90].)**

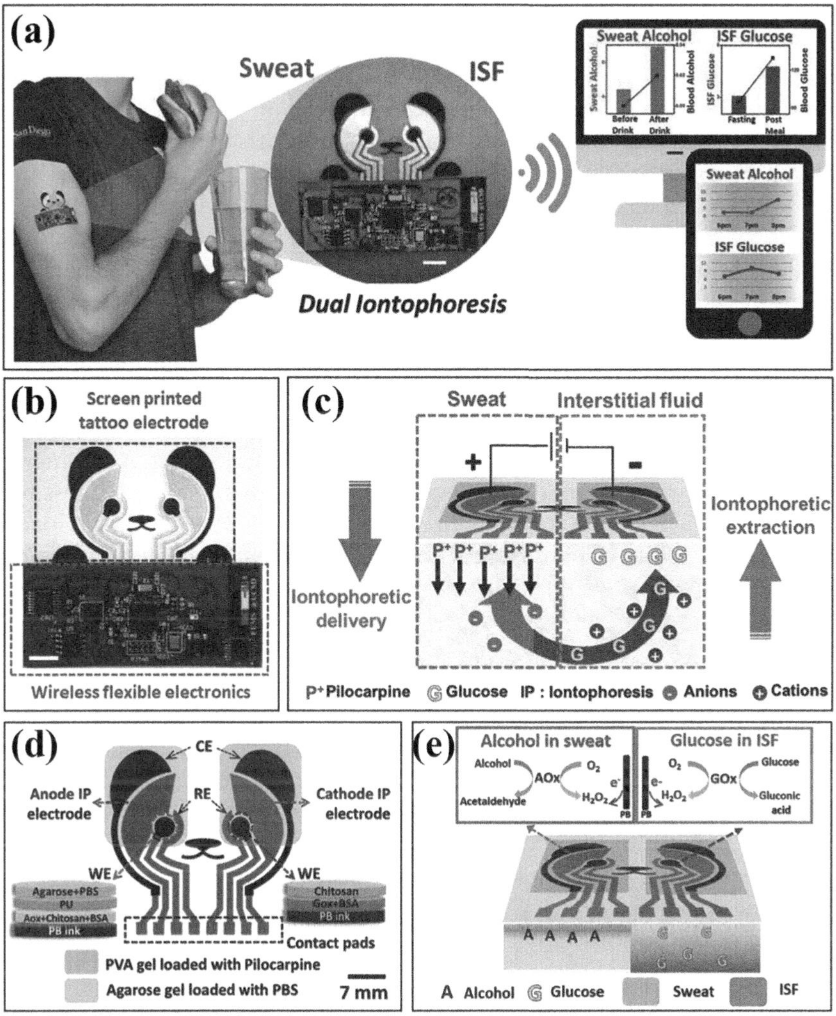

Figure 12.25 (a) Illustration of a wearable iontophoretic biosensor device integrated on a printed tattoo platform for glucohol (glucose + alcohol) detection on a human subject, featuring wireless real-time transmission of interstitial fluid (ISF) glucose and sweat alcohol levels. Scale bar: 7 mm. (b) Photograph of the screen-printed glucohol biosensor paired with a wireless flexible printed circuit board. Scale bar: 7 mm. (c) Diagram illustrating the iontophoretic process. The anode facilitates the generation of alcohol-laden sweat through the iontophoretic delivery of pilocarpine, while simultaneously, ISF glucose is sampled via reverse iontophoresis at the cathode. (d) Diagram of the glucohol biosensor, highlighting the "panda-like" electrode arrangement with iontophoretic gels and the layered structure of the working electrodes. Scale bar: 7 mm. (e) Diagram of the biosensing process. Amperometric detection identifies alcohol in the generated sweat and glucose in the collected ISF, using AO_x and GO_x, respectively, along with the measurement of hydrogen peroxide produced from these enzymatic reactions.

(Adapted from Kim et al. [91].)

and alcohol concentrations in real-time, showing strong correlations with conventional blood testing methods during in vivo assessments [91]. These advancements mark significant progress in healthcare technology, creating new possibilities for effective disease management, especially for diabetes and alcohol dependence, and ultimately making personalized health monitoring more accessible and convenient.

Other than molecular substrate or binder-based biosensors, crown ethers, as shown in Figure 12.26, are also used to detect alkali metal ions, such as sodium and potassium, by developing biosensors anchored on nanodiamonds functionalized with crown ether structures [92]. It shows unique optical properties of nitrogen-vacancy (NV) centers within nanodiamonds, which enable highly sensitive detection of specific ions due to changes in photoluminescence correlated with ion concentration.

Crown ether sensors can accurately differentiate among various alkali ions, which is crucial for medical diagnostics and bioimaging. Assessing the sensors' performance in diverse experimental setups reveals new possibilities for employing these functionalized nanomaterials in real-time biological monitoring, highlighting the significant implications of this technology for healthcare [92].

12.5 CHALLENGES TO OVERCOME

On one hand, molecular electronics have shown tremendous promise for the advancement of nanoelectronics, through creating efficient and powerful miniature-sized electronic devices; on the other hand, it faces several challenges. Prominent issues include a lack of reliable performance due to slower switching speed, stability issues, and practical implementation in real-time devices.

Some of the prominent challenges are as follows:

1. Stability and Reliability: Unlike conventional silicon devices, molecular electronic devices lack stability and reliability because of their unique molecular structures. Hence, it plays as a major hurdle to create durable molecular devices. Issues of stability and reliability can be addressed by developing stable molecular structures and by using techniques that can shield the structures from changes in the working atmosphere.
2. Scalability: The next challenge to be addressed is scalability, which needs further research and development in achieving self-assembly techniques and molecular engineering to create molecular devices with consistent performance by achieving tighter control over the placement and orientation of molecules. This will pave the way for mass production of molecular devices, resulting in commercial products.

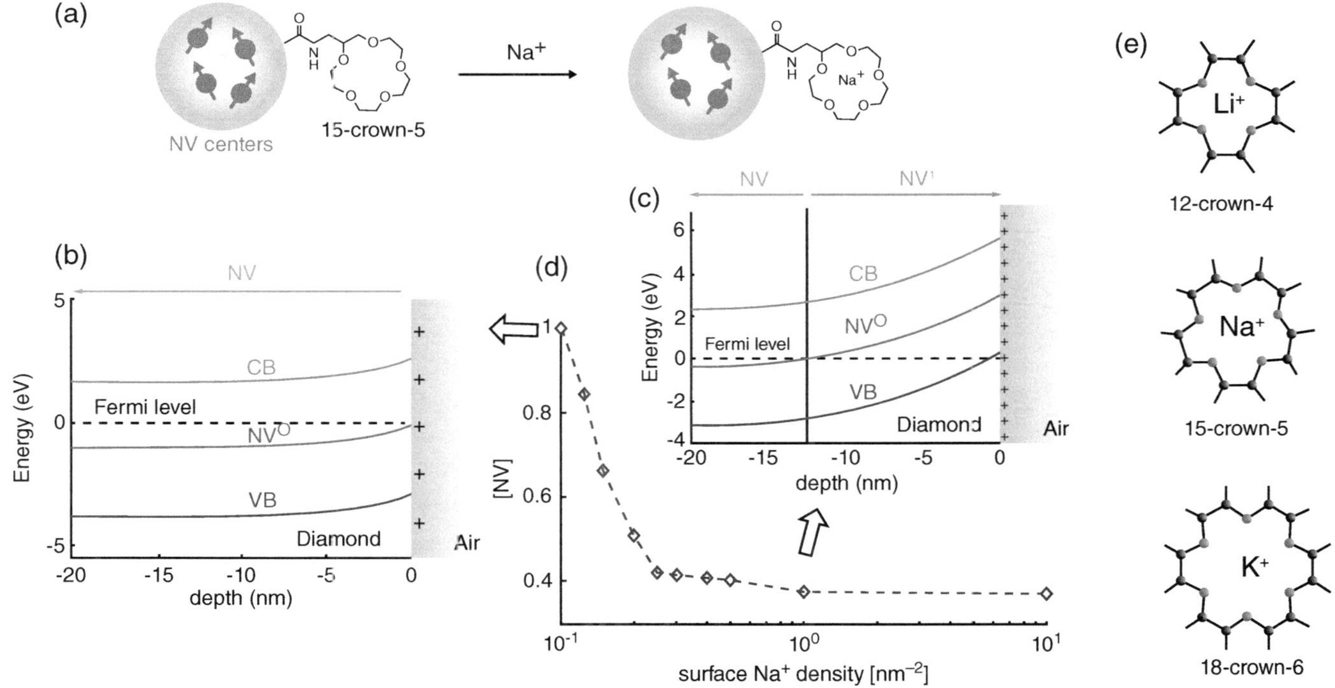

Figure 12.26 (a) Schematics of the ion sensor. NV center-containing NDs are first coated with a crown ether. Upon their introduction, sodium cations will be trapped in the 15-crown-5 compounds. (b) Energy band schematic of the diamond-air interface. Positive charges on the diamond surface lead to upward bending of the conduction band (CB) and valence band (VB) as well as the $NV^{-/0}$ transition level. The dashed line indicates the Fermi level. NV^0 is the dominant NV charge state when the $NV^{-/0}$ transition level is above the Fermi level, while NV^- dominates in the opposite case. The surface density of Na^+ is taken to be $0.1 nm^{-2}$ here. (c) Energy band structure when the surface density of Na^+ is $1 nm^{-2}$. (d) Simulated fraction of negatively charged NV states versus surface Na^+ density. (e) Generalization into other types of crown ethers, which can selectively bind certain cations and form complexes.

(Adapted from Li et al. [92].)

3. Integration and energy efficiency: The next challenge is integrating molecular structures with substrates or external circuits. A lack of proper interfacing between the molecular systems and conventional electronic devices is a huge gap to be addressed to have sustainable, effective charge transport. The current-day molecular electronic devices lack efficiency due to a lack of proper integration, resulting in high contact resistance.
4. Standardization: High-speed applications need a proper charge transport mechanism, and the molecular systems lack efficiency due to the distinct nature of each molecular structure. Creating a standard molecular design with desired electronic properties is important to improve charge transport. Unlike the conventional semiconductor industry, there are no established industry standards and fabrication techniques in molecular electronics. This is a bane in developing a unified approach to molecular electronics.
5. Controlling the Quantum Effects: Controlling the quantum effects, such as interference, tunneling, and quantization of energy levels at the molecular level, is a key to achieving stable performance.
6. Characterization and Probing Techniques: Studying the properties of individual molecules and molecular devices is essential to evaluate and optimize the molecular systems for practical use. Advancements in scanning probe microscopy, single-molecule spectroscopy, and other nanoscale characterization techniques are helping overcome these limitations.
7. Cost: The next challenge to be addressed is the high cost arising from complex synthesis techniques of molecular components and time consumption. This is another major impediment to establishing the commercial viability of molecular electronics.

12.6 FUTURE PROSPECTS

For molecular electronics to compete with traditional electronics, it is essential to promote broad interdisciplinary collaboration and research among the scientific community and establish unified approaches and standards for molecular material synthesis, fabrication, and measurement techniques, as well as innovations in molecular engineering and interface engineering, such as creating molecular junctions that promote charge transport and enhance conductivity; demonstrating superior energy efficiency; developing more cost-effective synthetic routes; and enabling scalable production methods. These efforts could help address this issue and will make this field suitable for applications in ultralow power devices, flexible electronics, quantum computing, advanced sensors, nano processors, and high-density data storage. Overcoming the challenges relies on the unified approach from the scientific community, and despite all these challenges, molecular electronics holds immense potential, and the days are not far from seeing commercial molecular electronic devices in reality.

12.7 CONCLUSION

In summary, molecular electronics represents a promising frontier in the field of electronics, with contributions from countless articles and studies. Key advancements in molecular conductance, self-assembly, molecular switches, and quantum effects have brought us closer to building functional molecular devices. However, significant challenges remain in integrating these devices into scalable systems that can rival the performance of conventional electronics. As the field continues to evolve, it is likely that molecular electronics will play a pivotal role in the development of next-generation technologies, ranging from ultra-miniaturized circuits to flexible, high-performance devices.

REFERENCES

1. D. K. Aswal, S. Lenfant, D. Guerin, J. V. Yakhmi, and D. Vuillaume, "Self assembled monolayers on silicon for molecular electronics," *Analytica Chimica Acta*, vol. 568, no. 1–2, pp. 84–108, May 2006, https://doi.org/10.1016/j.aca.2005.10.027
2. P. T. Mathew and F. Fang, "Advances in molecular electronics: a brief review," *Engineering*, vol. 4, no. 6, pp. 760–771, Dec. 2018, https://doi.org/10.1016/j.eng.2018.11.001
3. B. L. Feringa, "The art of building small: from molecular switches to molecular motors," *The Journal of Organic Chemistry*, vol. 72, no. 18, pp. 6635–6652, Aug. 2007, https://doi.org/10.1021/jo070394d
4. D. K. James, J. M. Tour, Molecular wires. In: *Molecular Wires and Electronics. Topics in Current Chemistry*, Decola, L., De Cola, L. (eds.), vol 257. Springer, Berlin, Heidelberg, 2005. https://doi.org/10.1007/b136066
5. X. D. Cui, "Reproducible measurement of single-molecule conductivity," *Science*, vol. 294, no. 5542, pp. 571–574, Oct. 2001, https://doi.org/10.1126/science.1064354
6. L. B. K. Phillip, C. Lavigne, and A. Aspuru-Guzik, "Toward quantum computing with molecular electronics," *Journal of Chemical Theory and Computation*, vol. 18, no. 6, pp. 3318–3326, May 2022, https://doi.org/10.1021/acs.jctc.2c00162
7. F. Kilibarda et al., "Single-molecule doping: conductance changed by transition metal centers in salen molecules," *Advanced Electronic Materials*, vol. 7, no. 10, Jul. 2021, https://doi.org/10.1002/aelm.202100252
8. P. Wu, B. Dharmadhikari, P. Patra and X. Xiong, [2] Rotaxane as a switch for molecular electronic memory application: a molecular dynamics study, *Journal of Molecular Graphics & Modelling*, vol. 2022, 108163 —108176, 2022, https://doi.org/10.1016/j.jmgm.2022.108163

9. N. A. Zimbovskaya and M. R. Pederson, "Electron transport through molecular junctions," *Physics Reports*, vol. 509, no. 1, pp. 1–87, Dec. 2011, https://doi.org/10.1016/j.physrep.2011.08.002
10. S. Han, X. Liang, I. Razdolski, Y. Bai, H. Li, and D. Lei, "Optical and charge transport characteristics of photoswitching plasmonic molecular systems," *Progress in Quantum Electronics*, vol. 95, p. 100517, May 2024, https://doi.org/10.1016/j.pquantelec.2024.100517
11. N. Chen et al., "Electronic logic gates from three-segment nanowires featuring two p–n heterojunctions," *NPG Asia Materials*, vol. 5, no. 8, pp. e59–e59, Aug. 2013, https://doi.org/10.1038/am.2013.36
12. R. Gupta, J. A. Fereiro, A. Bayat, A. Pritam, M. Zharnikov, and P. C. Mondal, "Nanoscale molecular rectifiers," *Nature Reviews Chemistry*, vol. 7, no. 2, pp. 106–122, Jan. 2023, https://doi.org/10.1038/s41570-022-00457-8
13. L. Venkataraman, J. E. Klare, C. Nuckolls, M. S. Hybertsen, and M. L. Steigerwald, "Dependence of single-molecule junction conductance on molecular conformation," *Nature*, vol. 442, no. 7105, pp. 904–907, Aug. 2006, https://doi.org/10.1038/nature05037
14. S. Gunasekaran, J. E. Greenwald, and L. Venkataraman, "Visualizing quantum interference in molecular junctions," *Nano Letters*, vol. 20, no. 4, pp. 2843–2848, Mar. 2020, https://doi.org/10.1021/acs.nanolett.0c00605
15. M. Famili et al., "Self-assembled molecular-electronic films controlled by room temperature quantum interference," *Chem*, vol. 5, no. 2, pp. 474–484, Feb. 2019, https://doi.org/10.1016/j.chempr.2018.12.008
16. Y. Deng et al., "Photo-responsive functional materials based on light-driven molecular motors," *Light: Science & Applications*, vol. 13, no. 1, Mar. 2024, https://doi.org/10.1038/s41377-024-01391-8
17. H. Mao, G. J. Pažėra, R. M. Young, M. D. Krzyaniak, and M. R. Wasielewski, "Quantum gate operations on a spectrally addressable photogenerated molecular electron spin-qubit pair," *Journal of the American Chemical Society*, vol. 145, no. 11, pp. 6585–6593, Mar. 2023, https://doi.org/10.1021/jacs.3c01243
18. I. Hnid et al., "Molecular junctions for terahertz switches and detectors," *Nano Letters*, vol. 24, no. 8, pp. 2553–2560, Feb. 2024, https://doi.org/10.1021/acs.nanolett.3c04602
19. R. Tuovinen and Y. Pavlyukh, "Electroluminescence rectification and high harmonic generation in molecular junctions," *Nano Letters*, vol. 24, no. 29, pp. 9096–9103, Jul. 2024, https://doi.org/10.1021/acs.nanolett.4c02609
20. S. Pecorario et al., "Stable and solution-processable cumulenic sp-carbon wires: a new paradigm for organic electronics," *Advanced Materials*, vol. 34, no. 15, Feb. 2022, https://doi.org/10.1002/adma.202110468
21. A. von Hippel, "Molecular engineering," *Science*, vol. 123, no. 3191, pp. 315–317, Feb. 1956, https://doi.org/10.1126/science.123.3191.315
22. R. P. Feynman, "There's plenty of room at the bottom," *Engineering and Science*, vol. 23, no. 5, pp. 22–36, Feb. 1960, Available: https://calteches.library.caltech.edu/1976/
23. H. Choi and C. C. M. Mody, "The long history of molecular electronics," *Social Studies of Science*, vol. 39, no. 1, pp. 11–50, Feb. 2009, https://doi.org/10.1177/0306312708097288
24. B. Mann and H. Kuhn, "Tunneling through fatty acid salt monolayers," *Journal of Applied Physics*, vol. 42, no. 11, pp. 4398–4405, Oct. 1971, https://doi.org/10.1063/1.1659785

25. A. Aviram and M. A. Ratner, "Molecular rectifiers," *Chemical Physics Letters*, vol. 29, no. 2, pp. 277–283, Nov. 1974, https://doi.org/10.1016/0009-2614(74)85031-1
26. M. Ratner, "A brief history of molecular electronics," *Nature Nanotechnology*, vol. 8, no. 6, p. 378–381, Jun. 2013, https://doi.org/10.1038/nnano.2013.110
27. L. Sun, Y. A. Diaz-Fernandez, T. A. Gschneidtner, F. Westerlund, S. Lara-Avila, and K. Moth-Poulsen, "Single-molecule electronics: from chemical design to functional devices," *Chemical Society Reviews*, vol. 43, no. 21, pp. 7378–7411, 2014.
28. M. A. Reed, C. Zhou, C. J. Muller, T. P. Burgin, and J. M. Tour, "Conductance of a molecular junction," *Science*, vol. 278, no. 5336, pp. 252–254, Oct. 1997, https://doi.org/10.1126/science.278.5336.252
29. S. Marqués-González and P. J. Low, "Molecular electronics: history and fundamentals," *Australian Journal of Chemistry*, vol. 69, no. 3, p. 244, 2016, https://doi.org/10.1071/ch15634
30. D. Kos et al., "Optical probes of molecules as nano-mechanical switches," *Nature Communications*, vol. 11, no. 1, Nov. 2020, https://doi.org/10.1038/s41467-020-19703-y
31. T. Harashima et al., "Single-molecule junction spontaneously restored by DNA zipper," *Nature Communications*, vol. 12, no. 1, Oct. 2021, https://doi.org/10.1038/s41467-021-25943-3
32. M. Ghasemi et al., "A new full adder cell for molecular electronics," *International Journal of VLSI Design & Communication Systems*, 2012. https://doi.org/10.5121/vlsic.2011.2401
33. B. Capozzi et al., "Single-molecule diodes with high rectification ratios through environmental control," *Nature Nanotechnology*, vol. 10, no. 6, pp. 522–527, Jun. 2015, https://doi.org/10.1038/nnano.2015.97
34. S. Sherif, G. Rubio-Bollinger, E. Pinilla-Cienfuegos, E. Coronado, J. C. Cuevas, and N. Agraït, "Current rectification in a single molecule diode: the role of electrode coupling," *Nanotechnology*, vol. 26, no. 29, pp. 291001–291001, Jul. 2015, https://doi.org/10.1088/0957-4484/26/29/291001
35. J. Trasobares, D. Vuillaume, D. Théron, and N. Clément, "A 17 GHz molecular rectifier," *Nature Communications*, vol. 7, no. 1, Oct. 2016, https://doi.org/10.1038/ncomms12850
36. Z. Wang et al., "Electrochemically controlled rectification in symmetric single-molecule junctions," *Proceedings of the National Academy of Sciences*, vol. 119, no. 39, Sep. 2022, https://doi.org/10.1073/pnas.2122183119
37. W. Xu et al., "Supramolecular diodes with donor–acceptor interactions," *Journal of the American Chemical Society*, Feb. 2025, https://doi.org/10.1021/jacs.4c14656
38. T. T. Phuc, "On theory of single-molecule transistor," *Journal of Physics Conference Series*, vol. 187, pp. 012055–012055, Sep. 2009, https://doi.org/10.1088/1742-6596/187/1/012055
39. S. Heinze, J. Tersoff, R. Martel, V. Derycke, J. Appenzeller, and P. Avouris, "Carbon nanotubes as Schottky barrier transistors," *Physical Review Letters*, vol. 89, no. 10, Aug. 2002, https://doi.org/10.1103/physrevlett.89.106801
40. A. Javey, J. Guo, Q. Wang, M. Lundstrom, and H. Dai, "Ballistic carbon nanotube field-effect transistors," *Nature*, vol. 424, no. 6949, pp. 654–657, Aug. 2003, https://doi.org/10.1038/nature01797

41. T. W. Ebbesen, H. J. Lezec, H. Hiura, J. W. Bennett, H. F. Ghaemi, and T. Thio, "Electrical conductivity of individual carbon nanotubes," *Nature*, vol. 382, no. 6586, pp. 54–56, Jul. 1996, https://doi.org/10.1038/382054a0
42. X. Guo, Q. Zhu, L. Zhou, W. Yu, W. Lu, and W. Liang, "Evolution and universality of two-stage Kondo effect in single manganese phthalocyanine molecule transistors," *Nature Communications*, vol. 12, no. 1, Mar. 2021, https://doi.org/10.1038/s41467-021-21492-x
43. B. L. Feringa and W. R. Browne, *Molecular Switches*. John Wiley & Sons, 2011.
44. T. Ye et al. "Artificial sodium-selective ionic device based on crown-ether crystals with subnanometer pores," *Nature Communications*, vol. 12, p. 5231, 2021, https://doi.org/10.1038/s41467-021-25597-1
45. J. Riebe et al., "Molecular folding governs switchable singlet oxygen photoproduction in porphyrin-decorated bistable rotaxanes," *Communications Chemistry*, vol. 7, no. 1, Aug. 2024, https://doi.org/10.1038/s42004-024-01247-7
46. A. Hamedian, M. Vakili, S. A. Brandán, M. Akbari, A. Kanaani, and V. Darugar, "Theoretical study on the structure, spectroscopic, and current–voltage behavior of 11-cis and trans retinal isomers in rhodopsin," *Scientific Reports*, vol. 14, no. 1, May 2024, https://doi.org/10.1038/s41598-024-63249-8
47. H. Kang, Y. Yang, and B. Wei, "Synthetic molecular switches driven by DNA-modifying enzymes," *Nature Communications*, vol. 15, no. 1, May 2024, https://doi.org/10.1038/s41467-024-47742-2
48. B. Daly et al., "Molecular memory with downstream logic processing exemplified by switchable and self-indicating guest capture and release," *Nature Communications*, vol. 10, no. 1, Jan. 2019, https://doi.org/10.1038/s41467-018-07902-7
49. J. Trasobares et al., "Hybrid molecular graphene transistor as an operando and optoelectronic platform," *Nature Communications*, vol. 14, no. 1, Mar. 2023, https://doi.org/10.1038/s41467-023-36714-7
50. Z. Chen et al., "Quantum interference enhances the performance of single-molecule transistors," *Nature Nanotechnology*, vol. 19, pp. 986–992, 2024, https://doi.org/10.1038/s41565-024-01633-1
51. C. Yan et al., "From molecular electronics to molecular intelligence," *ACS Nano*, vol. 18, no. 42, pp. 28531–28556, Oct. 2024, https://doi.org/10.1021/acsnano.4c10389
52. J. Song, W. Lyu, K. Kawakami, and K. Ariga, "Bio-gel nanoarchitectonics in tissue engineering," *Nanoscale*, vol. 16, no. 28, pp. 13230–13246, Jan. 2024.
53. M. Lemmer et al., "Unsupervised vector-based classification of single-molecule charge transport data," *Nature Communications*, vol. 7, p. 12922, 2016, https://doi.org/10.1038/ncomms12922
54. C. Shang et al., "Low-probability events detection using unsupervised multi-prototype clustering for single-molecule electronics," *Nano Research*, vol. 18, no. 4, p. 94907276, Apr. 2025, https://doi.org/10.26599/nr.2025.94907276
55. M. Tsutsui and M. Taniguchi, "Single molecule electronics and devices," *Sensors*, vol. 12, no. 6, pp. 7259–7298, May 2012, https://doi.org/10.3390/s120607259
56. T. Li, V. K. Bandari, and O. G. Schmidt, "Molecular electronics: creating and bridging molecular junctions and promoting its commercialization," *Advanced Materials*, p. 2209088, Dec. 2022, https://doi.org/10.1002/adma.202209088

57. PubChem, "18-Crown-6," Nih.gov, 2025. https://pubchem.ncbi.nlm.nih.gov/compound/18-Crown-6#section=InChI (accessed Mar. 29, 2025).
58. G. W. Gokel, W. M. Leevy, and M. E. Weber, "Crown ethers: sensors for ions and molecular scaffolds for materials and biological models," *Chemical Reviews*, vol. 104, no. 5, pp. 2723–2750, Mar. 2004, https://doi.org/10.1021/cr020080k
59. R. A. Bartsch, "Effects of structural variation within proton-ionizable crown ethers upon the selectivity and efficiency of solvent extraction of alkali metal and alkaline earth cations," *Solvent Extraction and Ion Exchange*, vol. 7, no. 5, pp. 829–854, Jan. 1989, https://doi.org/10.1080/07360298908962340
60. B. P. Czech, D. A. Babb, B. Son, and R. A. Bartsch, "Functionalized 13-crown-4, 14-crown-4, 15-crown-4, and 16-crown-4 compounds: synthesis and lithium ion complexation," *The Journal of Organic Chemistry*, vol. 49, no. 25, pp. 4805–4810, Dec. 1984, https://doi.org/10.1021/jo00199a011
61. P. Wu, B. Dharmadhikari, P. Patra, and X. Xiong, "Rotaxane nanomachines in future molecular electronics," *Nanoscale Advances*, vol. 4, no. 17, pp. 3418–3461, 2022.
62. M. Feng et al., "Stable, reproducible nanorecording on rotaxane thin films," *Journal of the American Chemical Society*, vol. 127, no. 44, pp. 15338–15339, Oct. 2005, https://doi.org/10.1021/ja054836j
63. S. L. Broman and M. B. Nielsen, "Dihydroazulene: from controlling photochromism to molecular electronics devices," *Physical Chemistry Chemical Physics*, vol. 16, no. 39, pp. 21172–21182, Aug. 2014.
64. S. L. Broman et al., "Dihydroazulene photoswitch operating in sequential tunneling regime: synthesis and single-molecule junction studies," *Advanced Functional Materials*, vol. 22, no. 20, pp. 4249–4258, Jun. 2012, https://doi.org/10.1002/adfm.201200897
65. M. Morimoto and M. Irie, "Photochromism of diarylethene single crystals: crystal structures and photochromic performance," *Chemical Communications*, no. 31, p. 3895, Jan. 2005.
66. S. L. Gilat, S. H. Kawai, and J.-M. Lehn, "Light-triggered electrical and optical switching devices," *Journal of the Chemical Society, Chemical Communications*, no. 18, p. 1439, Jan. 1993.
67. S. Fan, Y. Lam, L. He, and J. Xin, "Synthesis and photochromism of catechol-containing symmetrical azobenzene compounds," *Royal Society Open Science*, vol. 9, no. 6, Jun. 2022, https://doi.org/10.1098/rsos.211894
68. X. Zhang, L. Hou, and P. Samorì, "Coupling carbon nanomaterials with photochromic molecules for the generation of optically responsive materials," *Nature Communications*, vol. 7, p. 11118, Apr. 2016, https://doi.org/10.1038/ncomms11118
69. K. Matsuda and M. Irie, "Diarylethene as a photoswitching unit," *Journal of Photochemistry and Photobiology C: Photochemistry Reviews*, vol. 5, no. 2, pp. 169–182, Oct. 2004, https://doi.org/10.1016/j.jphotochemrev.2004.07.003
70. C. Jia et al., "Covalently bonded single-molecule junctions with stable and reversible photoswitched conductivity," *Science*, vol. 352, no. 6292, pp. 1443–1445, Jun. 2016, https://doi.org/10.1126/science.aaf6298
71. A. Kumar et al., "Reversible photo-switching of single azobenzene molecules in controlled nanoscale environments," *Nano Letters*, vol. 8, no. 6, pp. 1644–1648, Apr. 2008.

72. S. Lara-Avila et al., "Bianthrone in a single-molecule junction: conductance switching with a bistable molecule facilitated by image charge effects," *The Journal of Physical Chemistry C*, vol. 114, no. 48, pp. 20686–20695, Oct. 2010, https://doi.org/10.1021/jp1060667
73. D. M. Guldi, H. Nishihara, and L. Venkataraman, "Molecular wires," *Chemical Society Reviews*, vol. 44, no. 4, pp. 842–844, 2015.
74. P.J. Low and S. Marqués-González, *Molecular Wires: An Overview of the Building Blocks of Molecular Electronics*. Springer eBooks, pp. 87–116, Jan. 2016, https://doi.org/10.1007/978-981-10-0724-8_4
75. R. Asakawa et al., "Periodically twisted molecular wires based on a fused unit for efficient intramolecular hopping transport," *Journal of the American Chemical Society*, vol. 146, no. 33, pp. 23529–23536, Aug. 2024, https://doi.org/10.1021/jacs.4c07548
76. J. Terao et al., "Design principle for increasing charge mobility of π-conjugated polymers using regularly localized molecular orbitals," *Nature Communications*, vol. 4, no. 1, Apr. 2013. https://doi.org/10.1038/ncomms2707
77. Y. Ie et al., "Improving intramolecular hopping charge transport via periodical segmentation of π-conjugation in a molecule," *Journal of the American Chemical Society*, vol. 143, no. 2, pp. 599–603, Dec. 2020. 10.1021/jacs.0c10560
78. S. Hiroto, Y. Miyake, and H. Shinokubo, "Synthesis and functionalization of porphyrins through organometallic methodologies," *Chemical Reviews*, vol. 117, no. 4, pp. 2910–3043, Oct. 2016, https://doi.org/10.1021/acs.chemrev.6b00427
79. Z. Li, T. Park, J. Rawson, M. J. Therien, and E. Borguet, "Quasi-ohmic single molecule charge transport through highly conjugated meso-to-meso ethyne-bridged porphyrin wires," *Nano Letters*, vol. 12, no. 6, pp. 2722–2727, May 2012, https://doi.org/10.1021/nl2043216
80. N. Algethami, H. Sadeghi, S. Sangtarash, and C. J. Lambert, "The conductance of porphyrin-based molecular nanowires increases with length," *Nano Letters*, vol. 18, no. 7, pp. 4482–4486, Jun. 2018, https://doi.org/10.1021/acs.nanolett.8b01621
81. Y. Wada, T. Uda, M. Lutwyche, S. Kondo, and S. Heike, "A proposal of nanoscale devices based on atom/molecule switching," *Journal of Applied Physics*, vol. 74, no. 12, pp. 7321–7328, Dec. 1993, https://doi.org/10.1063/1.354999
82. Y. Chen et al., "Nanoscale molecular-switch crossbar circuits," *Nanotechnology*, vol. 14, no. 4, pp. 462–468, Apr. 2003, https://doi.org/10.1088/0957-4484/14/4/311
83. L. Jia Yen, A. R. Mat Isa, and K. Ahmad Dasuki, "Modeling and simulation of single-electron transistors," *Malaysian Journal of Fundamental and Applied Sciences*, vol. 1, no. 1, Jun. 2005, https://doi.org/10.11113/mjfas.v1n1.9
84. Y. Wada, "A prospect for single molecule information processing devices," *Pure and Applied Chemistry*, vol. 71, no. 11, pp. 2055–2066, Nov. 1999, https://doi.org/10.1351/pac199971112055
85. L. D. A. Siebbeles, F. C. Grozema, *Charge and Exciton Transport through Molecular Wires*. Wiley, 2011, pp. 334.
86. L. Zhou et al., "High-performance flexible organic light-emitting diodes using embedded silver network transparent electrodes," *ACS Nano*, vol. 8, no. 12, pp. 12796–12805, Dec. 2014, https://doi.org/10.1021/nn506034g

87. H. Li et al., "Biodegradable flexible electronic device with controlled drug release for cancer treatment," *ACS Applied Materials & Interfaces*, vol. 13, no. 18, pp. 21067–21075, Apr. 2021, https://doi.org/10.1021/acsami.1c04653
88. G. S. Zeng et al., "Realizing 17.5% efficiency flexible organic solar cells via atomic-level chemical welding of silver nanowire electrodes," *Journal of the American Chemical Society*, vol. 144, no. 19, pp. 8658–8668, Apr. 2022, https://doi.org/10.1021/jacs.2c01503
89. P. Sharma, D. Damien, K. Nagarajan, M. M. Shaijumon, and M. Hariharan, "Perylene-polyimide-based organic electrode materials for rechargeable lithium batteries," *The Journal of Physical Chemistry Letters*, vol. 4, no. 19, pp. 3192–3197, Sep. 2013, https://doi.org/10.1021/jz4017359
90. A. W. Muench, C. Friebe, B. Häupler, T. Janoschka, and U. S. Schubert, "Polymer-based organic batteries," *Chemical Reviews*, vol. 116, no. 16, pp. 9438–9484, Aug. 2016, https://doi.org/10.1021/acs.chemrev.6b00070
91. J. Kim et al., "Simultaneous monitoring of sweat and interstitial fluid using a single wearable biosensor platform," *Advanced Science*, vol. 5, no. 10, p. 1800880, Aug. 2018, https://doi.org/10.1002/advs.201800880
92. C. Li, S.-X. L. Luo, D. M. Kim, G. Wang, and P. Cappellaro, "Ion sensors with crown ether-functionalized nanodiamonds," arXiv.org, 2023. https://arxiv.org/abs/2301.03143 (accessed Mar. 29, 2025).

Chapter 13

Introduction to spintronics memristors

M. Sathvika, S. Srimaan, and K.P. Pradhan

13.1 ORIGIN OF SPINTRONICS: FROM MAGNETORESISTANCE TO MODERN SPIN-BASED DEVICES

Spintronics emerged in the late 20th century when researchers found that an electron's spin (a fundamental quantum characteristic) could enhance the charge in electronic processes. The discovery of the Giant Magnetoresistance (GMR) effect in multilayered ferromagnetics by Albert Fert and Peter Grünberg in 1988 [1, 2] established a key milestone in spintronics. Fert and Grünberg received the Nobel Prize in Physics in 2007 for their discovery that the resistance of an object could be influenced by the alignment of the magnetic field with respect to the electric field, ultimately marking the beginning of spintronic technology [3]. In addition to providing high-density storage capability, GMR technology transformed the way data is stored, with IBM introducing the use of GMR-based sensors in its hard-disk drive read-heads in 1997, thus creating a dramatic increase in data storage capacity [4].

Growing discoveries and research in the early 2000s have opened up new avenues of research for the field of spintronics beyond metals. The creation of "organic spin valves," as well as the demonstration of the injection of spin-polarized current into "semiconductors," has enabled a broader range of spintronic applications [5]. The use of Magnetic Tunnel Junctions (MTJs), which consist of two ferromagnetic materials separated by a nonconductive barrier, has also become increasingly embraced. MTJs are the foundations of Magnetoresistive Random Access Memory (MRAM), which combines both high speed and low power consumption [4]. Early in the 21st century, Freescale Semiconductor released the first commercially available MRAM products [6].

In the mid-2000s, spin-transfer torque (STT) was proposed as a technique to control magnetization via spin-polarized currents, rather than by magnetic fields [7]. This allowed for the development of STT-MRAM, which has a greater density and lower power consumption during switching. Later on, spin-orbit torque (SOT) was developed using materials with high levels of spin and orbital coupling properties (e.g., topological insulators), resulting

 DOI: 10.1201/9781003635123-13

in the ability to switch materials rapidly, while utilizing much lower amounts of energy [8–10]. These advances have made spintronic-based systems a viable alternative to standard complementary metal-oxide-semiconductors (CMOSs) for future computing architectures.

Modern spintronics research has branched into neuromorphic and quantum computing fields, amalgamating magnetism with nonvolatile memory principles. The Landau–Lifshitz–Gilbert (LLG) equation now underpins the nonlinear magnetization dynamics responsible for memristive behaviors in MTJ-based systems. The spintronic memristor, which can be adapted as a learning machine, is defined by the symmetry between spin dynamics and resistance variability.

TDK Corporation made the first spin-memristor prototype for neuromorphic hardware in 2024. This system used 100 times less processing power than digital processors for AI tasks [11, 12]. The device, created in collaboration with Tohoku University and the French Atomic Energy Commission (CEA), showed how magnetic memory and computation could work together [13].

Spintronics has changed from using magnetism as storage to using it as computation since the discovery of GMR. This has laid the scientific foundation for low-power, adaptive, intelligent electronics [1, 14].

13.2 FUNDAMENTALS OF SPINTRONICS

Spintronics leverages the quantum property of electron spin, in addition to charge, to enable new functionalities in electronic devices. An electron's spin can be either "up" or "down," and the net imbalance of these states in a material is termed *spin polarization* [5]. Spin-polarized currents are central to spintronic phenomena and device operation.

13.2.1 Magnetoresistance phenomena

Two key effects are foundational:

Giant Magneto-resistance (GMR): In metallic multilayers, the electrical resistance is contingent upon the relative orientation of the magnetization within the neighboring ferromagnetic layers. The GMR ratio is defined as follows [1, 15]:

$$\mathrm{GMR} = \frac{R_{\uparrow\downarrow} - R_{\uparrow\uparrow}}{R_{\uparrow\uparrow}} \tag{13.1}$$

where $R_{\uparrow\uparrow}$ and $R_{\uparrow\downarrow}$ are, respectively, the resistances for parallel and antiparallel magnetic alignments.

Tunneling Magneto Resistance (TMR): In MTJs, electrons move through an insulating barrier that separates two ferromagnets. The TMR ratio is given by the following relation [16]:

$$\mathrm{TMR} = \frac{R_{AP} - R_P}{R_P} \tag{13.2}$$

where R_P and R_{AP} are the resistances for parallel and antiparallel magnetization.

13.2.2 Spin torque mechanisms

Spin-Transfer Torque (STT): A ferromagnet can transfer angular momentum and apply a torque on the magnetization when a spin-polarized current flows through it. This phenomenon is described by the following relation [7]:

$$\frac{d\mathbf{M}}{dt} = -\gamma \mathbf{M} \times \mathbf{H}_{\text{eff}} + \alpha \mathbf{M} \times \frac{d\mathbf{M}}{dt} + \frac{\gamma \hbar}{2eM_s V} \mathbf{I}_s \times \mathbf{M} \tag{13.3}$$

where $\mathbf{M}$ is the magnetization, γ is the gyromagnetic ratio, α is the Gilbert damping, M_s is the saturation magnetization, V is the volume, and $\mathbf{I}_s$ is the spin current.

Spin-Orbit Torque (SOT): The spin Hall effect (SHE) in materials with strong spin–orbit coupling allows an in-plane charge current to produce a transverse spin current, which can change the magnetization of a nearby ferromagnet [17].

Spin Hall Effect (SHE): In materials with strong spin–orbit interaction, the SHE converts a longitudinal charge current into a transverse spin current [18].

13.2.3 Metal-based versus semiconductor spintronics

Ferromagnetic metals and alloys are used in metal-based spintronics (GMR, TMR, and STT-MRAM), providing fast switching and reliable room-temperature operation [1]. Although spin injection and coherence present tunable characteristics, they also show promise for spintronic devices with low dimensions [19].

Rare-Earth Alloys: Strong magnetic anisotropy and high coercivity alloys are utilized in permanent magnets and stable memory [20].

13.3 MEMRISTOR BASICS

In 1971, Leon Chua formally proposed the **memristor** (memory resistor) as the fourth essential circuit component, in addition to resistors, capacitors, and inductors [21]. Nonvolatile memory is embodied in a memristor's ability to maintain a resistance state based on the history of current or voltage that has passed through it.

13.3.1 Voltage–current pinched hysteresis

The memristor's **pinched hysteresis loop** in its voltage–current $(V - I)$ characteristics is a crucial experimental signature. The $V - I$ curve displays a pinched loop that passes through the origin when an AC voltage with a variable frequency is applied. This loop collapses as the frequency rises, which is a feature of memristive behavior [22].

13.3.2 Mathematical modeling

The following relation can be used to model the behavior of the memristor:

$$V(t) = (q(t)) \cdot I(t) \tag{13.4}$$

where $V(t)$ is the voltage, $I(t)$ is the current, and $M(q)$ is the memristance, which depends on the charge $q(t)$ that has passed through the device.

One common way to express state-dependent memristance is:

$$M(q) = \frac{d\phi}{dq} \tag{13.5}$$

where ϕ is the magnetic flux linkage related to the charge.

The dynamics of the state variable $q(t)$ (or other internal states) are usually given by the following differential equation:

$$\frac{dq}{dt} = f(q, V, t) \tag{13.6}$$

which models how the internal state changes in response to applied voltage or current.

13.3.3 Classification of memristor types

Memristors can be categorized based on their physical mechanisms:

- **Ionic Memristors:** Uses ion migration to change resistance (e.g., metal oxides) [22].
- **Electronic Memristors:** Use only electronic effects, such as charge trapping.
- **Spintronic Memristors:** Uses magnetic states and spin polarization to adjust resistance [23].

- **Ferroelectric Memristors:** Use nonvolatile ferroelectric polarization states.
- **Organic Memristors:** Uses organic materials with ionic or electronic conduction pathways [24].

13.4 INTRODUCTION TO SPINTRONIC MEMRISTORS

The continuous scaling of semiconductor technology and the increasing demand for nonvolatile, low-power, and highly dense information processing systems have driven research toward new computing paradigms. *Spintronics* and *memristive systems* are among the most revolutionary approaches. By using the electron's inherent spin and magnetic moment in addition to its charge, spintronics opens up new possibilities for logic operations and data storage [5]. In contrast, memristors provide nonvolatile resistance states that rely on the electrical excitation history, allowing for neuromorphic computing and energy-efficient memory [22]. These two fields have come together to create *spintronic memristors*, which combine memory-dependent resistive switching and magnetic spin control, opening the door to sophisticated in-memory and neuromorphic computation architectures.

13.4.1 Evolution from spintronics to spin-memristive systems

In 1988 [1, 15], Albert Fert and Peter Grünberg discovered the GMR effect, which laid the foundation of spintronics. This effect showed that the relative alignment of the magnetizations of neighboring ferromagnetic layers separated by a nonmagnetic spacer affects the electrical resistance of a magnetic multilayer structure. Resistance is low when the magnetic moments are aligned in parallel, and high when they are aligned antiparallel. This discovery, which earned the 2007 Nobel Prize in Physics, served as the basis for contemporary magnetic data storage technologies like MRAM and MTJs [25].

The existence of a fourth basic passive circuit element, the *memristor*, which relates charge and flux linkage [21], was theoretically proposed by Leon Chua in 1971 in a different field of study. In 2008, researchers at HP Labs achieved the first experimental demonstration of a titanium dioxide (TiO_2) thin-film device exhibiting resistive switching due to the drift of oxygen vacancies under an applied bias [22]. These gadgets created new opportunities for nonvolatile and brain-inspired electronics by "remembering" their previous states.

The resulting *spintronic memristors* show resistance modulation regulated by both ionic motion and magnetic state when spintronic and memristive concepts are combined. Multilevel resistive states and synaptic functionalities are made possible by these devices, which manipulate

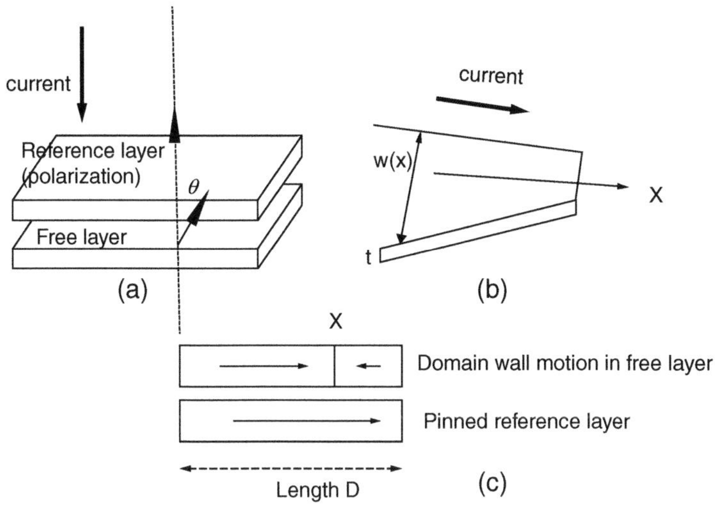

Figure 13.1 Schematic illustration of a spintronic memristor combining a magnetic tunnel junction (MTJ) with a resistive switching layer. The device exhibits magnetization-controlled and history-dependent resistance states. Spintronic memristors. (a) MTJ with spin-torque-induced magnetization switching. (b) Thin-film element with spin-torque-induced domain-wall motion. (c) Spin valve with spin-torque-induced domain-wall motion in the free.

(Wang et al. [28].)

magnetization states using the STT or SOT effects. Compact, energy-efficient, and nonvolatile components appropriate for logic-in-memory and neuromorphic computing are further made possible by the combination of MTJs and resistive switching layers.

Figure 13.1 shows a typical spintronic memristor configuration, which is made up of an oxide layer that provides resistive switching behavior integrated into a ferromagnet–insulator–ferromagnet stack (MTJ). While the oxide layer aids in ionic modulation of resistance, spin torque mechanisms can be used to switch the magnetization of the free layer. Thus, hybrid resistive and spin-based state control, which is essential to developing spin-based neuromorphic systems, is supported by the combined system.

13.4.2 Magnetic tunnel junction (MTJ) memristors

At the core of spintronic memristor technologies are MTJs, which combine analog resistive switching and magnetic memory phenomena in nanoengineered devices. MTJs are essential for applications in neuromorphic computing, spintronic logic, and STT magnetic random-access memory (STT-MRAM) [26, 27].

13.4.2.1 MTJ structure and operation

The two ferromagnetic layers that comprise an MTJ are separated by an ultrathin insulating barrier, typically made of magnesium oxide. While the magnetization of the bottom layer is fixed (pinned), the top layer's magnetization (free) can be altered by injecting spin-polarized current.

When the magnetizations are parallel, electrons tunnel more easily (low resistance, R_P). Antiparallel alignment (high resistance, R_{AP}) inhibits tunneling (Figure 13.2).

13.4.2.2 Current-induced magnetization switching

Switching the magnetization of the free layer is done by STT, which is defined by the Landau–Lifshitz–Gilbert–Slonczewski (LLGS) equation [15]:

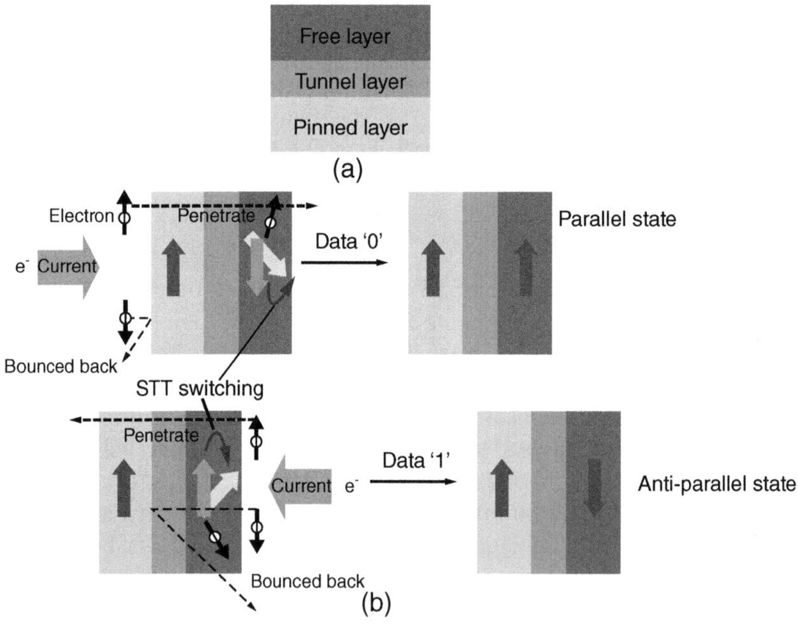

Figure 13.2 Schematic structure of a magnetic tunnel junction. The tunneling probability, and hence resistance, depends on the relative orientation of the magnetizations $\vec{M}_1$, $\vec{M}_2$ in the free and pinned layers. (a) Schematic structure of a magnetic tunnel junction (MTJ) comprising a free layer, tunnel barrier, and pinned layer. (b) Spin-transfer torque (STT) induced switching of the free-layer magnetization by spin-polarized current, producing the parallel (Data '0') and anti-parallel (Data '1') magnetic states used for information storage.

(Joshi et al. [26].)

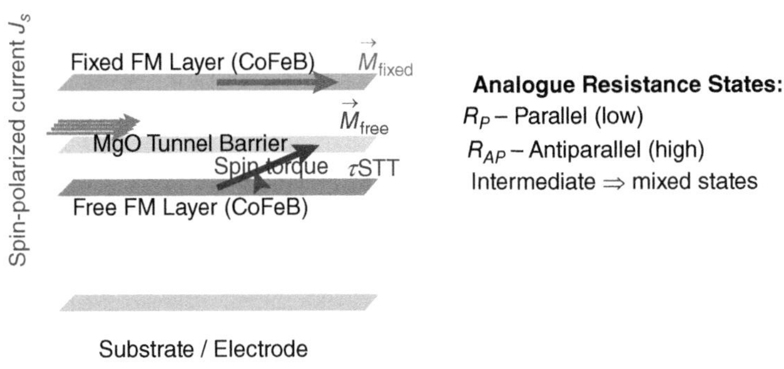

Figure 13.3 An example of an MTJ's current-induced magnetization switching. Analog resistance states and controlled rotation are made possible by the torque that spin-polarized current applies to the free layer.

(Zhang et al. [27].)

$$\frac{d\vec{m}}{dt} = -\gamma\left(\vec{m}\times\vec{H}_{eff}\right)+\alpha\left(\vec{m}\times\frac{d\vec{m}}{dt}\right)+\frac{\gamma\hbar\eta I}{2eM_sV}\left(\vec{m}\times\left(\vec{m}\times\vec{p}\right)\right) \tag{13.7}$$

$\vec{m}$ represents the magnetization vector, γ represents the gyromagnetic ratio, α represents the damping constant, η represents the spin-transfer efficiency, I represents the applied current, M_s represents the saturation magnetization, and V represents the volume of the free layer (Figure 13.3).

13.4.2.3 Multilevel resistance states

MTJs are not limited to binary switching. By altering the amplitude and duration of the current pulses, magnetization can be shifted to intermediate angles to create multilevel analog resistance. The magnetization angle and resistance in relation to θ:

$$R(\theta) = R_P + \left(R_{AP} - R_P\right)\frac{1-\cos\theta}{2} \tag{13.8}$$

where R_P and R_{AP} are parallel and antiparallel resistances, respectively [15] (Figure 13.4).

13.4.2.4 Experimental demonstrations and modeling

The memristive behavior of MTJs has been demonstrated through voltage-controlled switching experiments, hybrid MTJ-resistive devices, and perpendicular magnetic anisotropy stacks. The use of artificial synapses in neuromorphic circuits is made possible by the possibility of more than 60 programmable resistance states (Figure 13.5) [26, 27, 29].

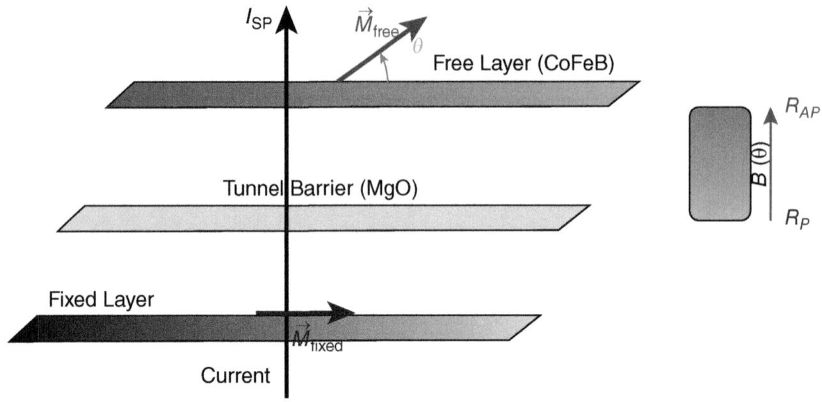

Figure 13.4 Analog tuning between R_P and R_{AP} is demonstrated by the resistance response of an MTJ memristor as the free layer's magnetization angle (θ) is changed [27].

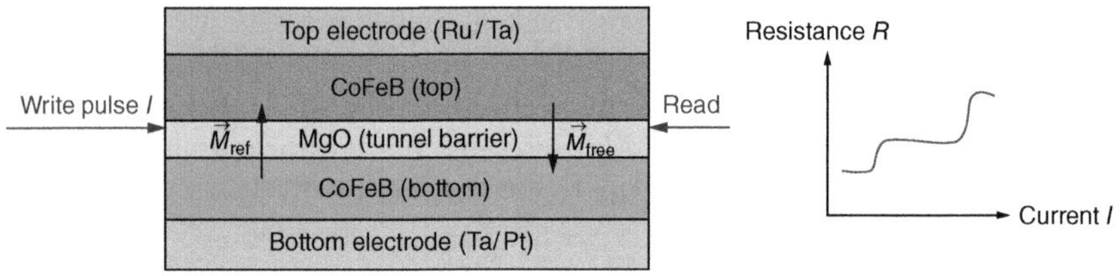

Figure 13.5 CoFeB/MgO/CoFeB perpendicular MTJ cell with experimental multilevel resistance switching.

(Reproduced from Zhang et al. [27].)

By combining the TMR equation, LLGS equation, and micromagnetic simulation of domain-wall motion, modeling is used to predict analog switching and hysteretic behavior.

13.4.3 Domain-wall memristors

A class of spintronic devices known as domain-wall memristors uses controlled displacement and pinning of magnetic domain walls in confined geometries, like racecourse memory elements and nanowires [30, 31], to modulate resistance states [30, 31]. These devices provide the possibility of neuromorphic architectures and high-density data storage by enabling multilevel, nonvolatile, and analog memory.

13.4.3.1 Domain-wall motion and pinning

Within a ferromagnetic material, a *magnetic domain wall* is a nanoscale boundary that divides areas (domains) of uniform but distinct magnetization. A nanowire's domain wall moves when an electric current or magnetic field is applied, moving the border between regions that are oppositely magnetized. With velocities as high as hundreds of m/s at room temperature, domain-wall motion dynamics are essential to the device's operation [30].

In order to stabilize the wall at specific locations, domain-wall pinning takes place at material inhomogeneities or engineered notches [32]. The device is appropriate for multilevel memory, as it contains multiple pinning sites that produce a distinct set of stable resistance states (Figure 13.6).

13.4.3.2 Device architectures: nanowires and racetrack memory

Ferromagnetic nanowires are most frequently used to create domain-wall devices. In the racecourse memory concept, information is stored as a pattern of domains and domain walls that are moved by a sequence of pulses of current. This enables three-dimensional, high-density solid-state storage without any moving mechanical components (Figure 13.7).

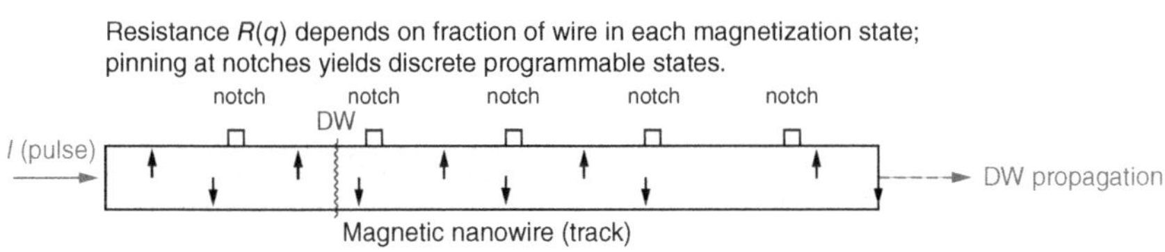

Figure 13.6 Domain-wall memristor schematic. Magnetic domain walls are pinned and contained by a nanowire with specially designed notches. Modifying the domain wall modifies the device resistance by altering the relative lengths of magnetized regions.

(Parkin et al. [30].)

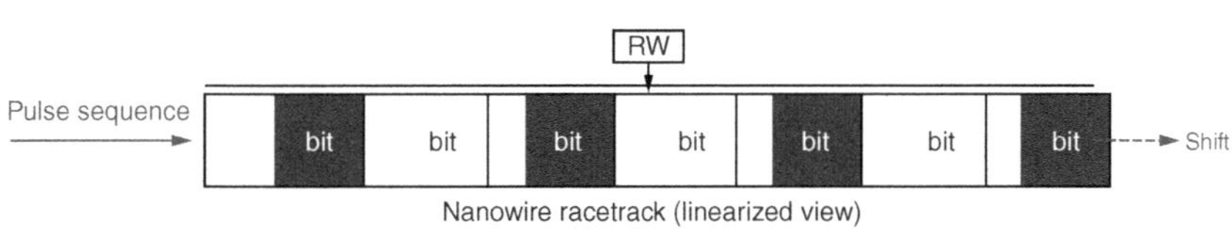

Figure 13.7 Racecourse memory architecture: Current pulses can move the magnetic regions (blue/white arrows) that are separated by domain walls (red) via the nanowire "track." One bit of information is encoded by each domain wall.

(Parkin et al. [30].)

13.4.3.3 Memristive switching via domain-wall dynamics

The percentage of the nanowire (or device) occupied by each magnetic domain configuration determines the resistance of a domain-wall memristor. The lengths of the "up" and "down" magnetic regions alter when an electrical current pulse shifts a domain wall to a new pinning location, resulting in a new resistance value. This mechanism enables programmable, nonvolatile multilevel memory.

13.4.3.4 Equations for domain-wall velocity and resistance change

The one-dimensional (1D) equation of motion (Thiele equation) for the wall position can be used to model domain-wall dynamics q [30]:

$$\dot{q}(t) = \frac{\gamma\Delta}{\alpha} H_{STT} \tag{13.9}$$

where α is the Gilbert damping parameter, γ is the gyromagnetic ratio, Δ is the domain-wall width, $\dot{q}(t)$ is the domain-wall velocity, and H_{STT} is the effective spin-torque field from current injection.

It is common practice to model the total device resistance $R(q)$ as a function of domain-wall position q as follows:

$$R(q) = R_{base} + (R_{AP} - R_P)\frac{q}{L} \tag{13.10}$$

where L is the total length of the nanowire, R_{base} is the intrinsic device resistance, and R_{AP} and R_P are the resistances for fully antiparallel and parallel configurations, respectively (Figure 13.8).

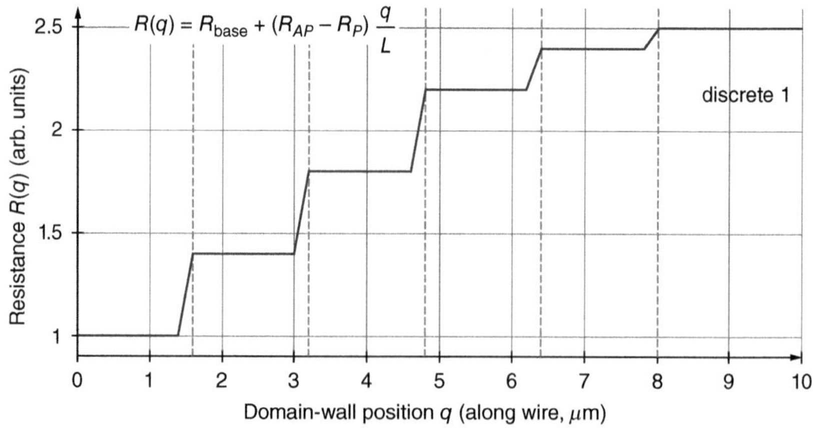

Figure 13.8 Device resistance in relation to q, the domain-wall position. Because each notch serves as a pinning site, distinct programmable resistance states are made possible [32].

13.4.4 Spin-wave memristors

Spin-wave memristors are a novel class of spintronic memory devices that utilize propagating magnons—quantized spin-wave excitations—as the physical carriers for storing and processing information. Unlike charge-based devices, magnonic systems leverage the low-dissipation propagation of spin waves in magnetic insulators, enabling nonvolatile, analog, and frequency-tunable memory for neuromorphic and microwave signal processing applications [33, 34].

13.4.4.1 Magnonic transport and spin-wave excitation

Spin waves are collective precessional motions of spins in magnetic lattices. Their propagation can be excited by applying microwave fields or spin currents generated using the SHE. The key spectrum governing spin-wave behavior is:

$$\omega(k) = \gamma\mu_0\left(H + Dk^2\right) \tag{13.11}$$

where ω is the magnon angular frequency, γ is the gyromagnetic ratio, μ_0 is the vacuum permeability, H is the bias field, k is the wavevector, and D is the exchange stiffness (Figure 13.9) [33].

13.4.4.2 Encoding memory states with spin waves

The device's cumulative history of magnon interference and spin-wave excitation determines its memristive state. The amplitude, phase, or constructive/destructive interference of propagating magnons can all be used to encode information. The following relation models history-dependent resistance:

$$V_{out}(t) = M_{SW}\left[\phi(t)\right]\cdot I_{in}(t) \tag{13.12}$$

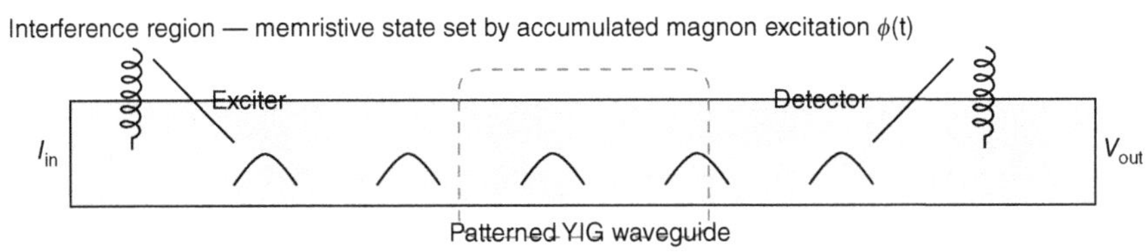

Figure 13.9 Schematic of a spin-wave memristor utilizing a patterned YIG waveguide with microwave antennas for excitation and detection. The output resistance may depend on magnon amplitude and interference pattern.

(Chumak et al. [33].)

where M_{SW} is the memristance function of accumulated magnon excitation $\phi(t)$ (the integral of magnon amplitude or phase over time), and V_{out} and I_{in} are the device's output and input signals.

13.4.4.3 Device implementations and modeling

Device platforms include the following:

- Patterned YIG (Yttrium Iron Garnet) waveguides.
- Magnonic crystals with superlattice structures or engineered grooves.
- Nanowire or microdisk magnetic resonators stimulated by spin current injection or antennas [34].

By adjusting the magnon amplitude, phase, and spatial arrangement within the device, experimental demonstrations have revealed programmable resistance states. Strong data storage with nonvolatile recall and multilevel analog states is produced by magnon scattering at engineered boundaries.

13.4.5 Topological spin-texture memristors

Topological spin-texture memristors are a sophisticated class of spintronic devices that achieve programmable resistance states by using nanoscale magnetic structures with nontrivial topology, such as *skyrmions* and *chiral domain walls*. These devices achieve robust, low-power, multilevel memristive behavior appropriate for neuromorphic computing and beyond [35–37] by utilizing the special stability and dynamic properties of topological spin textures.

Because of their nonzero topological charge, which shields them from constant deformation and collapse, magnetic skyrmions are whirl-like spin configurations. Skyrmions are perfect candidates for memory elements due to their nanoscale size, usually ranging from a few to tens of nanometers, and their manipulation capabilities with incredibly low current densities. Through variations in device resistance brought on by the shifting skyrmion population or position, the memristive response is determined by the dynamics of the skyrmions under applied spin currents (Figures 13.10 and 13.11).

When a skyrmion is driven by a spin current density $\mathbf{J}_s$, its velocity $\mathbf{v}_{sk}$ is determined by the following relation [37]:

$$\mathbf{v}_{sk} = \frac{\gamma}{2eM_s}\mathbf{J}_s \times \mathbf{n} \tag{13.13}$$

where e is the electron charge, M_s is the saturation magnetization, $\mathbf{n}$ is the unit vector that represents the topological charge direction of the skyrmion, and γ is the gyromagnetic ratio. According to this equation, the *skyrmion Hall effect* occurs when the skyrmion moves perpendicular to the spin current.

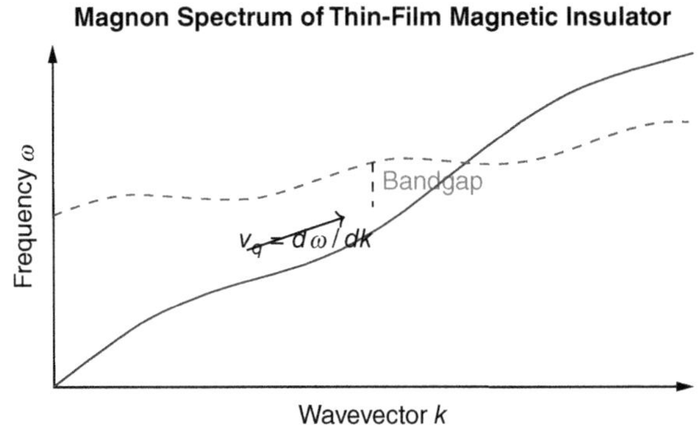

Figure 13.10 A thin-film magnetic insulator's magnon (spin-wave) spectrum. Geometry, field, and material properties can be used to adjust the band structure and group velocity.

(Chumak et al. [34].)

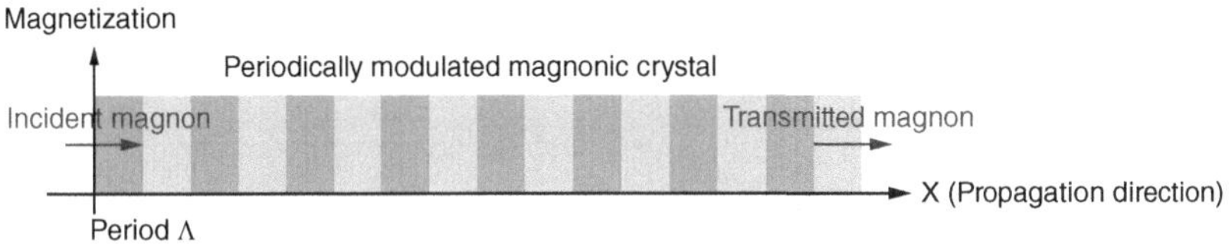

Figure 13.11 Periodically modulated magnonic crystals are employed in selective magnon transmission and bandgap engineering.

(Modified from Chumak et al. [34].)

Spin-polarized currents or electric fields can be used experimentally to nucleate, move, and annihilate skyrmions within confined nanostructures, allowing for programmable resistance states [35]. For the emulation of synaptic plasticity, gradual and reversible control of skyrmion density and configuration results in analog memristive behavior. Changes in the topological Hall effect and magnetoresistance brought about by the presence and location of skyrmions within the device cross-section give rise to the resistance modulation mechanism.

Another kind of topological spin texture that plays a role in memristive switching is chiral domain walls. Their motion under spin torque induces dynamic resistance changes, which affect local magnetic configurations. Engineered anisotropies and interface properties can be used to precisely control the robust, stable behavior of these domain walls [37].

Superior scalability, reduced energy consumption per switching event, and intrinsic stability to topological protection are some of the benefits that topological spin-texture memristors have over traditional memristors. These characteristics demonstrate their potential as foundational elements for upcoming hardware security primitives, neuromorphic systems, and more.

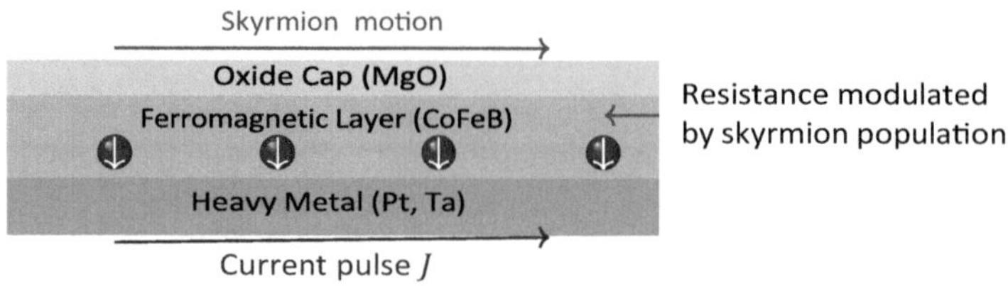

Figure 13.12 Schematic of a topological spin-texture memristor that uses current pulses to control skyrmions. Multilevel nonvolatile memory states are made possible by the modulation of the device resistance by the skyrmion population.

(Modified from Song et al. [36].)

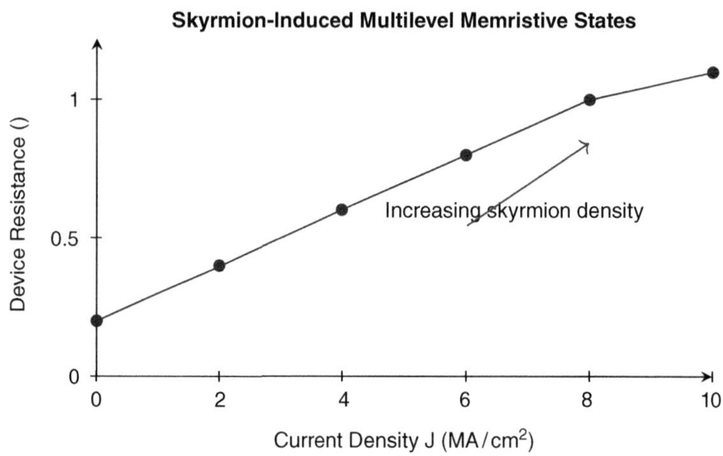

Figure 13.13 Experimental results showing device resistance modulation due to skyrmion dynamics, indicating multilevel memristive states.

(Reproduced from Luo et al. [35].)

There are still issues with accurately controlling skyrmion nucleation and detection, as well as integrating these devices at scale (Figures 13.12 and 13.13) [37].

13.4.6 Nanomagnet ensemble memristors

In the field of spintronic devices, nanomagnet ensemble memristors are a quickly emerging option for extremely dense, low-power, and durable memory architectures. These devices are made up of arrays of interacting nanoscale magnets that can change states through thermal activation, STT, or external fields. The behavior of these magnets as a whole results in rich memristive dynamics and emergent computational properties [38–40].

The physics of dipolar and exchange interactions provides the basis for the working principle. Collective switching events, noise-assisted transitions, and stochastic resonance phenomena are the results of interactions between each nanomagnet in the ensemble and its neighboring magnets as well as the applied field and current. Memristive transfer functions appropriate for reservoir computing, binary and multilevel memory, and probabilistic logic can be realized through the interaction of deterministic and stochastic mechanisms, which is enhanced by the array's size and geometry.

The coupled stochastic LLG equations, which capture the dynamic evolution of the magnetization $\vec{m}_i$ of each nanomagnet i, are a key framework for modeling these systems:

$$\frac{d\vec{m}_i}{dt} = -\gamma\left[\vec{m}_i \times \vec{H}_{eff,i}\right] + \alpha\left[\vec{m}_i \times \frac{d\vec{m}_i}{dt}\right] + \vec{\tau}_{STT,i} + \vec{\xi}_i(t) \tag{13.14}$$

where α represents Gilbert damping, γ represents the gyromagnetic ratio, $\vec{H}_{eff,i}$ represents the effective field (including dipolar interactions and anisotropy), $\vec{\tau}_{STT,i}$ represents the spin-transfer torque, and $\vec{\xi}_i(t)$ represents a stochastic thermal noise term [41].

The ensemble magnetization state, which is represented as the statistical distribution of individual nanomagnet orientations and their collective transitions, controls the device's overall resistive response. In real-world applications, the output can be made stochastic by using random thermal activation events for probabilistic bit operations or analog by adjusting the nanomagnet configuration, which allows for high-resolution weight storage.

In such architectures, experiments have shown reservoir computing, stochastic magnetic memory, and high-density logic arrays. For example, in physical neural networks, probabilistic bits (p-bits) are used to support noise-assisted inference and fast annealing [39, 40]. They are implemented as stochastic nanomagnets powered by low-energy currents. An ensemble memristor array with indicated directionality and current pathways for collective state transitions is shown in Figure 13.14. Nonvolatility, high packing density, intrinsic stochasticity for machine learning, and incredibly low power dissipation are the main benefits of nanomagnet ensemble memristors. The precise control of inter-magnet interactions, reproducible operation across large arrays, and scaling device fabrication to wafer-level processes—all of which are active research areas in modern spintronics and nanoelectronics—remain major challenges.

13.4.7 State-space dynamics in spintronic memristors

To fully utilize the computational and memory capabilities of spintronic memristors, it is essential to comprehend their state-space dynamics. Stable, oscillatory, stochastic, and chaotic collective behaviors are caused

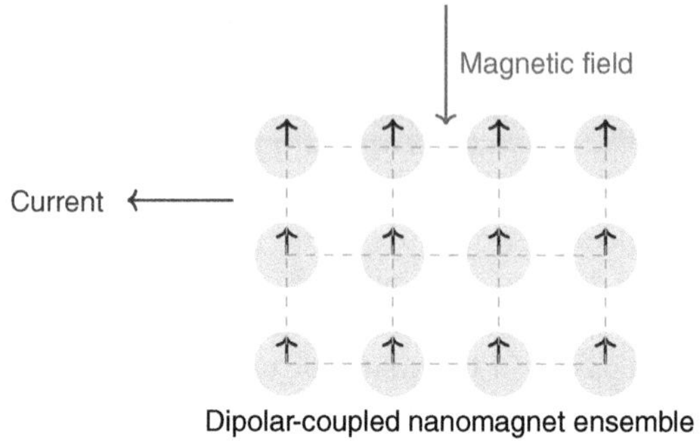

Figure 13.14 Schematic of a nanomagnet ensemble memristor: a collection of dipolar-field-coupled interacting magnets. Multilevel or probabilistic memory states are produced by stochastic collective switching under current or field.

(The figure was taken from Kamsari et al. [41].)

by nonlinear coupling, thermal effects, and intrinsic magnetization dynamics in nanoscale devices. In neuromorphic and probabilistic hardware, each regime has a direct effect on function, dependability, and adaptability [23, 39, 42].

13.4.7.1 Steady-state behavior

The device's magnetization stabilizes in steady-state operation, usually aligned parallel or antiparallel to the applied field or a reference layer. Reliability in binary memory encoding requires steady states, which are attained when damping and anisotropy effects balance STT and all driving forces. For nonvolatile digital logic, spintronic memristors such as domain-wall devices and MTJs take advantage of these stable points [23].

13.4.7.2 Oscillatory dynamics

Spintronic systems are capable of entering persistent oscillatory regimes when driven close to threshold or with periodic signals. Here, microwave-frequency spin-torque oscillators (STOs) are produced by the magnetization precessing around an effective field axis:

$$\frac{d\vec{m}}{dt} = -\gamma\vec{m}\times\vec{H}_{eff} + \alpha\vec{m}\times\frac{d\vec{m}}{dt} + \vec{\tau}_{STT} \tag{13.15}$$

These oscillators are utilized for neuromorphic spike generation, analog signal processing, and in-memory computation and frequency encoding for artificial neural networks [42]. A limit cycle represents a location of oscillatory trajectories in state space.

13.4.7.3 Stochastic switching

Probabilistic, noise-driven state transitions can be brought about by thermal fluctuations and inherent randomness (e.g., domain-wall depinning or nanomagnet switching). Because it allows for approximate inference, solves optimization problems, and introduces inherent randomness, this regime is essential for stochastic computing. This mechanism is physically realized in probabilistic bits (p-bits) using stochastic nanomagnets, where input current and temperature are used to adjust switching rates and probabilities [39]. The outcome is a broad distribution of switching trajectories, often visualized as a network of state-space paths weighted by transition likelihood.

13.4.7.4 Chaotic regimes

Nonlinear coupling between electrical and magnetic parameters, feedback loops, or multiple magnets can cause chaos in certain spintronic memristive circuits. Complex reservoir computing, in which time-series data is projected into high-dimensional space for pattern recognition, classification, or signal prediction, can make use of chaotic dynamics. Sensitive dependence on initial conditions and the appearance of odd attractors in the device phase diagrams are characteristics of such regimes (Figure 13.15) [42].

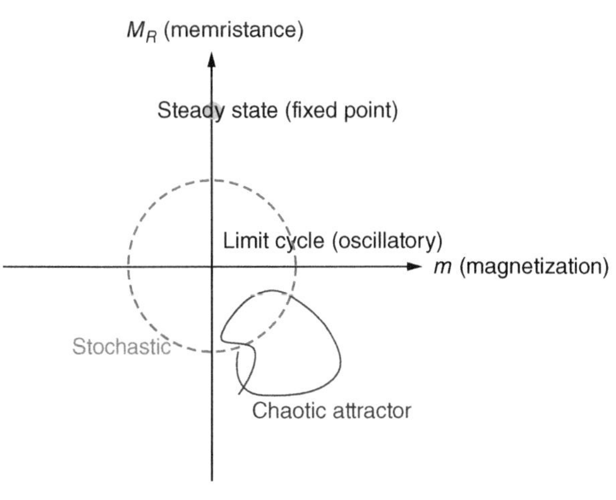

Figure 13.15 Phase diagrams and state-space trajectories in spintronic memristors. Steady states (fixed points) at the top; oscillatory regime (limit cycles) in the middle; stochastic switching events (diffuse phase space) at the bottom left; and a chaotic attractor as seen in a magnetization–memristance phase plot at the bottom right.

(Information modified from [23, 42].)

Table 13.1 Selected material properties and spintronic memristor device metrics

Material	*Spin polarization (%)*	*Curie temp (K)*	*Resistance-area (RA, $\Omega \cdot \mu m^2$)*	*TMR ratio (%)*
CoFeB/MgO	70–80	1000	5–20	200–600
Co_2MnSi	>90	> 900	5–50	~600
Bi_2Se_3	Surface only	–	100–500	—
CrI_3	Layer dep.	45	10–100	50–80
Fe_3GeTe_2	Layer dep.	200	20–200	60–100

13.5 MATERIALS AND FABRICATION

The scalability and performance of spintronic memristors are supported by the fabrication processes and materials landscape. Modern devices use a palette of carefully crafted layers, each intended to maximize resistive switching, magnetization dynamics, and spin transport. [4, 43, 44]. Based on recent IEEE and Nature reviews, Table 13.1 compares device performance metrics with exemplar material properties.

13.5.1 Key materials

In MTJs, magnesium oxide (MgO) is the preferred tunnel barrier. Strong spin filtering is made possible by MgO's crystalline structure and high symmetry, which results in high device endurance and large tunnel magnetoresistance (TMR) ratios. Maintaining optimal tunneling resistance and minimizing interfacial roughness require precise thickness control [4].

Heusler Alloys, ternary intermetallic compounds like Co_2MnSi and Co_2FeAl, show half-metallic behavior, providing almost 100% spin polarization at the Fermi level. Their robust magnetization and high Curie temperatures make them essential for room-temperature spintronic operation, and their electronic characteristics can be adjusted through epitaxial strain and stoichiometric control [43].

Topological Insulators, such as Bi_2Se_3 and Bi_2Te_3, have metallic, spin-polarized surface states and insulating bulk states that are shielded by time-reversal symmetry. These materials serve as the basis for SOT devices, in which spin-momentum locking on their surfaces is linked to effective spin current generation [44].

2D Magnets: New advances in van der Waals magnets, including $MnSe_2$, Fe_3GeTe_2, and CrI_3, pave the way for scalable, atomically thin spintronic architectures. Future quantum spintronic and multifunctional device applications are supported by their layer-dependent magnetic properties and ease of integration into heterostructures [19, 45].

13.5.2 Device fabrication techniques

The fabrication of spintronic memristors leverages the following:

- **Sputtering**: Physical vapor deposition for multilayer stacks guarantees high uniformity and scalable wafer processing, particularly for electrodes made of Heusler alloy and MgO barriers.
- **Molecular Beam Epitaxy (MBE)**: Applied to atomically precise layer growth, especially for engineered interfaces and topological insulators.
- **Lithography**: Device architectures, domain-wall tracks, and skyrmion racetracks can be patterned at the nanoscale using electron-beam or optical lithography.
- **CMOS Integration**: Direct compatibility with established silicon technology is made possible by flip-chip and back-end-of-line processes, which are essential for large-scale deployment and commercial viability [46].

13.5.3 Scalability and material engineering

Reducing dimensions while maintaining analog behavior and signal integrity is the main goal of material engineering for scalability. Increasing the density of a domain wall, skyrmion, or nanomagnet through interface tuning and defect engineering is one example of this. Exchange-biased stacks and synthetic antiferromagnets are used to stabilize small feature sizes. Crossbar scaling and new switching pathways are made possible by integrating 2D materials and functional oxides [19, 45].

13.6 APPLICATIONS

13.6.1 Multi-bit data storage and logic

13.6.1.1 Fundamentals and device structure

Multi-bit data storage using spintronic memristors represents one of the most promising applications for next-generation nonvolatile memory systems. Unlike conventional single-bit MTJ devices that can only store binary information (0 or 1), spintronic memristors enable the storage of multiple bits within a single device through continuous resistance state modulation. This capability is achieved through spin-torque-induced magnetization motion and domain-wall dynamics in ferromagnetic thin films (Figure 13.16).

The spintronic memristor for multi-bit storage consists of an MTJ structure or a spin-valve strip containing two ferromagnetic layers separated by a nonferromagnetic spacer. The reference layer maintains a fixed magnetization direction through exchange coupling with a pinned magnetic layer, while the free layer contains a magnetic domain wall that can be moved

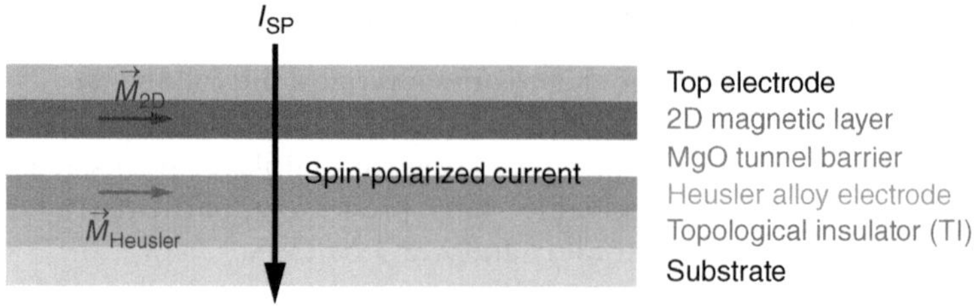

Figure 13.16 An example of a typical spintronic memristor layer stack that includes the topological insulator, Heusler alloy electrode, MgO tunnel barrier, and 2D magnet integration pathways.

(Chappert et al. [4].)

through applied current pulses. The resistance of the device is fundamentally determined by the position of this domain wall, as expressed by the following relation:

$$R(t) = R_L + (R_H - R_L)\frac{X(t)}{D} \tag{13.16}$$

where R_L and R_H represent the low and high resistance states, $X(t)$ represents the domain wall position at time t, and D represents the total length of the magnetic film (Figure 13.17).

13.6.1.2 Memristive switching behavior for multiple bits

This property enables precise control over the resistance states via modulation of applied current pulse width and amplitude. Multi-bit MTJ cells

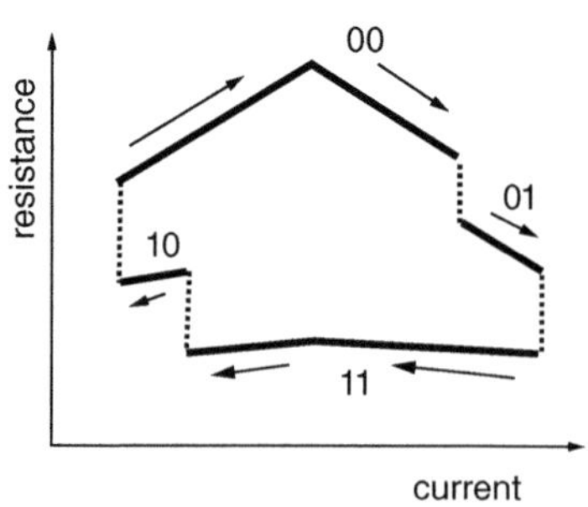

Figure 13.17 Multi-bit MTJ cell resistance versus current. This figure demonstrates the $R-I$ characteristic curves for a two-bit MTJ device showing four distinct states (00, 01, 10, and 11). One bit (hard bit) requires a higher current to switch, while the other bit (soft bit) requires a lower current.

(Wang and Chen [47].)

have been experimentally demonstrated, where two horizontal domains are introduced in the MTJ free layer to generate four different bits (00, 01, 10, and 11) or, alternatively, two MTJ devices are stacked vertically to achieve similar multilevel storage capacity [46].

The switching behavior exhibits both reversible and irreversible branches in the resistance–current $(R - I)$ characteristic curve. For example, reversible transitions can be achieved between the states (11 to 10) and (01 to 00), whereas direct transitions between some states, such as from 00 to 10, need sequential switching due to different critical current thresholds for individual bits. The presence of a hard bit (requiring higher current) and a soft bit (that requires lower current) allows for selective addressing through precise current pulse width modulation.

Experimental measurements demonstrate that the MTJ switching current magnitude is approximately inversely proportional to pulse width in the nanosecond region. This relationship allows the device to be operated at a crossover point where two MTJ sub-cells with different shape-induced anisotropies and effective volumes can be switched independently by controlling pulse duration. For instance, using current pulse widths of approximately 4 nanoseconds at around 900 μA enables discrimination between hard and soft bit switching, providing a practical mechanism for multi-bit manipulation (Figure 13.18).

13.6.1.3 Advantages for next-generation memory architectures

The multi-bit storage capability of spintronic memristors offers several significant advantages over conventional binary memory systems. First, the ability to store multiple bits per cell increases memory density without proportionally increasing the physical cell area. Second, the memristive nature of these devices enables gradual and continuous resistance changes rather than abrupt switching, which can reduce switching power and enhance reliability by reducing the impact of thermal noise. Third, spintronic memristors can be easily integrated on top of CMOS circuitry through standard back-end-of-line processes, similar to the successful integration of MTJs in commercially deployed MRAM devices.

13.6.1.4 Research challenges and future directions

Some current research challenges associated with multi-bit spintronic memristor storage include the stability of intermediate resistance states, which make for reliable sensing of several distinguishable levels while optimizing write-erase cycles for extended device life. The trade-off between the number of distinguishable states and device fabrication complexity remains a major consideration. Also, the development of robust error-correction and signal-processing circuits compatible with multi-bit readout mechanisms is essential for practical implementation in memory hierarchies.

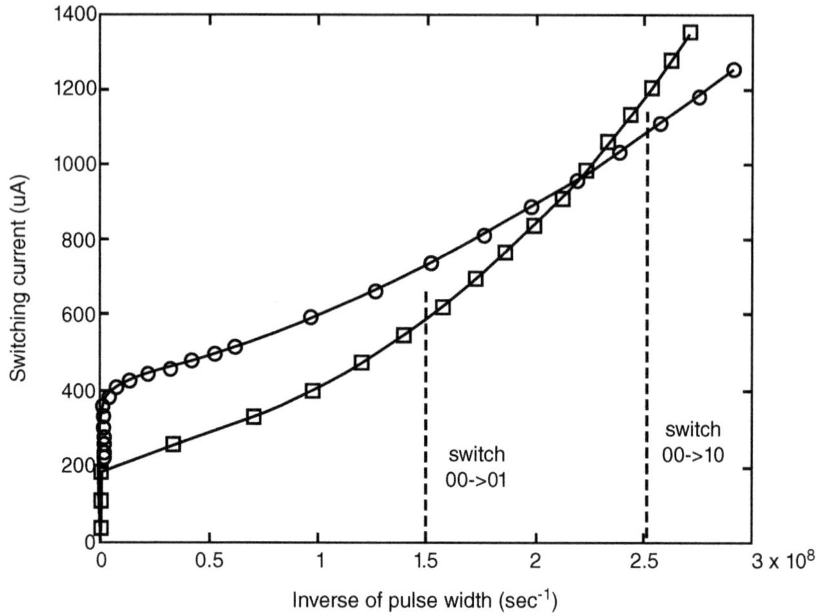

Figure 13.18 Switching current versus inverse pulse width for multi-bit MTJ sub-cells. The two curves correspond to MTJ sub-cells with different shape-induced anisotropies. The crossover point (around 900 μA and 4 ns pulse width) enables independent switching of hard and soft bits by controlling pulse duration.

(Wang and Chen [47].)

13.6.2 Application: on-chip temperature sensing

13.6.2.1 Temperature-dependent domain wall dynamics

Temperature sensing is another important application domain in which the spintronic memristors demonstrate superior performance characteristics compared with conventional on-chip temperature sensors. The physical basis of this application is rooted in the temperature-dependent mobility of domain walls in ferromagnetic materials. Indeed, thermal fluctuations directly influence the dynamics of domain wall motion via the stochastic LLG equation with spin torque terms [48].

The domain wall velocity is determined by the following relation:

$$v = \frac{dX}{dt} = \beta_0 \frac{\partial H}{\partial x} + \frac{\sigma_s(J)}{M_s w}\frac{dx}{dt} + \sqrt{\frac{2k_B T}{M_s \eta}}\xi_X(t) \tag{13.17}$$

where β_0 is the domain wall mobility coefficient, $\sigma_s(J)$ is the spin torque strength as a function of current density J, k_B is Boltzmann's constant, T is the temperature, M_s is the saturation magnetization, and $\xi_X(t)$ are thermal fluctuation fields. From this follows the important observation that the domain wall velocity around the critical current density J_{cr} has significant temperature sensitivity, which can be used to establish a temperature-sensitive operating regime in the device.

13.6.2.2 Sensor design and operation principle

For temperature sensing applications, the spintronic memristor structure consists of a long spin-valve strip with two ferromagnetic layers (reference and free layers) separated by a thin nonmagnetic spacer. The magnetization of the reference layer is fixed through exchange coupling, while the free layer is divided by a domain wall into two segments with opposite magnetization directions. When a constant-magnitude voltage pulse is applied to the device, the resulting current drives domain-wall motion. The resistance change during this excitation period is directly proportional to the domain-wall displacement, which in turn depends critically on temperature (Figure 13.19).

Its principle of operation utilizes the positive feedback between resistance and driving strength, a unique property of memristive devices. For a constant voltage pulse excitation:

$$R(T) = R_H - (R_H - R_L)\frac{X(T, t_{pulse})}{D} \tag{13.18}$$

When the temperature is raised, larger thermal fluctuations increase domain-wall mobility and produce greater domain-wall displacement and thus lower device resistance. This lower resistance in turn produces a higher driving current density through the device, which in turn further increases domain wall velocity due to the positive feedback mechanism amplifying the temperature sensitivity beyond what would be achieved through thermal effects alone.

13.6.2.3 Performance characteristics and advantages

Various experimental implementations have demonstrated spintronic memristor temperature sensors with operating ranges of 300–400 K, covering the normal operating temperature range of semiconductor chips. Key performance metrics include the following:

- Operating temperature range from 300 to 400 K.
- Power supply voltage required: 0.3 V (significantly lower than conventional on-chip sensors).
- Power consumption: 150 μW.
- Temperature sensitivity: 0.5 Ω/K (excellent linearity).
- Sensor area: 300 nm × 20 nm (nanoscale integration).

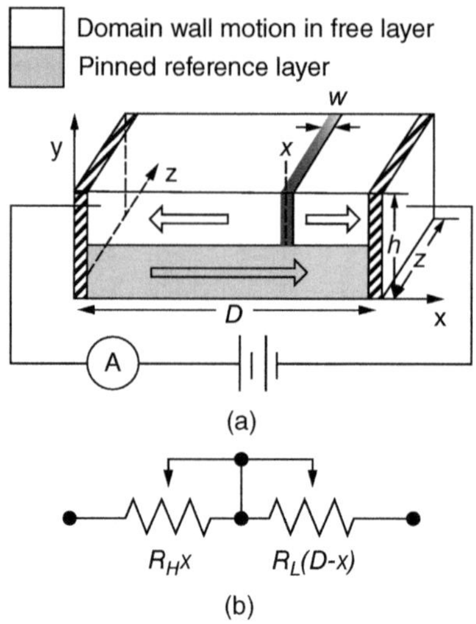

Figure 13.19 Spin-valve magnetic domain-wall memristor device structure. The device consists of a reference layer (with fixed magnetization) and a free layer (divided by a domain wall with opposite magnetization segments). Domain wall position $X(t)$ controls the device resistance according to $R(t) = R_H - (R_H - R_L)X(t)/D$. Temperature affects domain wall mobility through thermal fluctuations, and it enables temperature sensing. (a) Schematic of a domain-wall-motion-based spintronic device showing the free layer, pinned reference layer, domain wall position (x), device dimensions $(D, w, h, \text{and} z)$, and current measurement setup. (b) Equivalent electrical resistance model in which the total resistance is represented by two variable resistances, $R_H x \text{and} R_L(D-x)$, determined by the domain wall location.

(Wang et al. [48].)

Compared to oscillation-ring-based temperature sensors, which consume significant dynamic power, spintronic memristor temperature sensors have demonstrated superior power efficiency. Subthreshold voltage operation capability is especially valuable on advanced technology nodes, where conventional sensors struggle. The nanoscale feature size allows high-density integration directly on-chip, enabling distributed thermal monitoring across large computing systems without significant area overhead (Figure 13.20).

13.6.2.4 Integration and future research perspectives

Positive feedback between resistance and driving strength can be tuned through material design and geometric optimization. Varying the domain-wall thickness by manipulating magnetism depending on the material properties, the operating temperature range can be tuned for specific application requirements. Increasing the magnetoresistance ratio is possible by using

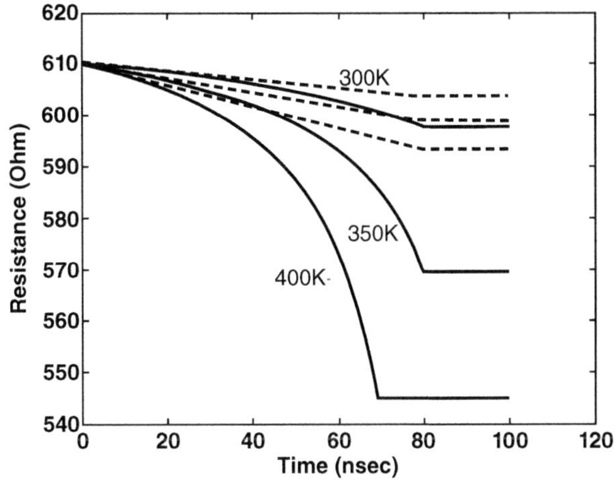

Figure 13.20 Spintronic memristor resistance as a function of time at different temperatures (300, 350, and 400 K) for constant-magnitude voltage pulse driving. Higher temperatures result in greater domain wall displacement and lower final resistance, while showing linear temperature sensitivity of 0.5 Ω/K. Solid curves show a memristor with positive feedback; dashed curves show a non-memristive device without feedback.

(Wang et al. [28].)

high-GMR or TMR stacks, which can further improve the sensing resolution and margin. Challenges in current research include achieving stable operations against process variation, managing effects of thermal noise on sensing accuracy, and developing reliable readout circuitry that can distinguish the ability to discern small resistance changes across the entire operating range. A combination of temperature sensors with other spintronic memory and logic elements on the same chip opens possibilities for closed-loop thermal management systems with minimal overhead.

13.6.3 Application: biomedical sensing and disease detection

13.6.3.1 Spintronic sensors for magnetic field detection

Biomedical sensing applications leverage the exceptional magnetic field sensitivity of spintronic memristors for noninvasive disease detection and diagnostic systems. Unlike conventional biomedical sensors, which often require invasive procedures or suffer from poor sensitivity, spintronic memristor-based sensors can detect very weak magnetic fields generated by biological markers with high specificity and sensitivity [49].

The most important merit of spintronic memristors in biomedical sensing comes from their capability to detect and store information about magnetic field perturbations caused by magnetic nanoparticles used as biomarkers. When magnetic nanoparticles bind to specific biomolecules, such as DNA sequences or disease antigens, on the sensor surface, they generate localized magnetic fields that influence domain-wall motion in the spintronic memristor. The relation between the normalized domain wall velocity and the normalized current density under different magnetic field strengths is:

$$v_{norm}\left(J_{norm}, B_{applied}\right) = f\left(J_{norm} - J_{cr,norm}, B_{applied}, T\right) \tag{13.19}$$

The domain wall velocity becomes highly sensitive to the applied magnetic field amplitude near the critical current density region.

13.6.3.1.1 Deoxyribonucleic acid (DNA) hybridization detection system

One well-characterized application is the detection of DNA hybridization using magnetic nanoparticle labels. In this configuration, single-stranded DNA receptors (probes) are immobilized on the sensor surface. Unknown DNA fragments (targets) are labeled with high-moment magnetic nanoparticles through biotin–streptavidin chemistry. When target DNA hybridizes with complementary probe sequences, the bound magnetic nanoparticles generate magnetic fields at the sensor location. These magnetic perturbations modulate the domain wall dynamics in the underlying spintronic memristor.

The sensing mechanism operates as follows: a constant current pulse is applied to the spintronic memristor at a current density slightly below the critical threshold. Without magnetic nanoparticles, the domain wall remains relatively stationary due to insufficient driving force. However, when magnetic nanoparticles are present, their magnetic fields add to the spin torque excitation, collectively exceeding the critical threshold and enabling domain wall motion. The final resistance state, determined by domain wall displacement, is directly proportional to the number of bound nanoparticles (Figure 13.21).

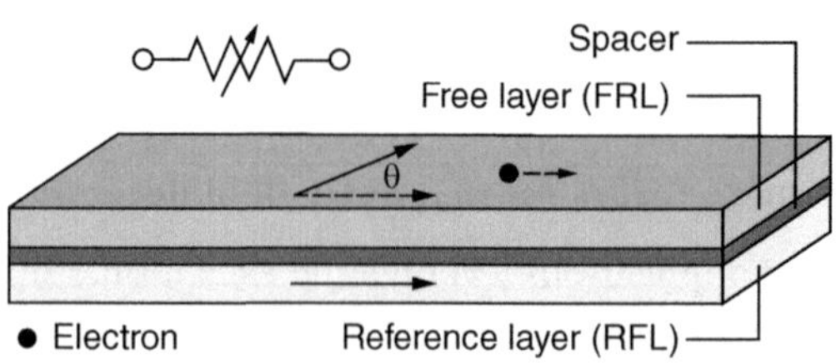

Figure 13.21 GMR spin valve sensor structure used in biomedical sensing. The device contains a reference layer (RFL) with fixed magnetization and a free layer (FRL) that responds to external magnetic fields from biomarkers.

(Chen et al. [49].)

13.6.3.2 Array architecture for multiplexed detection

Practical biomedical systems employ microarray architectures in which multiple spintronic memristor sensors are integrated on a single chip. A typical configuration includes 4 × 4 DNA spots, where each spot contains an array of spintronic memristor cells. To increase measurement throughput, both frequency division multiplexing (FDM) and time division multiplexing (TDM) are employed. Four DNA spots share one physical readout link using four distinct carrier frequencies for FDM, while these four FDM channels connect through a multiplexer to achieve TDM. The final output is processed off-chip using the Fourier transform and frequency spectrum analysis to extract individual spot signals.

The readout circuit includes a read current source, frequency divider, mixer, low-noise differential amplifier, programmable operational amplifier, and transmission gate-based multiplexer. The current source injects a small read current (typically 0.1 mA) through both the active sensor cell and a reference resistance to generate comparable voltage signals. The mixer multiplies the readout voltage by a carrier frequency generated from the master clock, enabling frequency-based signal extraction and filtering (Figure 13.22).

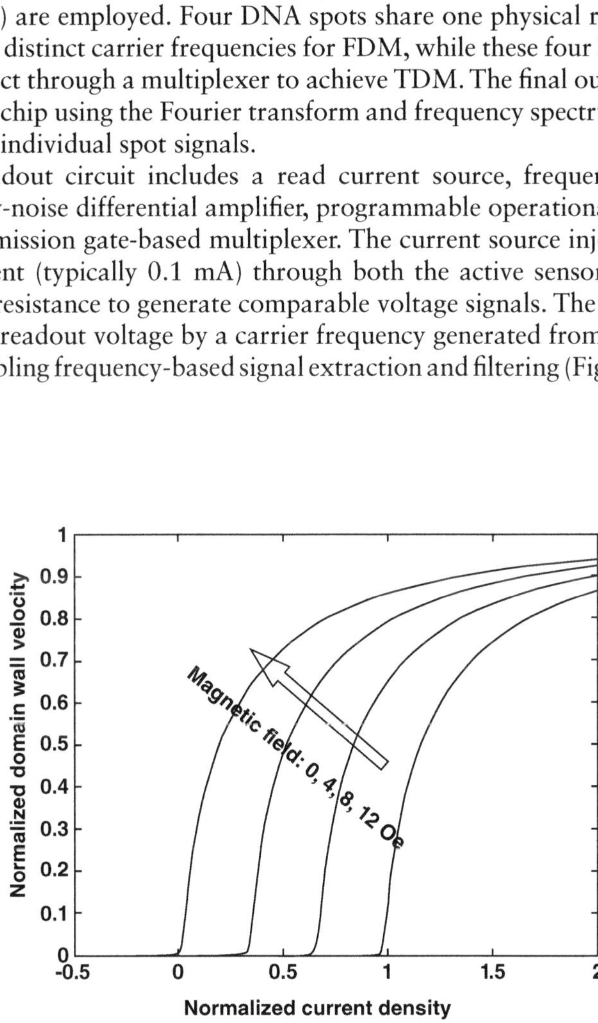

Figure 13.22 Normalized domain-wall velocity as a function of normalized current density under different magnetic fields at 300 K. The magnetic field represents the additional field due to bound magnetic nanoparticles on DNA targets. Enhanced sensitivity near the critical current density enables robust biomedical signal detection.

(Chen et al. [49].)

13.6.3.3 *Comparative advantages over traditional magnetoresistive sensors*

Spintronic memristors offer several advantages over conventional GMR and TMR sensors for biomedical applications. First, the magnetoresistance ratio can be substantially larger (exceeding 10,000 ohms or higher) compared to GMR sensors (typically 100–400 mΩ) or even TMR sensors (typically 1–5 kΩ). This directly translates to larger signal amplitudes and also improves the signal-to-noise ratio in readout circuits.

Second, spintronic memristors retain their resistance state after magnetic excitation is removed, thereby providing intrinsic data retention. This nonvolatile property enables simpler readout protocols and reduces the required dynamic range of analog-to-digital conversion. Third, the magnitude of the resistance change can be precisely controlled by adjusting the duration and magnitude of the applied voltage pulse, which enables signal amplitude optimization for various applications without needing to redesign the sensor hardware.

13.6.3.4 *Clinical and research applications*

The exceptionally high sensitivity and selectivity of spintronic memristor biomedical sensors make them ideal for diverse applications, including early disease detection (cancer biomarkers), pathogen identification, infectious disease diagnosis, and genetic screening. The nanoscale feature size and low power consumption enable portable diagnostic devices. Integration with microfluidic systems enables rapid, high-throughput screening of biological samples (Figure 13.23).

Current challenges to research include enhancing the specificity of biomarker capture and reducing nonspecific binding, surface functionalization of sensors, and development of robust calibration protocols for clinical use. Transitioning from research prototypes to FDA-approved diagnostic devices requires extensive validation studies, and operating procedure standardization is needed.

13.7 EMERGING TRENDS AND FUTURE OUTLOOK

Spintronic memristors are rapidly evolving toward new scientific and technological frontiers, driven by breakthroughs in quantum spin engineering, two-dimensional (2D) materials, organic spintronic devices, and multifield applications. This section outlines the main emerging trends, advanced use cases, and future challenges in the realm of spintronic memristive technologies [37, 50, 51].

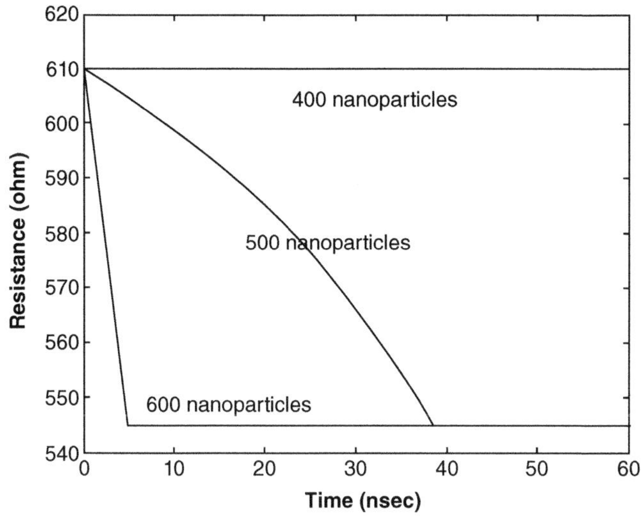

Figure 13.23 Spintronic memristor resistance as a function of time under different numbers of magnetic nanoparticles (400, 500, and 600) at 300 K. The different magnetic fields generated by varying nanoparticle numbers produce distinct resistance responses, enabling quantitative biomarker detection [49]. This demonstrates the sensor's capability for biomedical diagnostics.

13.7.1 Quantum spintronics and memristive qubits

Recent research is targeting the quantum limit of spintronic devices, utilizing single-electron spins and nonlocal spin transport for information processing. Quantum spintronic memristors leverage quantum coherence, tunneling, and entanglement to realize memristive qubits and elements for quantum neuromorphic processors [50]. These devices promise gate-tunable, nonvolatile memory together with quantum logic operations, as depicted in schematic crossbar circuits of spin qubits with memristive weighting elements (Figure 13.24).

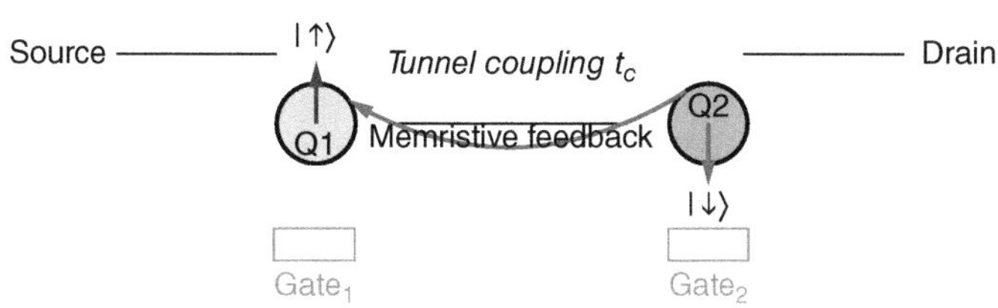

Figure 13.24 Quantum spintronic memristor concept with spin-qubit memory, tunnel coupling, and gate control.

(Adapted from Ghosh and Hoffmann [50].)

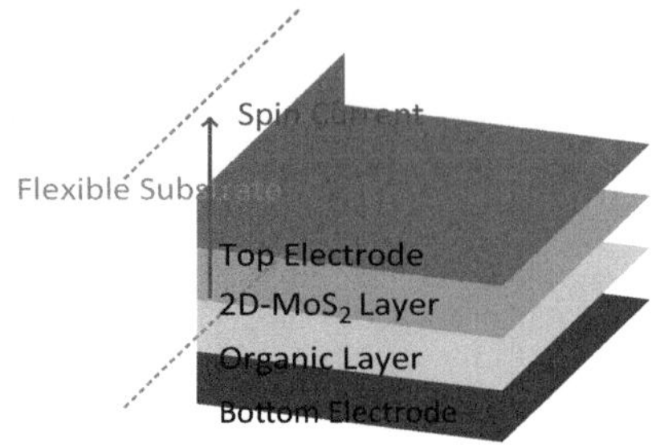

Figure 13.25 Atomically thin spintronic memristor integrated with a 2D-MoS_2/organic molecular cross-bar for flexible and wearable applications.

(Han et al. [45].)

13.7.2 Integration with 2D materials and organic spintronics

The combination of spintronic memristors with 2D magnetic materials (e.g., CrI_3, Fe_3GeTe_2, and MoS_2) and organic molecules delivers new pathways for low-power, flexible, and multifunctional electronics. 2D spintronic memristive junctions offer atomically thin devices, enhanced control over spin injection, and a large interfacial contact area for neuromorphic and wearable circuits. Organic spintronics introduces features such as biocompatibility, tunable interfaces, and high mechanical flexibility, supporting applications in the Internet of Things, brain-machine interfaces, and soft electronics (Figure 13.25) [37, 45].

13.7.3 Hardware security, wearables, and biomedical interfaces

Spintronic memristors offer unique physical fingerprinting, analog memory, and stochastic switching features that are highly attractive for hardware security primitives, true random number generators, physically unclonable functions (PUFs), and encrypted microcontroller design [51]. Their thin form factor and low energy requirements enable next-generation wearable sensors, neural stimulation systems, and low-profile biomedical implants, with ongoing progress in biocompatible encapsulation and skin-like device architectures.

13.7.4 Key challenges: endurance, scalability, and standardization

Despite rapid advances, there are essential challenges for the experimental characterization of spintronic and memristive devices:

- **Endurance**: Spintronic and memristive components must show billions of reliable switching cycles. Advancements in materials and interface control are thus crucial.

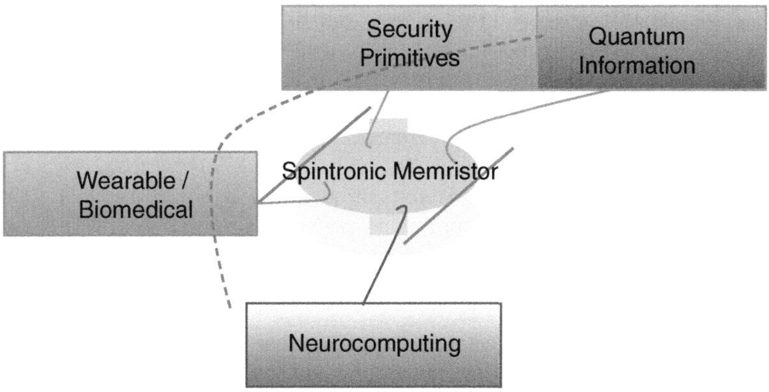

Figure 13.26 Emerging applications for spintronic memristors, including quantum information and its processing, neuromorphic computing, hardware security primitives, and wearable or biomedical electronics.

- **Scalability**: Continuous shrinkage of devices demands defect-free fabrication, controlled skyrmions/domains, and advanced wafer-level patterning methods.
- **Stochastic Reproducibility**: Utilizing (and controlling) random noise and variability in probabilistic inference is beneficial; nevertheless, this introduces a need for extensive statistical modeling and hardware evaluation [51].
- **Standardization**: To set global standards for device performance, error rates, reliability, and interoperability with memory/logic standards is essential for commercial use (Figure 13.26).

REFERENCES

1. A. Fert, "Origin, development, and future of spintronics (nobel lecture)," *Angew. Chem. Int. Ed.*, vol. 47, no. 32, pp. 5956–5967, 2008.
2. G. Binasch, P. Grünberg, F. Saurenbach, and W. Zinn, "Enhanced magnetoresistance in layered magnetic structures with antiferromagnetic interlayer exchange," *Phys. Rev. B*, vol. 39, pp. 4828–4830, 1989.
3. Nobel Committee for Physics, "The 2007 nobel prize in physics – popular information." https://www.nobelprize.org/prizes/physics/2007/popular-information/, 2007.
4. C. Chappert, A. Fert, and F. N. V. Dau, "The emergence of spin electronics in data storage," *Nat. Mater.*, vol. 6, pp. 813–823, 2007.
5. I. Žutić, J. Fabian, and S. D. Sarma, "Spintronics: Fundamentals and applications," *Rev. Mod. Phys.*, vol. 76, pp. 323–410, 2004.
6. Wikipedia, "Everspin technologies." https://en.wikipedia.org/wiki/Everspin_Technologies, 2024.
7. J. C. Slonczewski, "Current-driven excitation of magnetic multilayers," *J. Magn. Magn. Mater.*, vol. 159, pp. L1–L7, 1996.

8. Z. Han et al., "Spin-orbit torque in a three-dimensional topological insulator," *Phys. Rev. B*, vol. 97, no. 13, p. 134402, 2018.
9. Z. Han et al., "Lossless spin-orbit torque in antiferromagnetic topological materials," *Phys. Rev. Lett.*, vol. 132, p. 136701, 2024.
10. H. Yang, K. Garello, and S. S. P. Parkin, "Spin–orbit torques based on topological materials," *Accounts Mater. Res.*, vol. 3, pp. 987–1003, 2022.
11. Electronic Products News, "Spin-memristor slashes power use in AI." https://www.edn.com/spin-memristor-slashes-power-use-in-ai/, 2024.
12. TDK Corporation, "Tdk develops spin-memristor for neuromorphic devices." https://product.tdk.com/en/technews/2024/1024_spin-memristor.html, 2024. Accessed: 10 November 2025.
13. TDK Corporation and Collaborators, "Collaboration with Tohoku University and cea on spinmemristor neuromorphic hardware." https://www.tdk.com/en/news_center/press/20241002_01.html, 2025. Accessed: 10 November 2025.
14. Q. Shao, Z. Wang, Y. Zhou, S. Fukami, D. Querlioz, and L. O. Chua, "Spintronic memristors for computing," *npj Spintronics*, vol. 3, no. 1, p. 16, 2025.
15. P. Grünberg, "Enhanced magnetoresistance in layered magnetic structures with antiferromagnetic interlayer exchange," *Phys. Rev. Lett.*, vol. 57, no. 19, pp. 2442–2445, 1986.
16. J. S. Moodera, L. R. Kinder, T. M. Wong, and R. Meservey, "Large magnetoresistance at room temperature in ferromagnetic thin film tunnel junctions," *Phys. Rev. Lett.*, vol. 74, pp. 3273–3276, 1995.
17. I. M. Miron et al., "Perpendicular switching of a single ferromagnetic layer induced by in-plane current injection," *Nature*, vol. 476, pp. 189–193, 2011.
18. Y. K. Kato et al., "Observation of the spin hall effect in semiconductors," *Science*, vol. 306, pp. 1910–1913, 2004.
19. B. Huang et al., "Layer-dependent ferromagnetism in a van der Waals crystal down to the monolayer limit," *Nature*, vol. 546, pp. 270–273, 2017.
20. J. M. D. Coey, *Rare-Earth Iron Permanent Magnets*. Oxford University Press, 2011.
21. L. O. Chua, "Memristor – the missing circuit element," *IEEE Trans. Circuit Theory*, vol. 18, no. 5, pp. 507–519, 1971.
22. D. B. Strukov, G. S. Snider, D. R. Stewart, and R. S. Williams, "The missing memristor found," *Nature*, vol. 453, pp. 80–83, 2008.
23. J. Grollier, D. Querlioz, and M. D. Stiles, "Spintronic nanodevices for brain-inspired computing," *Proc. IEEE*, vol. 104, no. 10, pp. 2024–2039, 2016.
24. C. Ceri et al., "Conductive organic memristors for neuromorphic computing," *Adv. Mater.*, vol. 25, pp. 4895–4900, 2013.
25. S. Bhatti, R. Sbiaa, A. Hirohata, H. Ohno, S. Fukami, and S. N. Piramanayagam, "Spintronics based random access memory: A review," *Mater. Today*, vol. 20, no. 9, pp. 530–548.
26. V. K. Joshi, P. Barla, S. Bhat, and B. K. Kaushik, "From MTJ device to hybrid cmos/mtj circuits: A review," *IEEE Access*, vol. 8, pp. 194105–194146, 2020.
27. X. Zhang et al., "Spin-torque memristors based on perpendicular magnetic tunnel junctions for neuromorphic computing," *Adv. Sci.*, vol. 8, no. 10, p. 2004645, 2021.
28. X. Wang, Y. Chen, H. Xi, H. Li, and D. Dimitrov, "Spintronic memristor through spin-torque-induced magnetization motion," *IEEE Electron Device Lett.*, vol. 30, no. 3, pp. 294–297, 2009.

29. Z. Yu et al., "Heterogeneous memristive devices enabled by magnetic tunnel junction nanopillars surrounded by resistive silicon switches," *Adv. Electron. Mater.*, vol. 4, no. 3, p. 1700461, 2018.
30. S. S. P. Parkin, M. Hayashi, and L. Thomas, "Magnetic domain-wall racetrack memory," *Science*, vol. 320, pp. 190–194, 2008.
31. A. Bisig, M. P. Zimmermann, K. Zeissler, and et al., "Dynamic phase diagram of spin-torque switching in a nanomagnet," *Nat. Commun.*, vol. 4, p. 2328, 2013.
32. M. Hayashi, L. Thomas, C. Rettner, R. Moriya, and S. S. P. Parkin, "Current-controlled magnetic domain-wall nanowire shift register," *Science*, vol. 320, no. 5873, pp. 209–211, 2008.
33. A. Mahmoud et al., "Introduction to spin wave computing," *IEEE Trans. Magn.*, vol. 56, no. 1, pp. 65–101, 2020.
34. A. V. Chumak et al., "Advances in magnetics roadmap on spin-wave computing," *IEEE Trans. Magn.*, vol. 58, no. 5, pp. 1–46, 2022.
35. S. Luo, L. Kang, G. Liu, Y. Zhao, and X. Zhang, "Voltage-controlled skyrmion memristor for energy-efficient synapse applications," *IEEE Electron Device Lett.*, vol. 40, no. 4, pp. 626–629, 2019.
36. K. Song, Q. Shao, and W. Zhao, "Skyrmion-based synaptic devices for brain-inspired cognitive systems," *Nat. Electron.*, vol. 3, pp. 148–155, 2020.
37. Y. Zhou, P. Upadhyaya, and A. Hoffmann, "Topological spin textures: Basic physics and devices," *Adv. Mater.*, vol. 36, no. 1, p. 2007084, 2025.
38. M. K. Kamsari and S. Salahuddin, "Stochastic p-bits for probabilistic spin logic," *IEEE Electron Device Lett.*, vol. 38, no. 10, pp. 1457–1460, 2017.
39. M. K. Camsari, S. F. Junior, B. M. Sutton, and S. Datta, "Implementing p-bits with embedded stochastic nanomagnets," *IEEE J. Explor. Solid-State Comput. Devices Circuits*, vol. 5, pp. 18–22, 2019.
40. S. Daneshvar, F. H. Tahir, and M. K. Camsari, "Stochastic magnetic memory: From p-bits to reservoir computing," *IEEE Trans. Nanotechnol.*, vol. 19, no. 6, pp. 838–846, 2020.
41. M. K. Kamsari, W. Y. Zhang, and S. Datta, "Collective dynamics in nanomagnet ensemble memristors for stochastic computing," *IEEE Trans. Magn.*, vol. 57, no. 5, pp. 1–7, 2021.
42. B. Torresani, G. Malpuech, and J. Grollier, "Spintronic reservoir networks: Nonlinear dynamics and chaos," *IEEE Trans. Neural Netw. Learn. Syst.*, vol. 32, no. 8, pp. 3502–3513, 2021.
43. T. Graf, C. Felser, and S. S. P. Parkin, "Simple rules for the understanding of Heusler compounds," *Prog. Solid State Chem.*, vol. 39, pp. 1–50, 2011.
44. M. Z. Hasan and C. L. Kane, "Colloquium: Topological insulators," *Rev. Mod. Phys.*, vol. 82, pp. 3045–3067, 2010.
45. M. Han et al., "Wafer-scale integration of two-dimensional spintronic materials," *Nat. Electron.*, vol. 4, pp. 424–430, 2021.
46. Z. Li et al., "Cmos-compatible integration of spintronic memory," *IEEE Trans. Electron Devices*, vol. 64, pp. 4178–4185, 2017.
47. X. Wang and Y. Chen, "Spintronic memristor devices and application," *2010 Design, Automation & Test in Europe Conference & Exhibition (DATE 2010)*, 2010.
48. X. Wang, Y. Chen, Y. Gu, and H. Li, "Spintronic memristor temperature sensor," *IEEE Electron Device Lett.*, vol. 31, no. 1, pp. 20–22, 2010.

49. Y. Chen, X. Wang, Z. Sun, and H. Li, "The application of spintronic devices in magnetic bio-sensing," *Proceedings of the 2nd Asia Symposium on Quality Electronic Design (ASQED)*, pp. 230–235, 2010.
50. S. Ghosh and A. Hoffmann, "Quantum spintronics: Foundations and directions," *IEEE Trans. Magn.*, vol. 58, no. 6, pp. 1–10, 2022.
51. J. Grollier et al., "Spintronic memristors for advanced security architectures and wearable electronics," *Proc. IEEE*, vol. 108, no. 8, pp. 1388–1403, 2020.

Index

Pages in *italics* refer to figure and pages in **bold** refer to tables.

For Product Safety Concerns and Information please contact our EU representative GPSR@taylorandfrancis.com Taylor & Francis Verlag GmbH, Kaufingerstraße 24, 80331 München, Germany

Batch number: 10725800

Printed by Printforce, the Netherlands